Getting the Measure of the Stars

A fine example of an amateur astronomer's set-up for photometric measurement, as operated by Jack Ells of Crayford Astronomical Society. The telescope is fitted with a JEAP photometer, described in this book.

Getting the Measure of the Stars

W A Cooper

Open University, Milton Keynes

and

E N Walker

Herstmonceux, Sussex

Adam Hilger, Bristol and Philadelphia

British Library Cataloguing in Publication Data

Cooper, W. A.
Getting the measure of the stars.
1. Astronomy—Amateurs' manuals
I. Title II. Walker, E. N.
522

ISBN 0-85274-830-2

Library of Congress Cataloging-in-Publication Data

Cooper, W. A. (W. Alan)
Getting the measure of the stars

Bibliography: 3p.
Includes index.
1. Astrophysics—Amateurs' manuals. 2. Astrometry—Amateurs' manuals. I. Walker, E. N. II. Title.
QB461.C67 523.8′028 88-32057
ISBN 0-85274-830-2

Consultant Editor: **Professor A E Roy**, Glasgow University

Published under the Adam Hilger imprint by IOP Publishing Ltd
Techno House, Redcliffe Way, Bristol BS1 6NX, England
242 Cherry Street, Philadelphia, PA 19106, USA

Typeset by Mathematical Composition Setters Ltd
Salisbury, Wiltshire

Printed in Great Britain by J W Arrowsmith Ltd, Bristol

Contents

Preface

There are many excellent introductory books on astronomy, which outline the exciting questions to which astronomy holds a clue. You can see in them some of the wonderful images and radio maps, in full false colour, which in themselves are enough to stimulate the wish to go further. But how does one go further? Before you sign up for a degree in astrophysics or take out a mortgage for a secondhand NASA radio dish, try a couple of hours with this book. It is for anyone who is, or who might just be tempted to become, an amateur astronomer. Thanks to home computers that now includes armchair astronomers. Although many of the spectacular new results need professional astronomers and a spectacular budget, there are also many questions to which an amateur astronomer can contribute. It is on those topics that we have concentrated. There is, in fact, an increasing scope for amateur astronomers. Because spaceborne telescopes are so expensive, professional astronomers are obliged to concentrate on those problems at which such telescopes excel. That leaves many of the unsolved problems, especially those about variable stars, open to the amateur.

After an introductory chapter, the first part of this book (Chapters 2–8) explains how stars arrive at the delicately balanced state in which the rhythmic changes in luminosity characteristic of variable stars can arise spontaneously. There is a chapter on minor planets, the Cinderellas of astrophysics. The second part of the book explains how to make measurements on stars, and the third part suggests which might be the most interesting to concentrate on if you wish to make some measurements of your own.

You can, of course, find these things explained much more fully in textbooks of astrophysics. But they take a lot of physics and astrophysics for granted leaving, in our opinion, too big a gap separating them from the colourful introductory books. Try this book for bridging the gap. Keep the colourful pictures to hand, we couldn't afford them in this book. That should leave you enough for the first down payment on a secondhand NASA radiotelescope.

W A Cooper
E N Walker
June 1988

Foreword

Astronomy is often held up as the oldest of the sciences. Whether this is true, in the way that we understand a science today, is open to debate; what cannot be questioned is that, for thousands of years, mankind has had an awareness of and an interest in the night sky and the pageant continually unfolding there. Perhaps more important to everyday living, he has also had a practical concern with the seasonal cycles affecting the growing of his crops and the tides on his shores.

One might argue, with good cause, that astronomy was accessible to everyone well into the present century, with amateurs, albeit well heeled or well educated ones, playing a leading role in exploring the universe around us. One only has to look at the significant contribution of amateurs (in the strict sense) to the deliberations of the Royal Astronomical Society in the first half of its history to recognize this fact. Latterly, however, as more sophisticated and expensive equipment has been required to probe ever deeper into nature's secrets, one senses that the man in the street has been left behind. Not that it is a matter merely of cost but rather that, in science generally, both experiment and theory have of necessity got more complex and specialized to the point where full-time effort and large common-user facilities are apparently required to push the frontiers back still further.

In astronomy, where we are seeking to unravel the mysteries of a vast cosmos crammed full of exciting objects, the relentless march into ever deeper specialization by professional astronomers is fraught with danger. There is only a limited amount of money to fund a limited number of astronomers, which means that only a few of the most pressing questions can be tackled at any one time. As in most things, there are 'fashionable' topics in astronomy too, and it is very easy for whole areas, once thought to be fundamental, to be passed by as the hunt in full cry dashes off in another direction. Indeed, if we look on astronomical research as scaffolding reaching out into the Universe, it has to be conceded that, in places, the foundations and lower levels of what is now a very tall structure are less than secure.

It is against this background that the role of the dedicated amateur astronomer of the late twentieth century can be of considerable importance. Indeed, it has been recently said that we ought to be talking, rather, about paid and unpaid astronomers, with those who are privileged enough to be doing it full time encouraging and directing the efforts and energies of their colleagues who spend their daylight hours in other pursuits. Today's high-technology society makes this collaboration especially vital: not only do we expect leisure to become an increasingly significant activity generally, but the hobbyist is becoming ever more sophisticated and computerized. Failure to spark an interest in astronomy in this large potential part-time workforce would be folly indeed on the part of the professional. With modest telescopes and ancillary equipment, the cost of which can be no more than that of a good hi-fi system, an army of amateurs can collectively maintain a watch on the skies that professionals could never hope to achieve.

This is not to say that one expects amateurs to jump into high red-shift quasar searches or infrared spectroscopy of protostars, but there are many areas, particularly of variable star research, where the careful amateur observer can make a powerful contribution. Part of the reason for this, it must be said, is that professionals have, to some degree, painted themselves into an awkward corner: they have latterly concentrated their resources in relatively few large (and expensive) facilities which have tended to be used predominantly on the fundamental questions of cosmology. The assumption seems to have been made, by way of example, that we know how most stars work and that only 'dotting the i's and crossing the t's' remains to be done. The rich variety of stellar objects that is to be found in our Galaxy alone ensures that this is not the case. Stars are not static and they evolve and if we are to understand their histories we need to study a whole host of periodic and secular changes on timescales from milliseconds to many years. It is particularly difficult for professionals to tackle programmes where observing runs of more than a few days are required unless they are fortunate to have fairly exclusive access to a telescope. This is where the amateur can come into his own; he doesn't have to apply to an allocation panel for his telescope time; he doesn't have to produce a paper every so often to keep his salary coming in; he can get on with the job of observing and keep at it until the job is done.

A primary aim of the present book is to motivate and encourage amateurs to turn their skills to photoelectric photometry. In the United States, this area is already flourishing and there are now a few pioneers making PEP observations in Britain. Professionals have a responsibility to nurture growth in the field: Alan Cooper and Norman Walker have accepted the challenge, and it is to be hoped that others will add their support. Already, at the instigation of the dynamic Crayford Manor House Astronomical Society, a bridgehead has been built and the Pro–Am Coordinating Committee formed to explore ways of bringing professionals and keen amateurs into contact. There is so much to be gained and nothing to be lost; the movement deserves to succeed.

David Stickland
Patrick Moore

Part I

Patterns, Structures, Problems

Chapter 1

Physics at all sizes

1.1 AMATEURS AND PROFESSIONALS

An amateur astronomer stays up half the night watching a single star he has watched dozens of times before. He gets very cold, oversleeps and has to miss breakfast to catch the train next morning. Is that just a strange sort of fun? Surely it would be pretentious to call it 'delving into the basic laws of the Universe'. And in any case how can the amateur astronomer compete with a professional handling—from the comfort of a purely electronic control room—hundreds of millions of pounds worth of telescope orbiting well above the absorption and distortion caused by the atmosphere?

A particle physicist works for years at a huge particle accelerator, the only one of its type in the world, matching up her results against computations of theories expressed in a mathematical language only a few people can understand. Surely that can truly be called fundamental research into the laws of nature.

Until recently these three imaginary people would have been unlikely to meet each other to combine their ideas, but now that is changing. Study of the Big Bang at the creation of the Universe forced a shotgun marriage between particle physics and astrophysics. In the Big Bang the whole Universe was filled with the particles which are now made painstakingly, and expensively, at accelerators. There is no distinction between the nuclear physics and the astrophysics of the Big Bang. Orbiting telescopes have broken down another barrier. They can record radiation across practically the whole range from radio to x rays, and gamma rays such as those given off by radioactive nuclei. No astronomer can now afford to confine his attention to a particular type of radiation—certainly not to the very narrow band which is visible to the unaided eye.

In astronomy, if not in particle physics, the barrier between amateurs and professionals is also breaking down, to the benefit of both. The great expense of orbiting telescopes works, in an odd way, in favour of the amateur astronomer, because it has concentrated the money in a few instruments which are therefore in great demand, and can only be used for observations which are expected to yield a rapid and dramatic pay-off. But to benefit from this, the amateur must, like the professionals, have some familiarity with astronomy across the whole spectrum, and with results across the whole range of bodies from protons to galaxies. Obviously no-one can be an expert in all those fields, but a wider view is needed in understanding the large scale evolution of matter in the Universe.

As a start we will look for any overall pattern in the sizes of objects in the Universe. This needs examples from all sizes of object, which means using names of objects described later in the book. Fortunately some useful words have come into wide use in recent years, thanks partly to science fiction. A *black hole*, which was an arcane mathematical idea a few decades ago, is now a buzz word for bestsellers. The basic nuclear particles come up from time to time in news broadcasts, with 'neutron' beating 'proton' into second place for the number of citations. Some astronomical names are admirably self-explanatory—white dwarf star for instance. The name is enough for the moment, without going deeper. We will need to size up galaxies so as to put them in their proper place, but as you picked up this book you probably know that galaxies are collections of hundreds of millions of stars, produced from thousands of clouds of dust and gas. The word galaxy and that image of them are enough for the moment. On the other hand, of the four forces of nature, only gravity is so far part of the general vocabulary. The everyday experience of the electromagnetic force is limited to motors and meters, whereas it really comes into its own at the atomic level. The electromagnetic force between a proton and an electron in a hydrogen atom for instance is enough to give an acceleration a thousand million million (10^{15}) times greater than the acceleration due to gravity at the surface of the Earth! The nucleus is held together by even stronger forces—which of course is why nuclear energy is so potent. There are, in fact, two distinct nuclear forces (making up the total of four forces) but we will not need to distinguish between them. In combination they store the nuclear energy, which powers most, but not all, stars.

1.2 FROM PROTONS TO GALAXIES

To see patterns we need pictures or diagrams such as graphs, so how can we crowd all the different types of object in the universe into one graph? It is no use using ordinary *linear scales* on graphs, in which each successive scale mark on an axis represents an addition of the same amount to the value. If the proton were plotted at the first division, no piece of paper would be big enough to get as far as a meteorite, let alone a galaxy. The answer is to use a scale in which each division represents the same multiplied factor, rather than the same added amount. Usually the factor is ten, so, for instance, if the first division is a metre, the third division is 1000 metres, a kilometre, not three metres as it would be on a linear scale. Such a scale is called logarithmic. They may be used on one or both scales of a diagram. In one way it is a pity to have to use them, because they do distort the relation between sizes, but clearly an astronomer has little choice. To get used to the effect of a logarithmic scale, look at figure 1.1 which displays the size and weight of various creatures, from beetles to whales. Figure 1.1(*a*) has an ordinary, linear scale and figure 1.1(*b*), which has the constant factor, has a logarithmic scale. Both figures show, in their own way, two separate facts.

(1) The bigger animals weigh more. In fact weight increases roughly as the cube of the size because all creatures have much the same density—we all more or less float on water.

(2) Generally speaking, insects are smaller than mammals. The exceptions are rare enough to make good stories ('Honestly, I saw flies as big as rats out there ...'). The difference in size corresponds to a difference in structure—the totally different methods of respiration in the two cases. All through nature we find that differences in scale give different structures, and astronomy is no exception.

Now let us contract the scale so as to show a much wider range of sizes, down to nuclear particles and up to astronomical bodies. In doing so we must stop using the word weight because that refers to the force at the surface of the Earth on a body, due to the force of gravity. There is not much sense in talking about the force that there would be on the Sun if it were at the surface of the Earth! All this means in practice is that, from now on, we will use the word kilogramme in its proper sense, as a unit of mass, instead of the colloquial sense which was used in figures 1.1(*a*) and 1.1(*b*) (note 1.2 expands on this point).

Figure 1.2 is a mass against radius plot for the major types of object in the Universe. Whereas before there was only one gap, now there are many gaps, each corresponding to a change of structure. For instance, there are no superprotons the size of golf balls, and there are no animals as big as a planet. There is a very fundamental reason for the fact that elementary particles do not extend beyond the bottom left of the diagram. They are held together by the *strong nuclear force*, which only has a range of 0.000 000 000 000 001 m, usually written as 10^{-15} m or 1 femtometre or 1 fm, or if you prefer, a thousandth of a millionth of a millionth of a metre. The force does not operate beyond that distance, so clearly it is only of importance for nuclear particles. Bodies the size of creatures such as those in figure 1.1, on the other hand, are held together by electromagnetic forces. Astronomical bodies are, of course, held together by gravity. At the most fundamental level, therefore, the different groupings correspond to the dominance of different forces in their structures.

Although the points on figure 1.2 show the same diagonal trend as in figure 1.1 the pattern is more complicated. The bodies certainly do not all have the same density, but there is nevertheless a trend created by the fact that all the points lie above a diagonal line; a straight line on this

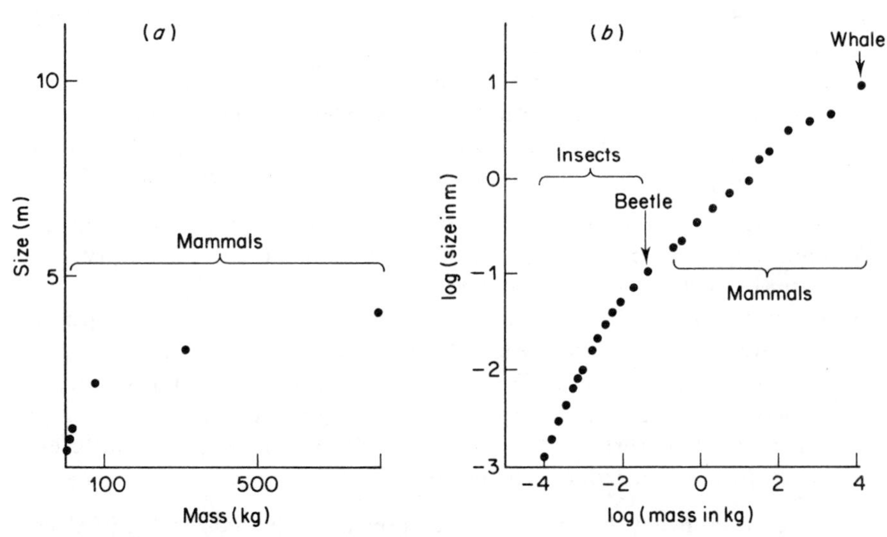

Figure 1.1 A diagram of all known animals. In trying to get familiar with bodies outside the Solar System we often have to make do with size, temperature and mass because that may be all that is known about them. How would we make out trying to find out about animals in the same way? They are all much the same temperature (apart from bacteria) so that does not help much. They cover quite a range of sizes and masses. If we plot one against another there is clearly a pattern, which is not surprising because they are all much the same density, so the mass is roughly speaking proportional to the cube of the size. So figure 1.1(*a*) is rather straightforward. It reveals a distinction between the insects and most mammals. Figure 1.1(*b*) shows the same numbers but plotted on logarithmic scales, which are very useful in astronomy. They are even useful here because they spread out the data so that it is easier to see, and also the distinction between the insects and the animals shows up better. Notice especially how different the two plots look at first sight. We will have to use logarithmic scales to show the much wider range of sizes and masses of astronomical bodies. On such a scale 1 means $10^1 = 10$, -1 means $10^{-1} = 0.1$, so 0 means 1 (since $\log 1 = 0$).

logarithmic plot. In other words, for each radius there is a maximum mass. At that mass and radius gravity is so strong that matter collapses in on itself. Electromagnetic radiation such as light is also constrained within this collapsed body—a *black hole*. We could not, as far as we know, observe any structure within a black hole. If more matter is added to the black hole, it becomes more massive, but it also becomes bigger, moving up to the point on the diagonal line corresponding to its mass. Thus the area below the diagonal line is simply empty. It is not that it contains bodies which are unobservable, it is empty. A black hole of a given mass has a given size. Stars and galaxies are not so simple. There are stars of the same mass but

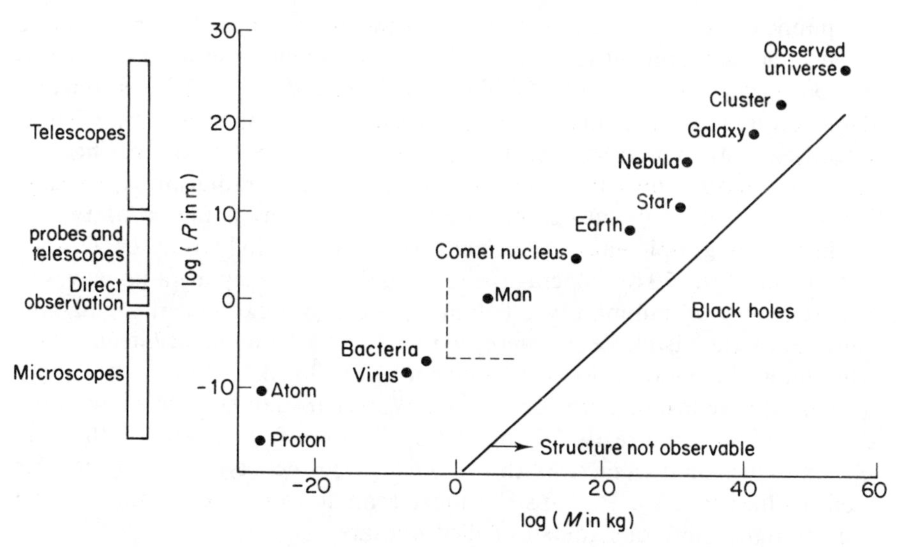

Figure 1.2 A diagram of all known bodies in the Universe. Again we look only at size and mass, but now the scales have to be very much compressed to get everything on the same plot. Radius is plotted vertically as in figure 1.1 and we will continue this practice, especially in Chapter 4. The area covered by figure 1.1 is shown by the broken lines. The gaps between different types of body now show up very clearly. There is still a definite pattern in the points, again because each type of body has a restricted range of typical densities. At stellar masses we have to plot two points, one for a typical star (at high density) and one for a typical cloud of matter (at low density) from which stars form. There is a big empty area at the top left corresponding to matter which is too diffuse to form any recognizable body. More surprisingly, there is a big empty area at the bottom right labelled black holes. This area corresponds to densities so high that the matter is not observable, because no light can escape. The diameter of a black hole of mass M is $2GM/(c^2)$, so a clearly defined straight line separates the observable from the non-observable bodies.

quite different sizes, corresponding to different internal structures. A request for a star of mass 2×10^{30} kg could, for instance, be satisfied by either the Sun or a white dwarf of a million times greater density. Much of this book will be about this diversity of ways in which matter in stars can respond to the force of gravity. Galaxies have a less individual nature. The differences are minor and rather subtle and correspondingly more difficult to explain.

1.3 FROM RADIO TO GAMMA RAYS

An amateur astronomer or self-employed inventor has difficulty getting an idea for a new instrument into the development stage. But if a scientist in the public eye brings it to the attention of the military, then it is much more likely to get government funding, and be able to exploit the scientific uses as a byproduct. This was true of Galileo, who borrowed a Dutch invention and demonstrated its power first to the defence authorities and then to his fellow scientists. Astronomy was tied to purely optical observations for the next 300 years. To see how restricted the visible range of radiation is, we may borrow a term from music, in which octave means a factor of two in frequency. A grand piano spans eight octaves. But light only covers one octave out of the 30 or so octaves spanning the frequency of radio waves to the frequency of gamma rays. The first people to break out of the narrow confines of the visible range were two amateur astronomers, Grote Reber who built the first radio telescope and Karl Jansky. Radio astronomy developed very fast after the 2nd World War, helped directly by ex-wartime equipment, and indirectly by the reflected glow of approval for the part played by wartime science. With radio telescopes new types of object were seen in the deep sky. Perhaps the most dramatic were very compact and exceptionally luminous galaxies called quasars.

We can expect to find new objects every time a new range of radiation becomes accessible to astronomers. By 1968 all thirty octaves had been exploited to some extent by astronomers, thanks to satellite-borne telescopes whose capabilities are increasing all the time. There is currently a queue of telescopes waiting to be launched. So will there be an end to astronomy when all the different types of object have been discovered? That seems to be a valid question since figure 1.2 suggests that there are discrete types of object, not a continuous spread (see Martin Harwit's '*Cosmic Discovery*' for a detailed discussion). There is an exactly analogous question posed in particle physics—'How powerful an accelerator is needed to produce all the possible fundamental particles?'. Because particles show a very clear pattern in their properties it is possible to give a definite answer, based on current theories, of about 10^{15} electron volts. To attempt to answer the question for astronomical objects we must have a more systematic and quantitative way of deciding when a range of wavelengths has been explored sufficiently well to say that no type of object has been missed. Let us say for simplicity that we need to be able to detect any object bigger than 100 km in size, within 250 light years of the Earth. Such a volume would contain several thousand ordinary stars, so this is a sensible figure to take, even though rather arbitrary. (Harwit gives more sophisticated methods.) Now we have a definite question to work on. What are the characteristics of telescopes which could detect such objects and to what extent have telescopes approached these 'ultimate' goals?

To begin with we can display the range of wavelengths of practical interest as in the second column of figure 1.3. Each range has a name, also given in the second column, with some ambiguity about just where to change from one name to another. Now, anyone who has a telescope would

T (K)	Segments of the spectrum	Sensitivity	Angular resolution	Time resolution	Spectral resolution
1	Radio				
	1 mm				
100	Infrared				
	1 μm				
5000	Visible				
	0.4 μm				
50000	Ultraviolet				
	0.01 μm				
10^7	X-ray				
	1nm				
	Gamma				
	Neutrino				
	Cosmic rays				
	Full scale Corresponds to	10 Photon m^{-2} s^{-1}	1 μarc sec	10 μs	$\frac{\Delta\lambda}{\lambda} = 10^{-6}$

Figure 1.3 Telescopes for the whole range of radiation with the shorter wavelengths omitted, in general, from hotter bodies. Early telescopes were naturally used for visible wavelengths. The bigger they were, the better the sensitivity and resolution of small bodies they could achieve. Later spectrometers were added, and the spectral resolution has also steadily improved. Two more subtle points are hinted at in subsequent diagrams. Firstly, the dotted areas under angular resolution indicate the use of long baseline interferometry, which allows a great increase in resolution, though without a corresponding increase in sensitivity. Secondly, spectral resolution is plotted vertically as an indication that it operates differently from the other parameters—improving spectral resolution usually decreases sensitivity and angular resolution. The next four diagrams aim to give an idea of how telescopes have allowed access to a greater range of radiation and have improved in accuracy over the years. The fraction of each cell which is shaded in indicates schematically the degree to which the practical limits have been achieved. They are set partly by the instruments, but also by the sheer difficulty of handling vast amounts of data. Imagine how Galileo's or Herschel's telescope would fare on such a diagram!

like to have a bigger one, so as to have higher sensitivity and so be able to see a given type of object out to greater distances. But we can be quite quantitative about a desirable sensitivity, given our assumption that we would like to be able to see any object bigger than 100 km in radius out to a distance of 250 light years. The answer follows from working out the rate at which photons are received from a star, or from any other object. (If you are familiar with the idea of photons, skip the next paragraph.)

The waves from a source of radiation are broken up into separate packets of waves, like this:

Each of these packets is called a *photon*. This applies to all forms of radiation, from gamma rays and x rays down to radio waves. The precise way to picture these packets has been a talking point since they were discovered, but the above picture is perfectly adequate for us. It is also essential; no modern astronomer could design a detector without some such picture in mind, as we will see repeatedly. This picture immediately raises some more questions. Are all the packets the same? How much energy is in each packet? The answers are as simple as they could be. For a given frequency they are all the same, and the energy in each packet is proportional to the frequency. (That means inversely proportional to wavelength, so for instance each radio photon has much less energy than each x-ray photon.) Correspondingly it is difficult to detect the photon nature of radio waves, whereas individual photons of light or x rays can be counted with modern electronics—a very important point for Part II of this book. It is quite easy to calculate the number of photons given off by a hot object, whether a headlamp bulb or a star. Forgetting about precise shapes, and assuming every body is a sphere, the result is as follows.

An object at a temperature T and radius R emits about $10^{15}R^2T^3$ photons per second. At a distance of 250 light years a telescope of, say, 1 square metre area collects 1 in 10^{33} of these photons because the surface area of a sphere of radius 250 light years is about 10^{33} square metres, as every pocket calculator knows. This gives us something to aim at. For a body of radius 100 km (10^5 m) at a temperature of 1000 K, a 1 square metre telescope would receive 10 photons per second—just enough to detect, given very good conditions. This is where the first column in the figure is important. A body at a temperature T will emit radiation at all wavelengths, but most strongly in the region of a wavelength $0.000\,29/T$ m. Thus for each temperature there is a wavelength which is most favourable for observation, because the photons are most abundant. We can use this to work out the photon rate that must be observable if the hypothetical 100 km body is not to be missed. These target sensitivities are shown in figure 1.3. A temperature of 1000 K corresponds to a peak wavelength of 2.9

micrometres, in the infrared part of the spectrum, so that is where you will see the target of 10 photons per second. With arithmetic such as this the sensitivity needed to detect any body at a given distance and temperature can be worked out.

The formula is for an idealized surface, but is a good approximation for a star. It would badly overestimate the radiation from, for instance, an asteroid with a surface of polished gold. (If you know of such an asteroid please write to the authors in confidence.) It would underestimate the radiation from a star which had, superimposed on the thermal radiation, a very strong spectral line. It fails completely of course for a black hole which gives out no radiation at all from anything which could be called its surface. Incidentally, a black hole would normally betray its presence by its violent effect on any surrounding matter, which is thereby heated to incandescence.

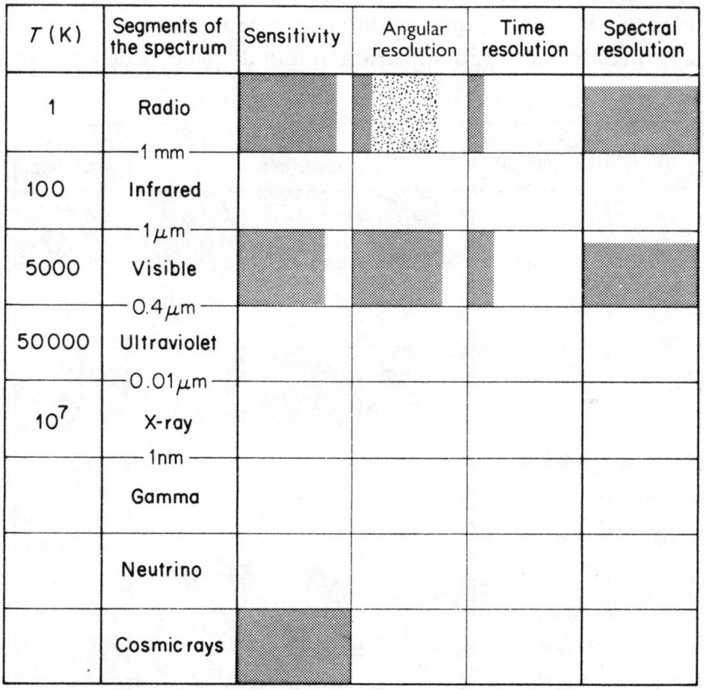

T (K)	Segments of the spectrum	Sensitivity	Angular resolution	Time resolution	Spectral resolution
1	Radio				
	1 mm				
100	Infrared				
	1 μm				
5000	Visible				
	0.4 μm				
50000	Ultraviolet				
	0.01 μm				
10^7	X-ray				
	1 nm				
	Gamma				
	Neutrino				
	Cosmic rays				

Figure 1.4 1960. 150 years after Herschel the first telescopes for radiation outside the visible range were built—the radio telescopes of Jansky and Weber. The completion of the 250 foot diameter steerable dish at Jodrell Bank was a milestone in the development of radio telescopes. It was immediately in the headlines—not for strictly astronomical reasons but because it received signals from Sputnik, launched on 4 October 1957! Soon, satellites were carrying small telescopes into orbit.

Detecting a few photons may be enough to record the presence of an object, but it will not tell us what it is. Good angular resolution is also needed to find out, for instance, how big the object is. The best resolution that looks feasible within the foreseeable future is about a millionth of an arcsecond. The resolution of a 15 cm telescope in visible light is, at best, about an arcsecond. A million times better than that would resolve the hypothetical 100 km object at a distance of 1 light year. It might be, for instance, a spot on Alpha Centauri—like a sunspot. That would count as a new object, since there is only very indirect evidence that such starspots exist. It is a missing link in stellar structure models. Thinking of bigger distances, such resolution would answer the mystery of the power sources at the centres of quasars and other active galaxies. A micro-arcsecond is therefore taken as a reasonable ultimate target in the fourth column of figure 1.3. Why not try for even better resolution? It sounds as though it would be nice to have even higher resolution, but it can, and already does, cause problems in storing and handling so much data.

To get some idea of the composition of a body needs a detector which can

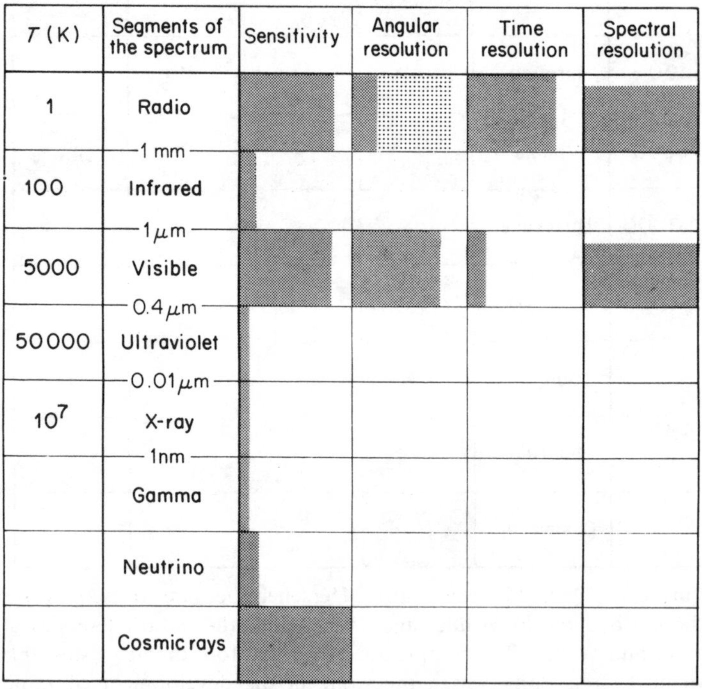

T (K)	Segments of the spectrum	Sensitivity	Angular resolution	Time resolution	Spectral resolution
1	Radio				
	1 mm				
100	Infrared				
	1 μm				
5000	Visible				
	0.4 μm				
50 000	Ultraviolet				
	0.01 μm				
10^7	X-ray				
	1nm				
	Gamma				
	Neutrino				
	Cosmic rays				

Figure 1.5 1970. In 1970 the entire range of radiation was accessible for the first time, thanks to telescopes borne on satellites. In most of the range, however, the telescopes were still very small and the detectors rudimentary.

identify the radiation produced by different atoms and molecules. Two examples of such distinctive radiation are familiar in everyday life, the yellow light of sodium and the red light of neon. Molecules have distinctive radio lines of their own which can be tuned in and identified in just the same way as identifying a radio station by its frequency. In fact it is less likely to find two molecules at the same frequency than to find two radio stations interfering with each other. Provided the stations are not at precisely the same frequency, they may be distinguished by improving the selectivity of the receiver. If a radio receiver tuned to 10 MHz is selective enough to reject signals outside a band only 1 kHz wide it may be said to have a *selectivity* of 10 000. In other wavelength ranges, the words *spectral resolution* are used, rather than selectivity, but it is quantified in the same way. A value of 10 000 is taken as a target across the whole range in figure 1.3, although this is already comfortably exceeded in some detectors.

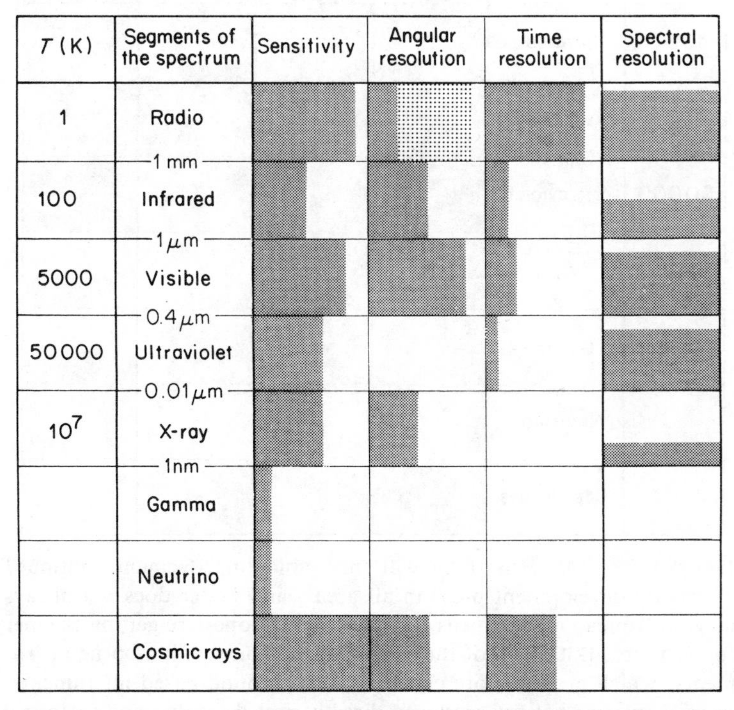

Figure 1.6 1988. Thanks to military needs more than astronomical research, satellite-borne telescopes rapidly increase in power. There is also a steady improvement in radio telescopes, due largely to grouping them together into interferometer networks. The spatial resolution of such a network greatly exceeds the spatial resolution of optical telescopes. Time resolution is now becoming important as rapid phenomena such as pulsars and flares are discovered and studied.

There is one more type of resolution which is becoming increasingly important. Observations at short wavelengths, especially in the x-ray region, reveal hot regions, which are sometimes violently active. To follow these rapid changes needs good time resolution. Pulsars were discovered when the time resolution of radio observations was improved from minutes to seconds. Even faster changes, down to a millisecond, have been discovered in some x-ray sources, and in a few pulsars which are thought to be in a second stage of development, having gained angular momentum. That is the time resolution which a modern telescope must aim at.

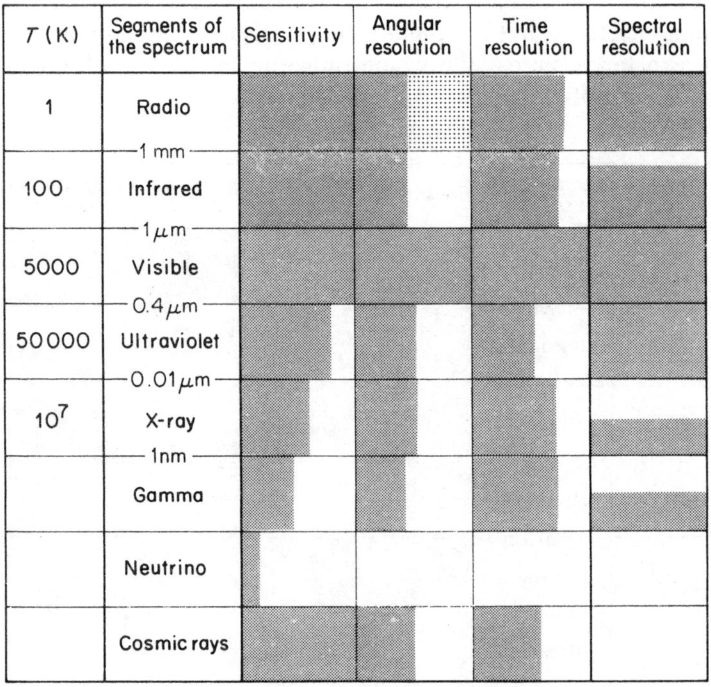

T (K)	Segments of the spectrum	Sensitivity	Angular resolution	Time resolution	Spectral resolution
1	Radio				
	1 mm				
100	Infrared				
	1 μm				
5000	Visible				
	0.4 μm				
50000	Ultraviolet				
	0.01 μm				
10^7	X-ray				
	1nm				
	Gamma				
	Neutrino				
	Cosmic rays				

Figure 1.7 2000. How long will this rapid improvement continue? There are development plans in all areas, and better does not always inevitably mean more expensive—optical telescopes are getting cheaper for instance. But much of the range depends on satellite-borne instruments, which are far more expensive than ground-based instruments. Astronomy has not yet produced directly useful results, and there is a limit to how much will be spent to satisfy curiosity, even in astronomy.

We can now use figure 1.3 to set out the degree to which these goals of a perfect observation have been attained. Each square is filled in by an amount roughly indicating, on a logarithmic scale, the degree of attainment of the goals chosen above. Figure 1.4 shows the situation in 1957, with

Jodrell Bank; figure 1.5 the first observations at short wavelengths so that the whole range of radiation is accessible. In 1978, the International Ultraviolet Explorer, one of the most successful orbiting telescopes—and so on. Figure 1.6 shows the current situation.

What will happen in the future? We can have a good guess because, thanks to the great cost of orbiting telescopes, they have to be planned years in advance. Both Europe and the USA have published the expected launches up to the year 2000. No doubt there will be delays due to unexpected difficulties, such as the succession of accidents in 1986, including the explosion of the Shuttle Challenger, but the programme will proceed. The capabilities of the projected instruments are indicated in figure 1.7.

1.4 UNSOLVED PROBLEMS

Astronomy tries to answer two questions: what objects exist, and how does each type evolve. Eventually perhaps we will have found examples of all the types of object that exist and have calculated all physical models consistent with known properties of matter. We will then be able to answer a very fundamental question—do all the types of object that could exist in fact occur. This is a question that astronomers face at the moment over the case of black holes. It seems almost certain that they are a theoretical possibility. But whether nature has actually made any is not at all certain. If they do exist they have a very important role to play in the Universe since the energy released in making a black hole from a star is comparable to all the energy emitted from the star during its lifetime. Moreover, they would very slowly become increasingly common since they would be the only final resting place for matter. Another major question of the same type is whether very massive stars—of hundreds or thousands of solar masses—exist now, or did exist in the early stages of the Universe. A population of early massive short-lived stars might have produced much of the background radiation which is at present ascribed to the Big Bang, so it is a question mark over the very heart of cosmology.

There are equally deep mysteries much closer to home. Obviously an 'object' called a *planetary system* exists, and we would recognize an example other than our own. But how does a protostellar cloud evolve into a planetary system? We need to find other examples. The search for them is at such an early stage at the moment that any detection of cool matter near a star raises an exciting possibility that a breakthrough is at hand. A disc of matter, as observed round the nearby star β Pictoris, is more exciting because its shape clearly distinguishes it from the plumes and irregular clouds which are also found associated with stars, but at a very early stage in their formation. The low temperature of the disc is indicated by the infrared radiation observed, but its shape is only seen in the visible light which it reflects. Despite this double signature, it is not known if the material is in the form of dust, or if it has condensed into planets. Much better resolution would be needed to detect individual planets. That would attack the problem of their existence. To attack the second type of problem, that of evolution, would need the recognition of several examples of planetary systems of different and known ages so as to sort out an evolutionary sequence. However, the minor planets and comets in our Solar System, left over from an earlier stage of evolution, can yield some clues, as described in Chapter 9. They are good objects for amateur observations. Many comets are discovered by amateurs, and photometric evidence bearing on the evolution of the asteroid family comes from amateur astronomers as well as from the big observatories.

There may be new types of star to be discovered. After all, a completely new type, pulsars, was discovered as recently as 1967 after hundreds of years of telescopic observation of stars. Certainly there is much to be discovered about the way stars interact with each other in binary pairs. The more they are observed, the more evident it becomes that simple ideas of such systems do not work. Sometimes the interaction produces novae that are bright in the visible region, so this is a fruitful field for the amateur, as will be described in Chapters 8 and 15. We will be concerned most of all with unsolved problems about the evolution of stars. This is governed by the energy generation deep inside the star, which we cannot (with one exception) observe. The picture of the core of a star therefore comes almost exclusively from computation. However there are so many discrepancies between the calculated and predicted behaviour of stars that stellar evolution is bound to be a growth area in astrophysics for many years to come.

Variable stars are an essential bridge between the calculated and observable characteristics of stars, so much of this book will be devoted to them. To obtain useful data on variable stars requires long series of observations, which is where the amateur astronomer wins. But a good observer must know what to look for and, in the case of a star, have a mental picture of what it looks like inside. Such pictures, or models, are usually hidden away in professional journals or, worse, in computer programs. We will do our best to make some of them more accessible. When an astronomer measures the variation of light or other radiation from the surface of a star, it is not the surface that is of interest. Most oscillations are generated in an intermediate layer, below the surface but outside the core. Thus, in studying the light curve, it is the properties of the intermediate layer that are being revealed. The modes of oscillation of layers of gas are thus important and are discussed in Chapter 6. Since the characteristics of oscillation depend on the rate of energy flow from the core and the evolved chemical composition of the material in the intermediate layer, variable stars test virtually all of stellar theory, directly or indirectly. It is this relation to unsolved problems that makes them so important, of course.

There are many other important unsolved problems of which we will say little because they are not directly accessible to amateur observations. For instance,

What is the unseen matter in galaxies?
How are spiral and elliptical galaxies and quasars related?
How many different types of molecules are formed in clouds of interstellar matter?
How are binary stars formed?
How do pulsars generate radio waves?
etc.

These problems are not within the realm of amateur observations—yet. It will be interesting to see what advances are made in their solution as more space telescopes are put into operation over the next decade. No astronomer, whether amateur or professional, can afford to neglect advances in other types of observation. They are all inter-related, and observations from space telescopes impinge on all topics in astronomy. Moreover, the various problems are interlinked. For instance, while planets are being formed the young star is developing and the two processes need to be understood together. The evolution of myriads of stars add together to shape the evolution of a galaxy. Eventually all the evolutionary models will merge and combine. But a lot of supernovae will have come and gone before that is achieved.

Note 1.1 Linear and logarithmic scales

Because astronomy has to deal with large numbers it is useful to have a way of squashing up the scales on a graph. The standard way is to use a logarithmic scale on one or both axes. On such a scale, as we will use it, each major division corresponds to a factor of ten. On a linear scale, on the other hand, each division corresponds to the same increment, i.e. added not multiplied. Thus logarithms, as we will use them, are simply powers of ten. For instance, the fact that $10^2 = 100$ can be turned round to say $\log(100) = 2$. Similarly, $\log(1000) = 3$ and so on. Since the square root of 10 is 3.124 we can also write $\log(3.124) = 0.5$. This can be extended to cover all the numbers. A straight line through the origin on a linear scale graph remains a straight line when plotted on a logarithmic graph. One of the nice features of logarithmic graphs is that any quantities related by a simple power with no additive constant (e.g. $y = x^n$) give a straight line of slope equal to the power n. On a linear scale graph they give different curves which are not so easy to identify as different slopes.

Note 1.2 Weight and mass

In the strict language of mechanics, weight is a force; the force on a body due to gravity. If a body has mass m, the force on it is mg where g is the acceleration caused by gravity, which is the same for all bodies. g on Earth has a value of about 9.81 metres per second per second, but it varies slightly from place to place on the Earth. It has totally different values on other planets, which is why mass is easier to use than weight. The units of mass are kilogrammes. The units of weight are newtons. A 10 kilogramme mass has a weight on Earth of 9.81 newtons. However, the term newton is never used in colloquial speech so an ambiguity is created. Spring balances measure force, but are marked in units of mass, kilogrammes, assuming the value of g is 9.81. This is good to within a few tenths of a per cent all over the surface of the Earth and so is perfectly satisfactory for weighing out vegetables. For an astronomer, however, it is clearly essential to distinguish between mass and weight. The term mass comes naturally to an astronomer, except when dropping a mirror on his toe, in which case he might use the word weight!

Chapter 2

Looking at the stars

2.1 THEIR TEMPERATURE AND LUMINOSITY

How can one make sense of the thousands of stars seen by the unaided eye, and the millions revealed by telescopes? Which are young, which are old? Are there different types of star, or is age the only difference? The first attempts at classification were by position. We can imagine early man, who had the best seeing conditions of any astronomer in the dark and then totally unpolluted skies of Africa, seeing the Milky Way with unexcelled clarity, as a major feature in the sky. Later, groupings into constellations were recognized and gradually standardized. But since these are based only on angular positions, without regard to distance (which is notoriously difficult to measure), the stars in a constellation are often not related to each other. The constellations are very convenient markers on the map of the sky, but have little physical significance. Tighter groupings of stars, however, often are physically related, and a telescope can give evidence of the nature of the relationship. In the Pleiades, photographs show wisps of gas linking the stars. In the Hyades careful measurements reveal a group of stars all moving together. In the case of the globular clusters of stars, many thousands of stars densely packed together in a spherical volume of the sky, we can scarcely doubt that the members share a common origin. (All the globular clusters look circular, so we can assume that they are spheres and not discs of stars.) A small telescope can resolve the stars in a globular cluster, but it was not until a powerful telescope was available, the 60 inch at Mount Wilson, that they were studied systematically. This was done by *Harlow Shapley* who used them as markers of the shape of the halo of our Galaxy. At about the same time *Henry Norris Russell* was looking for a classification of stars which could reveal something about their properties and origin. Clusters of stars provided samples which, having a common origin, probably had the same age—apart from second generation stars. They also sidestepped the problem of measuring distance since, even if the actual distance was not very surely known, one could at least assume that all the members of the cluster were at much the same distance. Russell in the United States and Hertzsprung in Denmark did parallel but independent work—this was just before the days of big international conferences for exchanging recent results—and their joint invention was the *Hertzsprung–Russell diagram*. Incidentally, the three men came to astronomy in totally different ways. Harlow Shapley went to Mount Wilson from being a crime reporter on newspapers in Kansas and Missouri. Hertzsprung was a chemist, and took up astronomy as a hobby. Russell followed a conventional academic career at Cambridge and Princeton leading to astronomical research—which makes him the odd man out of the three!

Hertzsprung and Russell classified stars according to two values which can be measured without too much difficulty, the luminosity and the spectral class. They were measuring the luminosity in visible wavelengths, using photographic plates, but were restricted to stars for which the distance was known so that the absolute luminosity could be calculated. Since the aim was to classify stars one must clearly use the luminosity of the star itself, not the apparent luminosity which it happens to have from the viewpoint of the Earth. If the distance of the star is known, one can calculate the tiny fraction of the total light output which is being picked up by the telescope mirror, and hence deduce the absolute luminosity from the apparent luminosity. That fraction is of course very tiny, but it is simply the mirror area divided by the area of a sphere of radius equal to the distance of the star. *Spectral class* is a way of describing the appearance of the spectrum of a star, judged by the number of lines and which elements produce them. The different classes are designated by a code letter in the series O, B, A, F, G, K, M, N which is, in fact, a sequence of decreasing temperature. Their original results are shown in figure 2.1. Despite the scatter, caused mainly by uncertainties in distance, a diagonal strip of stars stands out and has earned the name main sequence. Above it is a group of stars now called the red giants.

Rather than linger over the exact meaning of the parameters which Hertzsprung and Russell used, let us go straight to the more fundamental parameters which perhaps they would have used, had they been measurable. The visual range of wavelengths is a narrow and rather arbitrary slice out of the full range of wavelengths. A more fundamental quantity, which is easier to interpret, is the total luminosity, added up over all wavelengths. Something approximating this can be measured with the wide-range heat detector called a bolometer, but it needs to be used above the atmosphere to receive all the radiation. The bolometric luminosity then falls only a little short of the total luminosity. This is a total radiated power, so it can best be expressed in watts. The total luminosity of the Sun, for instance, is about 3.8×10^{26} watts. A red giant such as Betelgeuse in Orion emits over ten thousand times as much radiation—over 3×10^{30} W. Alternatively, we can use this solar luminosity as a unit.

Their second parameter was spectral class, but it was really being used as a measure of temperature. The relation between spectral class and temperature is shown in figure 2.2, and is complicated by the fact that the relation is

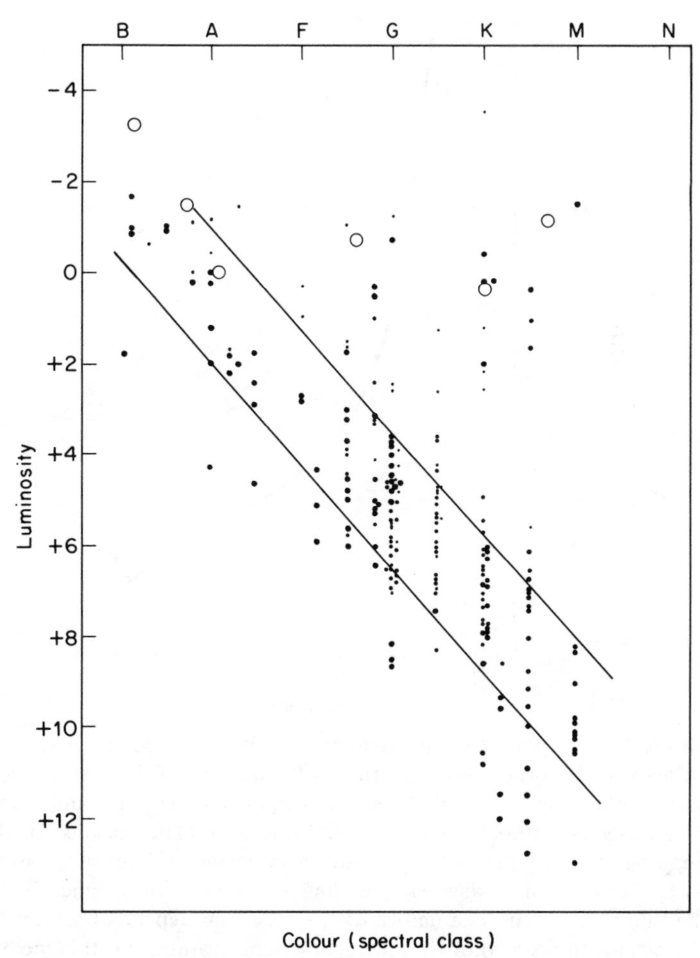

Figure 2.1 The first published Hertzsprung–Russell diagram which appeared in the journal '*Nature*' in 1914—a Dane and an American writing to a British journal. 'Spectral class' is a way of describing the appearance of stellar spectra according to the lines present. It is related closely (but not linearly) to temperature, and there are standardized procedures for assigning it. The problem for Hertzsprung and Russell was to choose stars with measured distances, so as to determine absolute luminosity, the vertical coordinate. Such a plot would be meaningless if apparent luminosity were used. Their distance estimates were in fact in error, but in a systematic way which does not mask the band of main sequence stars, with a scattering of red giants above them. Luminosity is plotted in 'magnitudes', a logarithmic scale explained in Chapter 10. (Reprinted by permission from *Nature* Vol. 93 p 252 © 1914 Macmillan Magazines Ltd.)

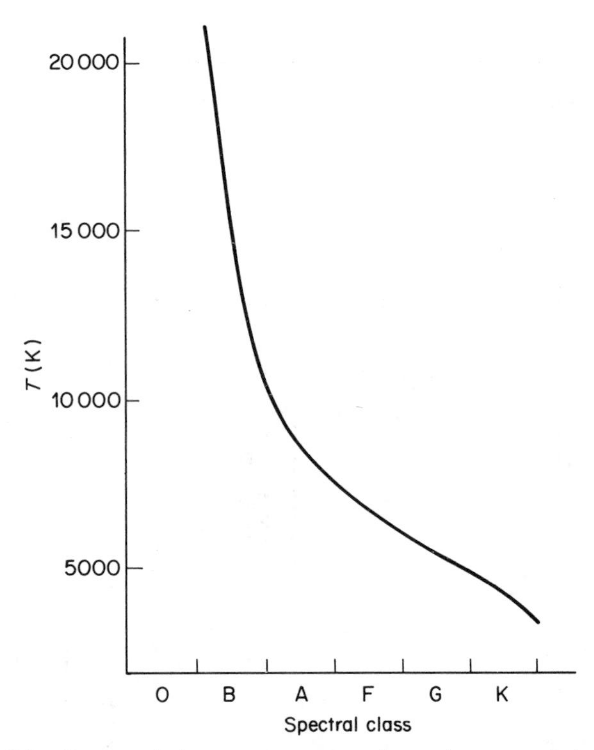

Figure 2.2 How to deduce a temperature from the spectral class. The appearance of a spectrum, in terms of the number of lines, their width and whether they are in emission or absorption, is classified and named by a series of letters O, B, A, F, G, K ,M, R or N (The reasons for this apparently random choice are no longer important.) Finer subdivisions within the class are indicated by adding a number—for instance G2 for the Sun's spectrum. The nature of the spectrum depends both on the chemical composition of the star and its temperature, but this method of classification emphasizes the temperature variation. The diagram shows the relation between spectral class and temperature.

different for different types of star. This does not matter too much if the aim of the Hertzsprung–Russell diagram is simply to classify stars, as was the case in their original work, since the different types are separated out by their different luminosities. In trying to understand the properties of stars, however, spectral class is clearly not a satisfactory parameter, and it is much better to use the fundamental physical quantity, temperature. If the complete spectrum of the star can be measured, the temperature can be deduced simply and reliably from the underlying shape of the spectrum, deliberately ignoring the spectral lines. Hertzsprung and Russell could not easily do this, because for most stars the spectrum extends well into the ultraviolet and infrared regions, to which the atmosphere is almost opaque. Even for the Sun, only a third of the radiation is in the visible region, and the fraction is even lower for most stars. Space telescopes can cover the whole range, and thereby produce much more complete information.

Figure 2.3 shows a modern HR diagram. There are, of course, many more measured stars than in figure 2.1 and a third group, below the main

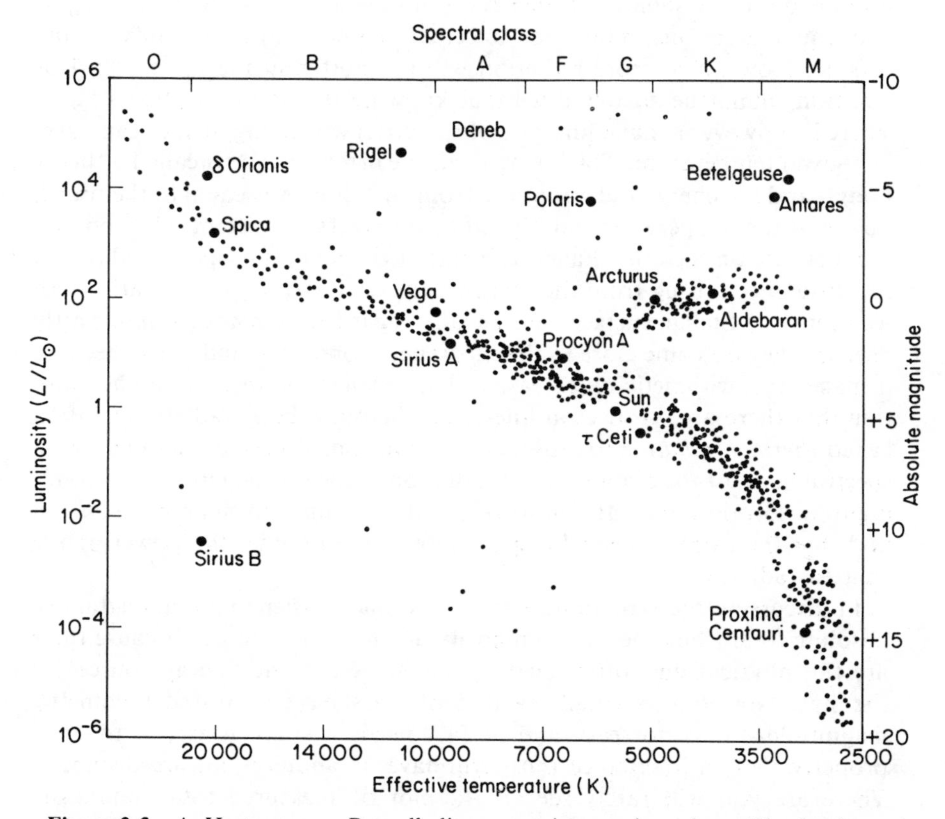

Figure 2.3 A Hertzsprung–Russell diagram using modern data. The horizontal coordinate is now temperature, but increasing to the left to retain the original pattern. The main sequence and the red giants are now very clearly separated, and a few white dwarfs appear, scattered below the main sequence. White dwarfs are all relatively dim but show a wide range of temperatures, up to values off the left-hand edge of the plot. Distances have been recalibrated upwards since 1914, moving the main sequence up slightly from its position in figure 2.1. The luminosity at a given temperature is proportional, for obvious reasons, to the surface area of the star, that is, to the square of the radius. Thus one can add to this plot lines joining points corresponding to a given radius. On a logarithmic plot these lines are straight, sloping downwards from left to right. From *Astronomy: the Cosmic Journey* 3rd edition, by William K Hartmann © 1985 Wadsworth, Inc. Reprinted by permission of the publisher.

sequence, is now apparent. These are the white dwarfs. We can now give a quantitative meaning to the names dwarf and giant, since there is a very simple relation between the total luminosity, temperature and radius.

Imagine two surfaces, both at the same temperature, but one twice the area of the other. Clearly the one with twice the area gives off twice as much radiation. The luminosity of two stars at the same temperature will be in proportion to the squares of their radii, if they are spherical. Now imagine two equal areas maintained at different temperatures; say 5000 K and 6000 K. How much more radiation will the hotter surface give off? This question cannot be answered without knowing the nature of the surfaces. There is, however a maximum rate of radiation which any surface can have, at a given temperature. This theoretical, idealized surface is called a 'black body', and the energy radiation rate from such bodies is equal to the fourth power of the temperature, multiplied by a universal constant. (The photon number rate on the other hand, as mentioned above, is proportional to T^3, the difference arising from the fact that the energy of a photon varies with wavelength.) Stellar surfaces are usually regarded as *black bodies*, partly because they do come close to the theoretical condition, and partly because it makes the arithmetic much easier. The idealization includes an assumption that there are no spectral lines, and the black body radiation is often called *thermal radiation* to distinguish it from emission of extra radiation in spectral lines, or the diminution of radiation at the wavelengths corresponding to absorption lines. If you work out the example given above, you will find that the extra thousand degrees more than doubles the power of the emitted radiation.

In discussing the evolution of stars it is much easier to use a diagram of absolute total ('bolometric') luminosity against temperature, because these are the physical quantities fixed by the structure and energy sources of the star. Now only a small fraction of stars have measured bolometric magnitudes—it is not easy and in fact needs a space telescope to do it properly. Only a fraction of those will have, in addition, known distances. Therefore you will rarely see an HR plot of measured total luminosity against temperature. It is much more normal to see visual magnitudes. The difference is not trivial. On the visual magnitudes plot, the main sequence dips down sharply on the right (i.e. low-temperature) side. This is because those stars give out mainly infrared. On a total luminosity plot the main sequence is practically straight. This difference will come up again in relation to the amplitude of red variable stars.

On a total luminosity plot (but not a visual luminosity plot) one can draw straight lines corresponding to given radii, which helps in interpretation. The reason that this is so easy on a total luminosity plot is that there is a simple relation between luminosity and radius for a given temperature. For two idealized spherical, black body stars with different radii and temperatures, the ratio of radiative powers is

$$L(1)/L(2) = R(1)^2 T(1)^4 / R(2)^2 T(2)^4.$$

Since we know the luminosity, radius and temperature of the Sun, it is easy to calculate the radius of any given star knowing its luminosity and temperature, and assuming it is a 'black body'. Clearly this means that we can draw a line on a total luminosity–temperature plot corresponding to a particular radius. All such lines have the same slope, which is a little less than the slope of the main sequence, corresponding to the fact that there is a modest increase in radius going up the sequence. The $10R_\odot$ line goes through the red giants such as Arcturus and the $100R_\odot$ line goes through the supergiants such as Polaris.

2.2 THEIR MASS, RADIUS AND TEMPERATURE

There is another parameter which increases up the main sequence, helping to increase the luminosity. There is no way of deducing what it is from the plot alone, though one might guess. It is the mass. Stars at the top of the main sequence are considerably more massive than those at the bottom.

The mass of the Sun can be accurately determined from the distance and period of the Earth, or any other planet. Measuring the mass of other stars would be equally easy if we could see planets in orbit around them. Since this is not yet possible, the best bet is to use binary stars in the same way. The relation needed is given in note 2.1. Even though many stars are binary few masses are well measured. The reason is not difficult to find. If the members are well enough separated to resolve, allowing the orbit to be plotted, the corresponding periods are uncomfortably long: hundreds or thousands of years. If the members are not resolved (which is much more common), the orbit cannot be plotted but the component of speed in the direction of the Earth can be measured by the Doppler shift of lines in the spectrum of either or both stars, provided the lines are clean and sharp. There are several ifs and buts already. The most awkward one is that only one component of the velocity can be measured. This does not, by itself, reveal the inclination of the orbit. As a result, only a lower limit to the mass can be deduced, unless the conditions are favourable in some way. For instance, if eclipses are observed, the Earth must be nearly in the plane of the orbit. Even if the masses of both stars can be obtained, the problems are not over. One of the main uses for the measurement is to compare the properties of the star (the position on the HR diagram for instance) with those predicted by calculation. The calculations, however, are for isolated stars, unaffected by a close neighbour.

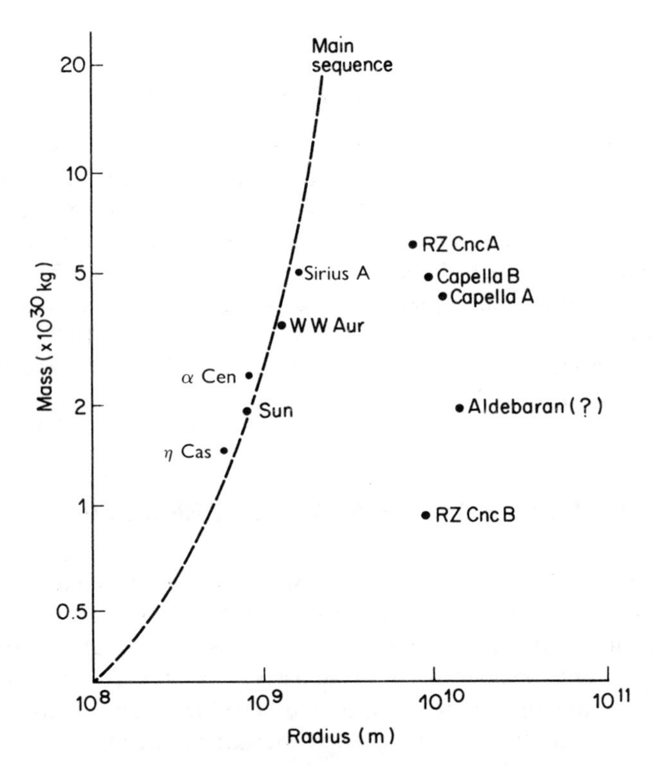

Figure 2.4 How much bigger are giant stars? To answer this we need to compare stars of the same mass, so we need a plot like figure 1.2, magnifying the small section covering stellar masses. The broken curve is the main sequence again. One point on that curve is well known—the Sun, whose mass is readily calculated from the way it swings us round in an orbit. A few other (binary) main sequence stars have well measured masses and radii, Sirius being a classic example, so we can regard this plot as a result of observations, connected together by calculation. Turning to red giants, their radii can be deduced from luminosity and temperature, and they are certainly bigger than any main sequence stars. Some of the very few that have measured masses are plotted on the diagram and show radii about 20–30 times greater than a main sequence star of similar mass. Supergiants are, as the name implies, bigger still, up to 1000 times the radius of the Sun.

On figure 2.4 some hard-won data on mass and radius are displayed. We have a big star and a small star with similar masses. So which way does the evolution go, for stars of this mass? Does a large low-density star condense, through the force of gravity, or does a small dense star heat up and puff out to enormous size? Remember that there is no direct way to measure the age of the stars. The answer had to be found through calculations, or predictions, of the way stars with a given composition would be expected to evolve, given the laws of physics and laboratory data on nuclear reactions. These reactions which drive the evolution of the star occur inside the star and the HR diagram only refers to the outside of the star. We need some new diagrams to link the two. It will need the next two chapters to set them up.

Note 2.1 The relation between period and radius

For a body in circular orbit, the inward acceleration is the square of the speed, divided by the radius of the orbit, v^2/r. This can be written in terms of the period P, using the obvious relation $P = 2\pi r/v$. The result is that the acceleration is $4\pi^2 r/P^2$. This must be equal to the gravitational force, divided by the mass of the body. Using Newton's law of gravitation, this is, using the notation shown in figure 2.5,

$$(GMm/s^2)/m.$$

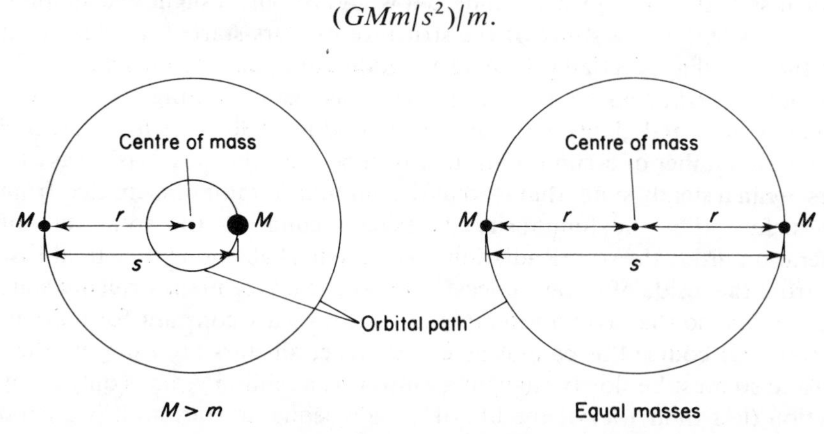

Figure 2.5 The relation between orbital size and period. Kepler's law, stated as 'the square of the period is proportional to the cube of the radius' leaves open the question of the point from which the radius should be measured. The answer is from the centre of mass, leading to the relation given in the text.

If m is very small compared with M, as in the case of a planet orbiting the Sun, then the centre of the orbit (which is the centre of mass, and therefore given by $mr = M(s - r)$) is practically in the centre of the more massive body, and so $r = s$. The equation then simplifies to

$$P = 2\pi r^{1.5}/(GM)^{0.5}$$

which is Kepler's law: that the square of the period is proportional to the cube of the radius. However, in the case of binary stars, the two bodies are of comparable mass, so this approximation is not valid. As figure 2.5 shows, r is appreciably less than s in these cases; in fact $r = 0.5s$ if the two masses are equal. This is because both bodies rotate round the centre of mass, so that in this case $r = s - r$, so $r = 0.5s$.

Chapter 3

The structure of stars—a first look at models

Although stars have been observed with great skill and steadily increasing sensitivity for thousands of years, it is only in this century that any real idea of their structure has emerged. Indeed it would be only a slight oversimplification to say that the study of the structure of stars started in 1926. That was the year that A S (later Sir Arthur) Eddington published his book '*The Internal Constitution of the Stars*'. He was really writing about main sequence stars; red giants were not understood until 30 years later. He had to make a number of assumptions, of which perhaps the most basic was that stars are in a steady state, that is, that the amount of radiation emitted from the surface (i.e. the luminosity) is exactly equal to the total rate of generation inside the star. Radiation takes a remarkably long time to diffuse out from the inside of a star, at least a million years for main sequence stars like the Sun, so that assumption implies that stars are constant for millions of years. Of course this cannot be exactly true; all stars are using up their fuel and so must be slowly changing. However, a million years is only a tiny fraction (less than 1%) of the life of a main sequence star, so it is a good assumption for them. It would not necessarily be a good assumption for variable stars such as Mira (*o* Ceti) whose luminosity changes by a factor of a hundred in less than a year. Eddington could not possibly have tackled such a case for two quite different reasons. First, the computation needed is far too complex for the mechanical computers then available. Secondly, Eddington did not know, in 1926, what the source of the energy was in stars. He could not therefore know, *a priori*, whether the variations were in the source of energy or in the insulation provided by the mantle of the star. He rightly expected, however, that the energy source must be in the hottest part of the star, near the centre, and that the long timelag in the diffusion of radiation would even out any variations except those near the surface. Any thermal insulator will damp out variations in the heat source. For instance, a few feet of earth damps out the variation between summer and winter temperatures, which of course is the reason why water pipes are buried and why wine is stored in the cellar. For the same reason, variations in luminosity cannot be caused by variations in the energy source, except in a very few, very special cases. Thus even for a star like Mira, the steady-state assumption is true for most of the bulk of the star, though certainly not for the surface layers.

The steady state imposes some very severe constraints on the structure of the star; or, to put it another way, as the star settles naturally into its stable, steady-state condition, it must take up a particular, specific structure. At every point there must be a *balance of forces* arising from thermal pressure and gravity, and any other causes. Secondly, the temperature gradient must everywhere be just such as to make heat flow outwards at precisely the rate which we observe at the surface. These conditions lead to no less than seven separate mathematical equations, and they are just sufficient, in principle, to work out the density and temperature throughout the star. The equations, should you wish to see them, are in note 3.1. These equations cannot be solved exactly and analytically. They can be tackled numerically but the difficulties in doing so are immense. First there is the composition problem. The *chemical composition* throughout the star must be known. We may choose some particular cases to work out—a 3 : 1 mixture of hydrogen to helium throughout might be thought suitable for instance—but how are we to know if the results are appropriate to a given star, whose internal composition cannot be measured? Secondly, there is the data problem. Suppose the composition is, magically, known. What is the density and heat conductivity of such a mixture? No chemical data book will be of much help. It may give experimental values up to a few thousand degrees, but the bulk of all stars is at far higher temperatures. The properties of matter at such temperatures must be found by calculation from basic physical principles. There is no direct way of knowing if the answers are right. The indirect way is to go ahead and do some calculations for well known stars, including the Sun, constraining the equations to give the measured surface temperature, and seeing if the calculated internal conditions are plausible and tally with any available evidence.

The internal temperature calculated from the degree of insulation is in the region $10–20 \times 10^6$ degrees and this is certainly plausible. If it were ten times higher it would not be plausible because *nuclear reactions* would then proceed so fast that stellar lifetimes would be far too short. To improve on this rather crude type of check on the model one needs to measure the rate of production of new particles by the nuclear reactions. These include, of course, new nuclei, especially the heavier nuclei produced by fusion. It would be very helpful if they could be detected when they are formed. Unfortunately, they stay put for a long time in or near the core, except in very rare cases where convection right through the star brings them up to the surface. Normally they do not appear until long after they were produced, and by then the star is in a different stage of evolution. Photons are also produced in many nuclear reactions, carrying away the surplus energy released. The energies involved are high—millions of electron volts rather than the few electron volts corresponding to light. Photons of such high energy have even shorter wavelengths than x rays and are called gamma rays. Gamma rays come from nuclear rearrangement, whereas x rays come from the rearrangement of electrons round the nucleus. Gamma rays travel at the same speed as light and interact with matter and their energy becomes degraded into lower and lower energy photons. None of the original gamma rays reach the surface, and all indications of their original nature—their energy distribution for instance—are totally lost. It looks hopeless to detect nuclear reactions in the Sun, still less any more distant star.

There the matter rested until 1968. In that year a way was devised of detecting another type of particle produced in some nuclear reactions. This is the *neutrino*, a particle without mass or charge and therefore able to traverse matter with very little probability of interaction. Most atoms simply do not notice it passing. Neutrinos can travel directly from the core of the Sun to the Earth. The problem is to find atoms which do notice them passing, so as to measure their rate. That is the experimental problem which was solved in 1968. The atom which can be used is chlorine, but we will not go into the formidable technical problems that had to be solved, for the very reason that they put this particular experiment for ever outside the scope of amateur work. The point for us is that the measured neutrino rate is a factor of two less than that given by the calculated rate for the nuclear reactions. That is a measure of the uncertainty in the knowledge of how stars work. Reducing that uncertainty is a job which, on the observational side at least, is shared by all astronomers, professional or amateur.

Beyond the main sequence there are a few stars that do allow direct observation of nuclear reactions, because they occur on the surface. These are the novae and 'bursters' which are very bright in x rays but can also be studied in visible light. We will return to them in Chapter 8. As mentioned above, there is also a handful of stars in which the products of nuclear reactions seem to be brought to the surface by convection very quickly. The parts of the star which are most amenable to study, however, are the layers just beneath the surface whose instability at certain stages of evolution lead to a wide range of rapid variations in luminosity, of which the best known are perhaps those in the Cepheid stars. They cannot be caused by variations in the nuclear energy source, which are very effectively damped out, and so cannot tell us directly about the nuclear reactions. Variable stars do however tell us about the insulating power of different layers in the star. In fact it is precisely because the insulating power varies that the instability arises. As the star evolves, changes in conditions in the outer layers can induce the variability. It is changes in the nuclear reactions which cause the main evolutionary stages—we could label different areas of the HR plot with different nuclear reactions. However, it is only the outer layers of the star which can be studied in each area, notably by the type of variability which arises. We might imagine a neutrino telescope which could be pointed at a particular star to study its core nuclear reactions. There would be a new plot, not surface temperature and surface luminosity, but core temperature and neutrino luminosity. Which two people will give their names to that plot?

Note 3.1 Equilibrium conditions

A description of the equilibrium conditions inside a star can be explained in seven equations, which split into two groups of three plus a single mass equation. The first group is concerned with the balance of forces across a shell such as that shown in figure 3.1. The following material is adapted from *Matter in the Universe Block 5, The Formation and Evolution of the Stars* copyright the Open University Press.

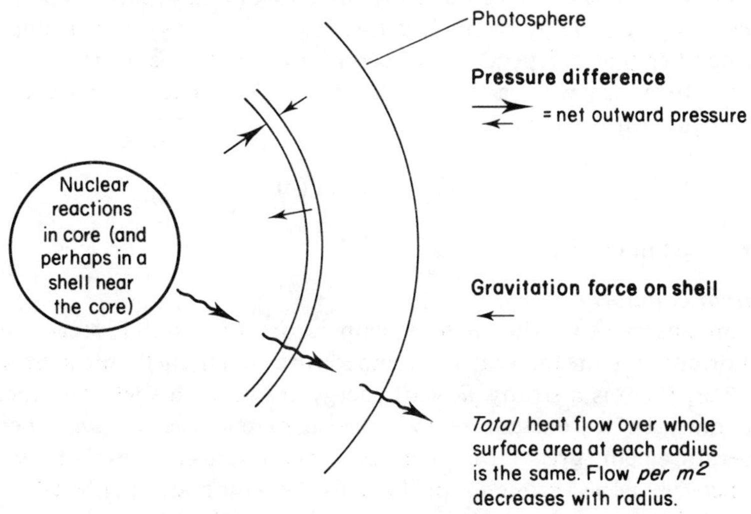

Figure 3.1 An idealized view of stellar structure. In steady equilibrium conditions, the net effect of the various forces must be exactly zero at each and every position in the star. This leads directly to seven powerful mathematical equations, which form the basis of stellar evolution equations. Even where there is convection or pulsation, meaning that conditions are definitely not steady, the same equations may give a good guide to average conditions.

Mechanical equilibrium

The gravitational force on any shell tends to pull it towards the centre of the star. But pressure increases with depth (also the result of gravity), and therefore the pressure below the shell is greater than the pressure above it. This pressure difference opposes the inward gravitational force on the shell, and in mechanical equilibrium the net force on the shell is zero. Therefore, this equation states that, for any shell, the pressure difference across it equals the inward gravitational force per unit area of shell.

$$\frac{dP}{dr} = -\frac{\rho MG}{r^2}$$

where P = pressure and $M(r)$ = mass within r.

Energy equilibrium

In main sequence stars, energy generation is confined to a core. Outside this core, the equation of energy equilibrium states that, for the temperature to remain steady, the rate of energy flow into a shell must equal the rate of energy flow out of the shell. Inside the core, the rate at which energy flows out of a given shell exceeds the rate at which it flows in, and this equation then states that this excess exactly equals the rate at which energy is being generated in the shell. This therefore introduces a new parameter, the rate of generation of energy by nuclear fusion per unit mass, ε. It is important to remember that ε depends on the composition, temperature and density, which in turn vary with radius, so we write $\varepsilon(r)$. Outside the core, $\varepsilon(r) = 0$, so the equation is simpler there.

$$\frac{dL}{dr} = 4\pi r^2 \rho \varepsilon(r)$$

where L = luminosity.

Thermal equilibrium

We can observe that the *surface* temperature of a star is steady, and at equilibrium this means that the temperature of all shells must be steady. However, there is a steady *flow* of energy across each shell, and therefore there must be a constant temperature difference across any shell, the temperature being greater on the inside. This equation states that the rate at which energy flows through a shell equals the temperature difference across the shell times an appropriate heat transport coefficient.

$$\frac{dT}{dr} = \frac{-3\varkappa \rho L}{16\pi acr^2 T^3}$$

where T = temperature, $\varkappa$ = opacity, and a and c are radiation constants.

These three equations must be satisfied at all radii in the star. They thus represent powerful constraints on the structure of a star. However, they cannot be solved by themselves because there are more unknown quantities than equations, so we need more equations.

The next three equations are concerned with the relations between different quantities at a point, rather than with striking a balance across each shell. However, under the assumption of spherical symmetry, these equations do not vary from point to point in any given shell, but only from one shell to another: in effect, they vary with r.

Equation of state

In a main sequence star, this is $p(r) = K\rho(r)T(r)/\bar{m}(r)$. The (r) is a continual reminder that we have a separate equation for each shell.

Energy generation rate

This equation relates the energy generation rate by nuclear fusion $\varepsilon(r)$ to other parameters:

$\varepsilon(r)$ rises very rapidly with temperature, and it also increases with density, and obviously depends on the composition.

$$\varepsilon = \varepsilon(\rho, T, \mu)$$

where μ = mean atomic weight, an indicator of chemical composition.

Rate of energy transport

This equation relates the rate of energy transport through a material to various properties of the material. For instance, the rate at which energy is transported by conduction through a material is proportional to the conductivity of the material. The transport by radiation depends on the opacity of the material, and so on.

$$\frac{dI}{dr} = \rho K(B - I)$$

where I is the luminosity in a particular range of wavelengths at a particular radius, and K and B are constants.

We have had three equations about the conditions of balance across each shell and three about the relations between quantities at a point. There is one further equation.

Cumulative mass

This equation states that the mass $M(r)$ equals the sum of the mass of all shells within r. The mass of a shell is its volume times the density $\rho(r)$. If R is the radius of the star, then $M(R)$ is the total mass of the star.

For each shell we thus have seven equations. With these equations we can, in principle, calculate the conditions throughout a star, using observational data whenever possible to constrain the outcome of the calculations. These data can include stellar radius, mass, luminosity, surface temperature and general constraints on composition.

Chapter 4

Evolving stars

4.1 STARS LIKE THE SUN

Measurements of the absolute luminosity and temperature of the Sun place it in the middle of the main sequence, first recognized as a band on the Hertzsprung–Russell diagram (figure 2.1). As we now know, the position of the band is governed by the nature of the nuclear reactions producing the power of the star. On the main sequence they are reactions in which hydrogen nuclei are fused together to produce helium nuclei. At the present stage of the evolution of the Universe, all newly formed stars have enough hydrogen to work in this way. When the Universe is many times older than at present this will presumably no longer be the case because most of the hydrogen will have been used up. There will no longer be a main sequence of young stars on the HR plot, at least not in its present position. The higher the luminosity of a star, the higher up the main sequence band it appears. But high luminosity is a consequence of high mass, so that luminosity and mass increase together up the main sequence. Sirius, Spica, Vega and Fomalhaut are examples of high-mass stars towards the top of the main sequence. The Sun is somewhat below the middle and Proxima Centauri, a dim red star which we would not see if it were not so close, is down at the bottom right.

While it is using hydrogen as a fuel, a star stays in much the same position on the plot. It does not move along the line as it evolves, but an early misunderstanding on this point has led to the confusing notation, still in common use, of 'early' and 'late' stars. These terms are traditionally used to refer to the temperature of main sequence stars, but in fact one cannot judge the age of a star from its temperature alone. A given star looks almost exactly the same all the time it is on the main sequence. No ageing is apparent. Even in the case of the Sun itself there is no known way of measuring the time it has spent on the main sequence, though lower limits can be put on that time by carbon dating of rocks and fossils, and from radioactive analysis of meteor material in the Solar System. The fossil evidence is particularly interesting because it indicates that several forms of plant and insect life have stayed much the same for at least 500 000 000 years, and if we accept the admittedly controversial identification of fossil bacteria, that time could be pushed right back to 2000 000 000 years ago. This means that the temperature of the Earth must have stayed the same to within a few degrees for that long. It is fair to conclude that the Sun has not evolved along the main sequence.

Assuming that other stars do not evolve along the main sequence either, it is possible to make sense of the Hertzsprung–Russell diagram for *clusters* of stars, which are presumably of the same age and initial composition. Whereas isolated globular clusters way out in the halo of the Galaxy must be old, open clusters in the disc of the Galaxy may be young, since star formation and recycling of matter is a continuous process here. Some clusters are clearly young because they still contain gas and dust. The Pleiades is an obvious example. The HR diagram for such young clusters shows points scattered along the main sequence. With the evidence from the

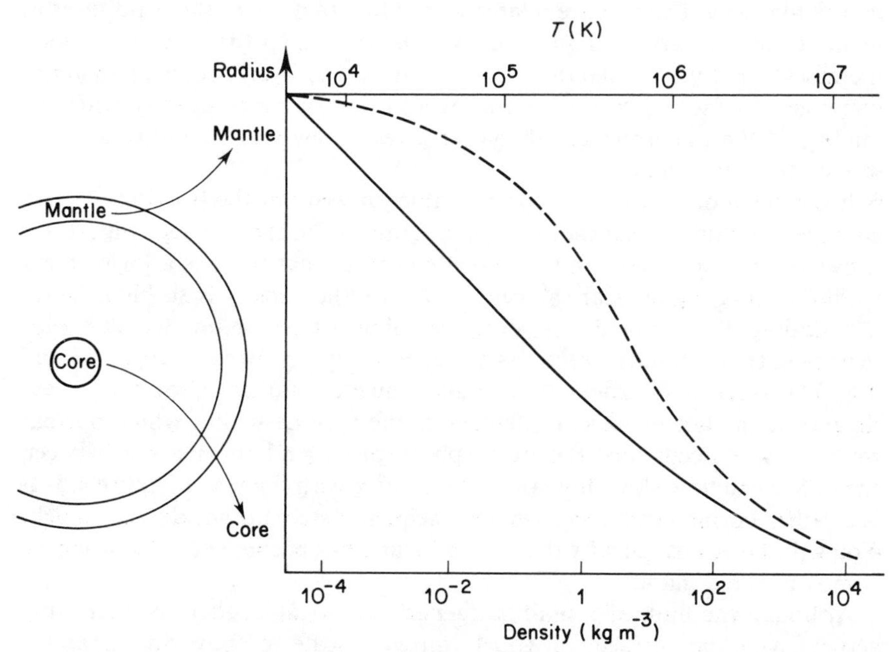

Figure 4.1 Radiation cools as it diffuses out. This is a calculated curve using opacities which are themselves computed from theory. The temperature (broken curve) falls steadily ('monotonically') from a stellar core to atmosphere as it must do where energy flow depends purely on temperature gradient. (There are no temperature inversions such as occur in the solar corona, and occasionally above some cities.)

Is it true that the Sun is denser than solid lead? The core, even though it is gaseous, is denser than lead (11 700 kg m^{-3}). The outer regions, however, are much more tenuous, giving an overall density of about 1400 kilogrammes per cubic metre, about 1.4 times that of water (1000 kg m^{-3}). (Try working it out—2×10^{30} kg spread through a sphere of radius 7×10^8m.)

Sun in mind, we assume that the position along the sequence depends on mass, not on age. Other clusters, showing stars away from the sequence, are interpreted as older clusters in which some stars have evolved away from the main sequence. The more massive stars, which evolve faster, move away before the less massive stars, lower down the sequence. There seems little doubt that this general picture is correct. Detailed modelling of particular clusters is, however, still a difficult and debatable matter.

If we accept the picture of a main sequence star as a quasi-stable and fairly straightforward structure we can introduce some simple diagrams which we will put to more testing use for the case of red giants. The way temperature varies with radius is shown in the top half of figure 4.1. The radius is plotted vertically, which is not traditional but you will see why it is done this way shortly. The lower part of figure 4.1 shows density falling off in a similar way. The two are related by the fact that under the conditions in main sequence stars their product is proportional to the pressure, which increases rapidly with depth into the star as the overburden of material increases. In fact, thinking of the overburden or more exactly what we might call the underburden of matter gives a new way of plotting radial distributions in stars.

It is often more convenient to plot values in terms of the fraction of mass contained within a given radius than in terms of the radius itself. Figure 4.2 shows the relation between the two. Remember that there is a lot of mass within a small radius near the centre because the density is so high, hence the sloping lines connecting corresponding values. Now we can plot density in terms of mass fractions as well as radius, shown in figures 4.3 and 4.4. The three main zones of a main sequence star are shown on these diagrams: the hot dense core defined as the volume within which nuclear reactions are occurring, the outer photosphere and the mantle between them. We can now show how these zones vary with time, as in figure 4.5. It is a rather boring graph because main sequence stars do not do very much. You will, however, see why this type of graph has become popular when we come on to red giants.

Although the bulk of a main sequence star is stable, there is interesting activity near the surface, on too localized a scale to show on the above diagrams. Sunspots are a modest example. In recent years there have been successful attempts to look on other stars for spots and flares similar to those on the Sun. Spots and flares demand very different techniques. *Sunspots* only change the luminosity of the Sun by about one part in a thousand, so that is not the way to look for them on stars. They also strengthen the calcium lines but recording their changes needs professional spectroscopy.

Flare stars are a different matter. They are dim red stars with emission lines, so a flare of say 10^{24} W might even double the luminosity, whereas the very same flare on the Sun which emits 3.8×10^{26} W would only cause a few

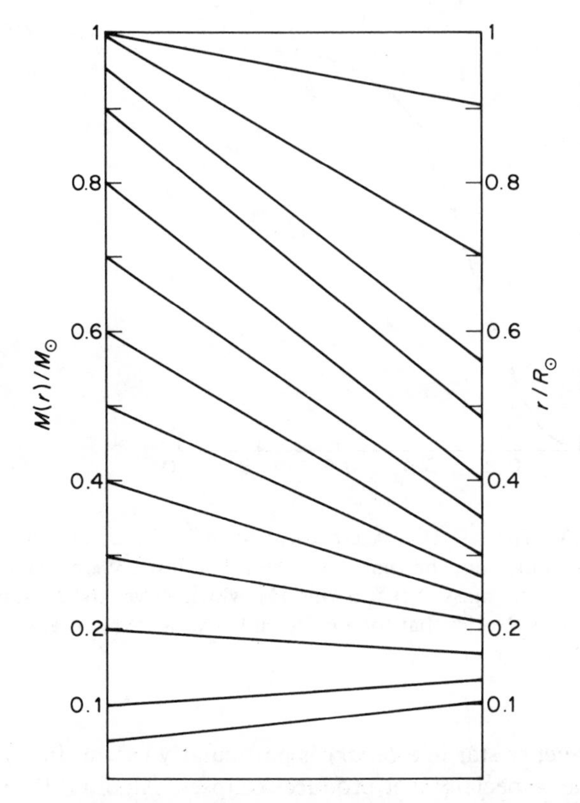

Figure 4.2 A 'ladder' diagram showing the relationship between fractional radius (r/R) and fractional mass (mass within a radius r/total mass within the outer radius R) for the Sun. Different stars will have different mass distributions and therefore different slopes to the 'rungs', but a common feature is the concentration of mass at the centre. Notice for instance how little mass is in the outer 20% of radius, and how much mass is within the central 10% of radius. The left-hand axis, the radial mass fraction, is often used in displaying stellar evolution calculations (see figure 4.10). The data are from Stromgren B 1965 *Stellar Structure* ed L H Aller and D B McLaughlin (Chicago University Press).

per cent change in luminosity, which is obviously more difficult to detect. Unfortunately flares only last a few minutes, and several days pass between outbursts, so it needs a lot of patience to see them. There are some exceptionally short flares, of a few seconds only, cause unknown. RS CVn stars also show flare activity but the cause is probably quite different because they are binaries and the usual cause of outbursts in binaries is matter transfer (or disc instability); see the section on cataclysmic variables.

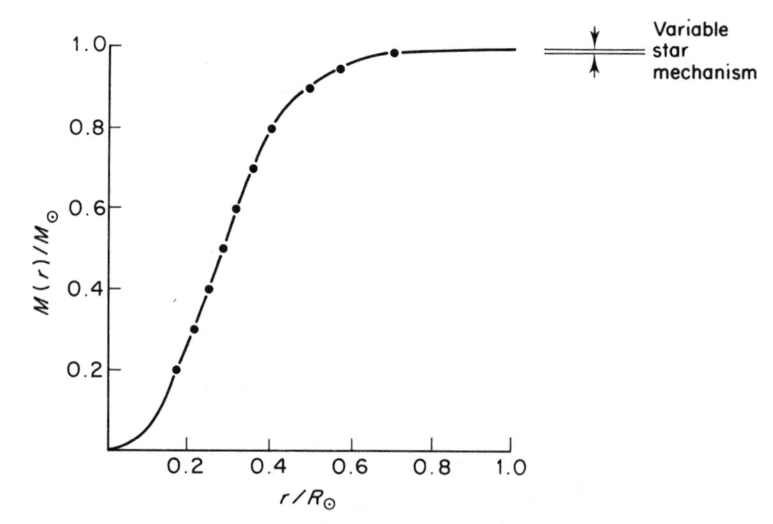

Figure 4.3 The last 1%. A graphical form of the data in figure 4.2. Travelling out from the centre of a star, the last 25% of radius only adds 1% to the mass. Yet it is this 1% which drives the pulsations in variable stars. Notice that for a uniform body, this type of graph would be a cubic curve.

A main sequence star in a *binary* is particularly useful since its orbit may be measurable, especially if it produces eclipses. Algol (β Per) is foremost among such binaries. Arab astronomers noted its variability and Goodricke interpreted it as an eclipsing binary in 1782. This was confirmed 107 years later still by spectroscopic measurements which revealed the motion of the two stars. Nowadays, one more century further on, one finds 485 spectroscopic binaries brighter than magnitude 6.3 listed in Sky Catalogue 2000.0. They are fruitful objects of study and we will return to them in Chapters 8 and 15. With the orbital speed known from the spectroscope, radii can be determined from the duration of the eclipses. The ratio of the masses, and perhaps the individual masses, can be determined. They therefore can provide valuable data on the evolution of stars, particularly if both are on the main sequence. At later stages, the evolution is likely to be significantly modified by matter transfer between the two. In the best cases, this transfer can actually be detected by its reflected light producing unusual features on the light curve between eclipses. Data on eclipsing binaries is always useful.

(You may wish at this point to look at note 4.1 which gives, very briefly and in tabular form, the way theories link observations to basic properties of stars.)

The end of the main sequence lifetime comes when a star can no longer

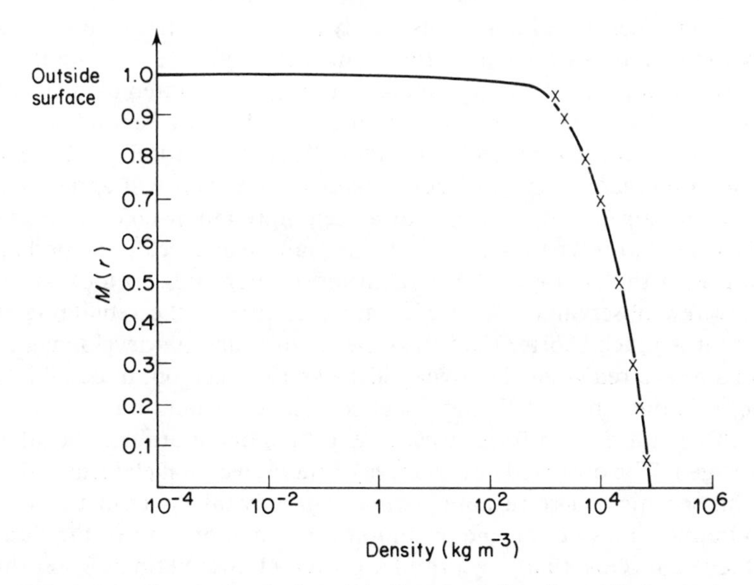

Figure 4.4 Density in the last 1%. Density falls off rapidly in the last 1% of mass. This diagram shows that it is valid to think of the outer part of the star as distinct from the inner part, and this is a useful picture for variability in stars which is confined to the outer part. Remember also however that the fall off would not be so sharp if plotted as a function of radius.

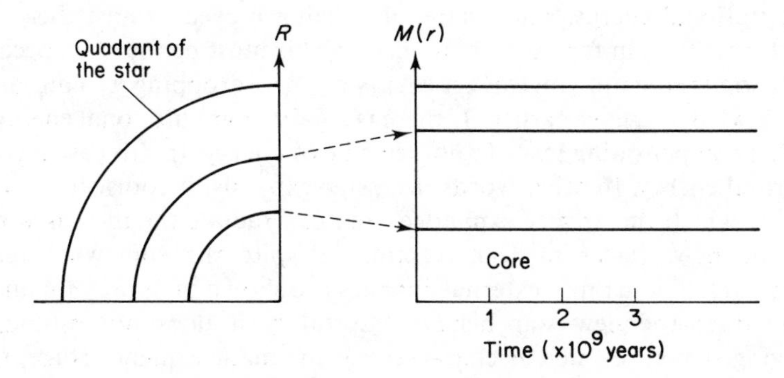

Figure 4.5 The uneventful life of a main sequence star. There is thought to be little change in structure during most of the main sequence lifetime of a star, as shown by the horizontal lines in this diagram. In fact there would be little point in drawing it if there were not change in the offing...

use protons effectively as a basis for fusion reactions in the core. It changes to a different fuel, *helium*, and this totally changes the structure. This is one indication of the delicate equilibrium which is maintained during the main sequence. As a fanciful analogy, imagine that if you wanted to change from gas to electrical heating in your house you had to rebuild to a new design—which turned out to be as big as Buckingham Palace. Unless you happen to be reading this in Buckingham Palace, you will agree that this would be an absurd consequence of a seemingly minor change. However, this is what happens to stars after the main sequence, and, perhaps in keeping with the analogy, makes them more interesting to look at. There are very few observable stars which are currently at the rebuilding stage, because it is much shorter than the other main stages in development. The rebuilt star is a red giant. Following ideas first worked out in detail by Fred Hoyle, it is thought that the new fuel consists of helium nuclei, which are fused three at a time to form *carbon*, and then one more may be added to give oxygen. The observed abundances of these two vital elements tally well with this scheme. These fusion reactions can only take over in a core which is a hundred times hotter and a million times denser than in the Sun, but they then run so fast that they produce power a hundred times faster than in the Sun. Only the most massive stars produce energy at this rate while they are still on the main sequence.

The basic restructuring of the star is violent and achieved in only a few million years after a few thousand million years on the main sequence for solar type stars. The core contracts and a new site of nuclear reactions is set up in a surrounding shell of 'unburnt' material, now hot enough for reactions to start. The mantle is forced outwards which must mean a change of gravitational energy, since material is being moved against the gravitational force. It is, in fact, a gain of energy. This must be the case, because if you lift something up, you can get energy back by dropping it. You can only get back what you gave earlier! If there is a gain of gravitational energy then there is a compensating loss of another type of energy. In this case it is a loss of thermal energy. In other words, as a star expands, it cools. It settles at a radius at which the greatly expanded area can radiate the increased power from the new, faster nuclear reactions despite the somewhat reduced temperature. These rapid external changes are shown in figures 4.6 and 4.7.

How does the new star behave? Certainly it does not establish an unvarying structure, such as characterized the main sequence stage, shown in the parallel lines of figure 4.5. *Red giants* undergo a progression of relatively minor internal changes. To push the Buckingham Palace analogy too far, after the conversion you never seem to get rid of the interior decorators. Some of the changes are apparent from the outside, and these are shown on the Hertzsprung–Russell diagram in figure 4.8. The timescale is indicated by the side of the trajectories in a form commonly used in papers and books. However, Hertzsprung and Russell did not intend their

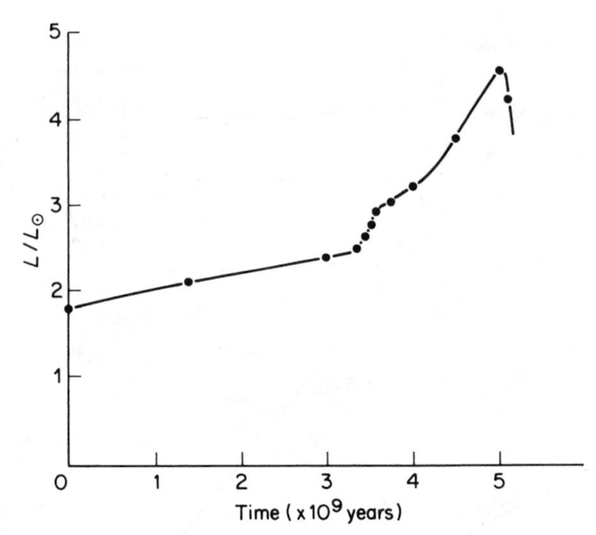

Figure 4.6 ...because the star has to react to the worsening hydrogen shortage in the core. If calculations are right, the adjustments cause an unsteady rise in luminosity, shown here, and...

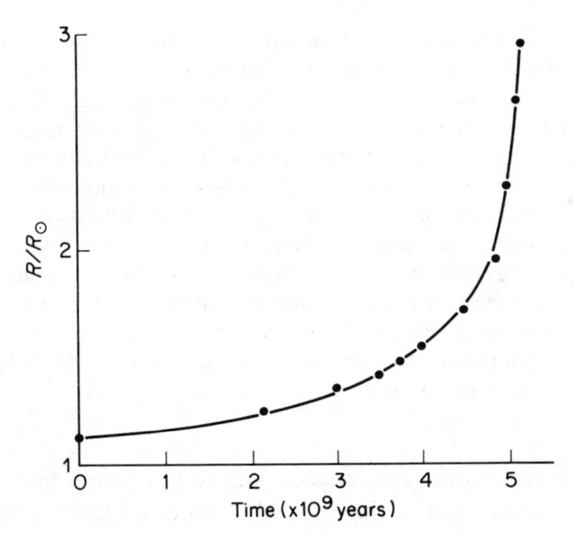

Figure 4.7 ... a radius which increases, slowly at first, then much faster. Compare this with figure 4.5. There is no contradiction; the star initially expands all its zones in proportion, so the fractional radii depicted in figure 4.5 do not betray the fact that the overall radius is slowly expanding.

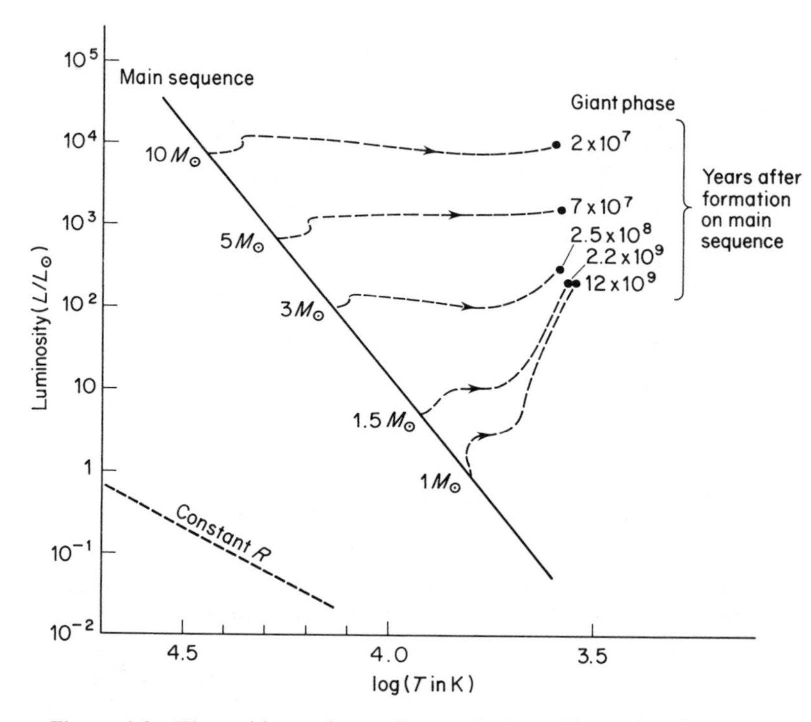

Figure 4.8 The evidence for stellar evolution. We obviously cannot watch a given star for 10^9 years and so draw figures 4.5–4.7 from direct observation. They come from pure calculation in fact. We can, however, look at different stars at different positions on the Hertzsprung–Russell diagram and interpret the positions as successive stages in evolution. The first attempts to do this were completely wrong because they assumed that stars cooled off and so drifted down and along the main sequence. Now, with better calculations, it is certain that they evolve off the main sequence, as indicated here. However there are too few measurements of mass and, just as important, of mass loss, to link a particular star with a particular trajectory. Moreover, there is no way of measuring the age of a star, so the values given on this diagram come purely from calculation.

diagram to show evolutionary timescales, and so you might find figures 4.6 and 4.7 easier to interpret. For the internal changes we have to rely heavily on calculations. Figure 4.10 sketches the way new zones appear and disappear during a red giant's life and it is an extension of figure 4.5. Remember that the plot gives no indication of the changes of the outer radius, since the diagram is in terms of mass. A change of total mass would not, in itself, show up either, though in practice it would cause a change of structure as well. Changes of total mass are quite common in stars. There is

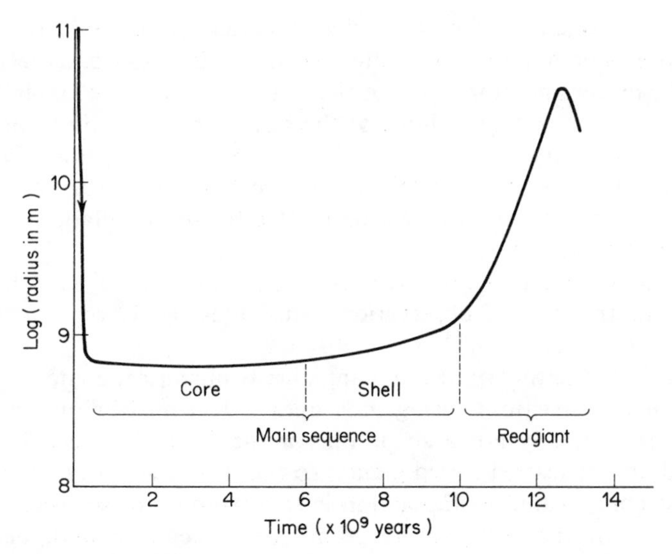

Figure 4.9 This diagram extends figure 4.7 to longer time scales. However, the calculations become progressively more uncertain since they involve progressively longer term predictions. They will improve as data improves.

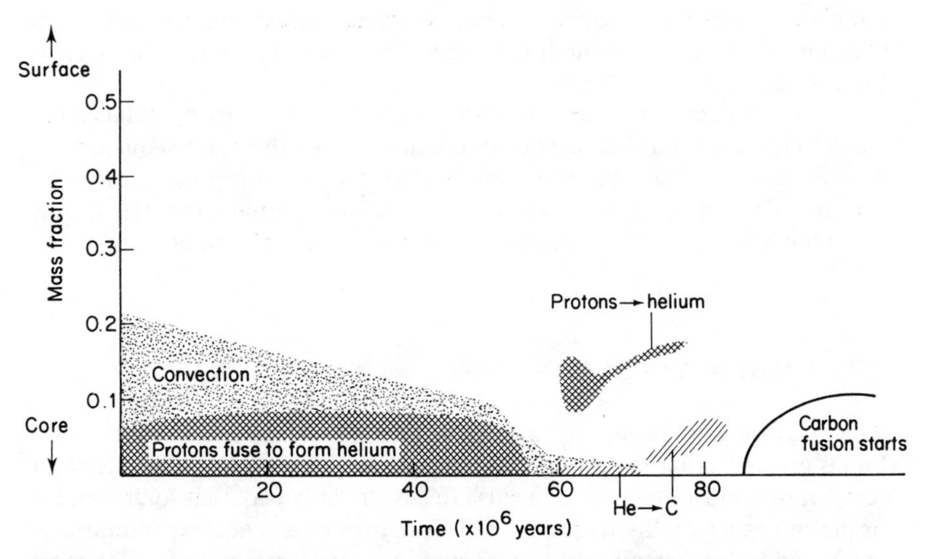

Figure 4.10 Evolution in a nutshell. This is where using the mass fraction variable pays off. It makes it easy to show the quite complicated arrangement of shells that builds up. The thickness of the shell on this plot is proportional to its mass and hence, in broad terms, to its importance in the star. This diagram is calculated for a star of 5 solar masses. What about different masses? Look at figure 4.11.

considerable exchange of mass in close binaries and even an isolated star loses appreciable amounts of matter blown off into space, as in the solar wind. The bigger and more tenuous the star, the greater the loss in this way.

Much of the now huge volume of the star is churned up by convection, which can easily take over where the pressure gradient is low. *Convection* carries both heat and new nuclei from the core up to the surface. We shall see later that the precise stratification needed for self-oscillation in variable stars cannot be set up where there is convection.

The study of red giants is currently at a fascinating stage, with a fertile interplay of theory and observation. This interplay is very briefly summarized in note 4.1.

The basis of the theory of red giant stars is numerical computation of a web of nuclear reactions, using data obtained in nuclear physics laboratories. Unfortunately not a single one of these reactions can be directly identified and measured in red giants, so the theory is perhaps the longest and most complicated extrapolation ever attempted in science. Even the endproducts of the red giant reactions are difficult to find, because in general the elements produced in stars do not surface until later stages of evolution, when convection has had a chance to bring them into the open from deep down in the core.

The radius of a star is normally inferred from its absolute luminosity and temperature, as was described in Chapter 2. However, in the case of nearby red giants direct measurement is feasible, using optical interferometers. At the time of writing only one star has been resolved in this way, *Betelgeuse*, but no doubt others will follow.

Whereas a main sequence star scarcely stirs from its starting point, a red giant follows a complicated path in the area above the main sequence. At various stages during this evolution it shakes itself into pulsations which vary its light output in ways which can reveal its internal structure, if only we know how to read the message. This is the topic of Chapter 6.

4.2 HIGH-MASS STARS

The pressure at the centre of a main sequence star more massive than the Sun is greater than that at the centre of the Sun. This means in turn that the temperature must be higher, because this is the only way that hydrogen gas or plasma can exert higher pressure. This in turn makes nuclear reactions go faster, and also extends the list of possible nuclear reactions. The more massive a star, the faster it evolves, as shown in figure 4.11. Nuclear reaction rates are very sensitive to temperature so the core temperature only needs to rise slowly with mass, as shown in figure 4.12. The outer layers of such stars are also hotter, partly because of the higher core temperature, but mainly because hot material is, generally, more *transparent* to radiation so the energy reaches the surface more easily. Massive stars are less well lagged than the Sun. Hot gas is less dense, so massive stars are more distended than the Sun. This effect far outweighs the fact that the small core is of high density, with the result that the *mean* density of stars decreases up the main sequence. Thus, higher mass goes with lower density because of the steady temperature increase up the main sequence. Incidentally these greater radii for the more massive stars mean reduced gravitational pull on the outer layers—they weigh less than if they were closer to the centre of the star. This means that the first sentence of this section is not as obvious as it looks. (Indeed if massive stars were extremely large, it would no longer be true.)

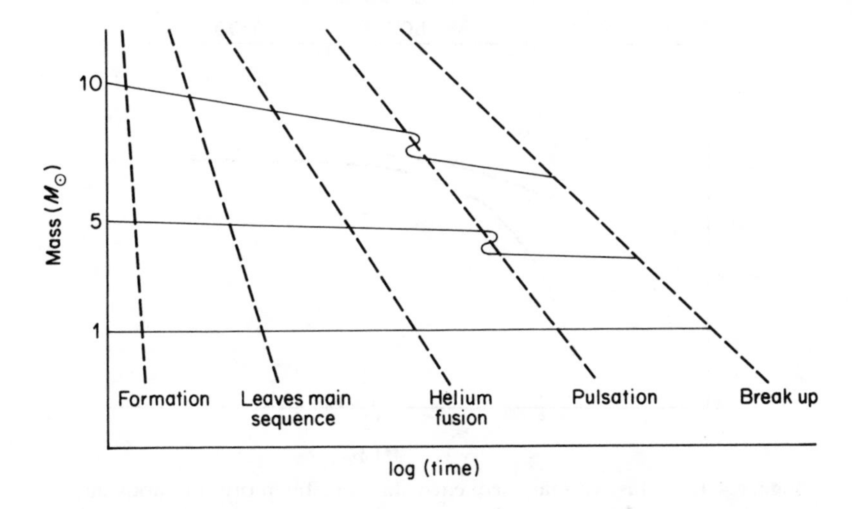

Figure 4.11 Does the evolution of a star depend on its mass? Stars between 0.2 solar masses and 10 solar masses go through broadly similar phases, but the more massive stars go through them much faster. Very light stars quietly subside into white dwarfs. Very massive stars probably explode as type II supernovae, producing a supernova remnant rather than a 'planetary' nebula, and a neutron star rather than a white dwarf.

What are the consequences of these differences for the amateur astronomer? First and foremost these hot, luminous stars are gratifyingly easy to see—Spica, Alkaid, Regulus and Vega are massive main sequence stars for instance. Also, being short lived, they are often still surrounded by the nebula from which they were formed, and in these cases the nebula will be lit up to produce very beautiful, and scientifically very interesting, emission nebulae, such as M42 in Orion and η Carina. Some of these stars have been found to be variable. Their periodic changes are far from dramatic, in fact they are difficult to detect. What makes them intriguing is that the mechanism must be quite different from that operating in the Cepheid variables and related stars, and nobody knows how they work. Appreciating this elegant, subtle but perhaps very significant difference may be one of the pay-offs for the hard work of learning about variable star mechanisms! The second brightest star in Canis Major, β CMa, is such a star and has given its name to the class. (They are also called β Cepheids, but this name can so easily cause confusion that we will not use it.) Measurements on such stars are described in §13.4.

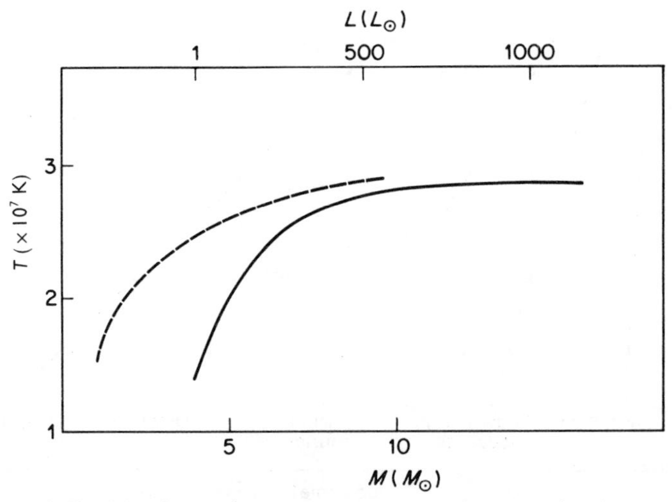

Figure 4.12 Massive main sequence stars are far more luminous and consequently much shorter lived than the Sun, but their high power output is achieved with relatively modest increase in core temperature. In fact, if the temperature axis is turned around, and luminosity plotted against it, it becomes apparent just how steep the dependence of output on temperature is for these reactions. Try to plot it on a linear scale if you are not convinced. This is typical of fusion reactions, including those in a hydrogen bomb which, when triggered by a rapid rise in temperature, go very fast indeed. In this light, the stability of stars should be seen as very remarkable. Judging only by hydrogen bombs one would have expected supernovae to be the norm and the Sun to be a special case!

Figure 4.13 shows how the density and temperature vary through the body of stars of three different masses. It also shows, as a contour diagram, the opacity of solar type material under the different conditions. The opacity will be important when we come to consider how variable stars work.

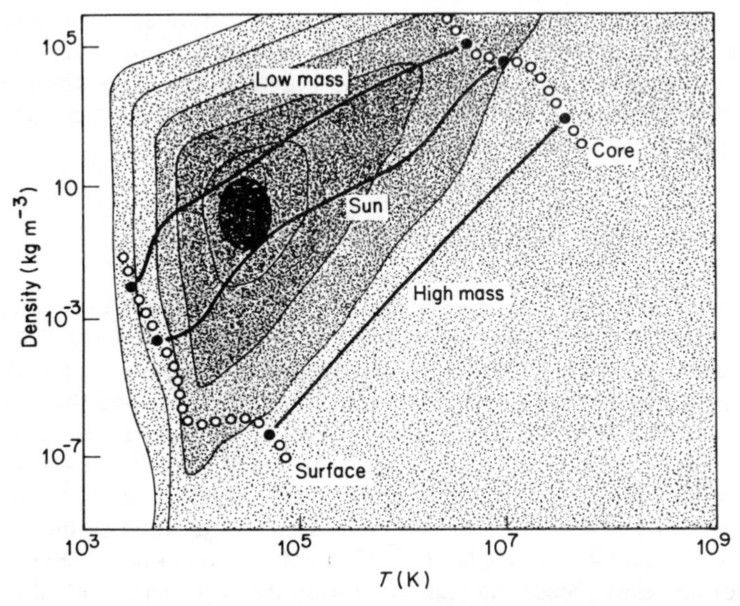

Figure 4.13 Journeys through the dark interior. The three lines show imaginary journeys through three different stars, from the core (top right) to the surface. The contours show the opacity—darkness or lack of transparency—of the material. Based on calculations by P Eggleton and J MacDonald, Institute of Astronomy, Cambridge.

4.3 LOW-MASS STARS

The Sun can serve as a prototype for all main sequence stars down to about 0.2 solar masses. These lower mass stars are cooler throughout, and are therefore more economical with their fuel and live longer but, according to calculations, will go through similar stages of evolution. However, it takes them longer than the present age of the Universe just to get off the main sequence, so there are as yet no examples to look at! For still lower masses, down to about a tenth of a solar mass, the red giant stage cannot occur and such stars are thought to produce white dwarfs immediately after leaving the main sequence. At still lower masses we have to ask 'when is a star not a star?' There must be a critical mass below which fusion reactions are not ignited or are too slow to be significant. Such stars, if that is the right word, are coming to be called 'brown dwarfs'. Are there distinct differences between planetary formation and stellar formation, or is the first a scaled down version of the second? In other words, should one consider whether *brown dwarfs* are the poor relations in stellar formation or the winners in planetary formation, even more 'successful' than Jupiter? Or does such a question not make sense? This is a question which amateur observations cannot help in solving. These bodies are extremely faint, and can only produce tiny gravitational effects on any neighbours. The observations demand large orbiting telescopes and astrometric precision which again can only be achieved in the perfect seeing of space.

4.4 WHAT MIXTURE OF STARS DO WE SEE?

As we have seen before, the restructuring stage after the main sequence is brief, so only a very small proportion of stars are in this stage at a given time. Correspondingly there is a gap, with few observed stars, on the Hertzsprung–Russell diagram between the main sequence and the region of the red giants. Although the surface is cooled by expansion there is a lot of energy being produced so when it starts to reach the surface—a journey which may take ten million years—the luminosity of all but the most massive stars increases. This is most easily seen on the Hertzsprung–Russell diagram in figure 4.8. These large, cool, highly luminous stars deserve their name of red giant. Since the high luminosity uses up energy so fast it is no surprise that calculations give the red giant stage a shorter life than the main sequence stage. Only very detailed calculations can say how much shorter, and the answer is about a factor of ten. But isn't this, at last, a prediction which should be directly verifiable, simply by counting red giants? Once a star can be plotted on a Hertzsprung–Russell diagram (which means knowing its absolute luminosity), there is no difficulty in distinguishing the red giants, so we can find the relative numbers on the plot. However, the sample of stars must be chosen carefully. We must take a given volume of space, say within 100 parsecs of the Earth, and plot all the stars contained within it—or at least be sure to include all the main sequence stars, even the faint ones at the bottom end of the diagonal strip. This is a very different matter from surveying a given area of the sky. When we look at the sky, we may well see *more* red giants than main sequence stars, but that is because the red giants are more luminous, so they can be seen out to greater distances. For instance, we could only see the Sun with the unaided eye out to a distance of about 10 parsecs (corresponding to an absolute magnitude of 5.5, about the limit for the unaided eye) but when it becomes a red giant we could see it out to about 100 parsecs. The ten fold increase in distance corresponds to a thousand fold increase in volume (the cube of the radius factor) so this makes a very big difference. Let us suppose for a moment that all red giants are a hundred times as luminous as their precursors. We can then see them ten times further away, so that we see them in a thousand times bigger volume. However, they only last a hundredth as long. Thus we expect on this rough calculation to see ten times as many red giants as main sequence stars in the sky as a whole. On the other hand, if we look at a

Hertzsprung–Russell diagram for a particular cluster (i.e. in a given volume) we only expect to see 1% of red giants.

This type of argument is important, so we shall show it graphically as well. Figure 4.14 shows the distances and volumes observable with the unaided eye for various luminosities. It shows again that the lower end of the main sequence can only be seen in a much smaller volume than the red giants. In practice, some parts of the volume which ought to be visible will be obscured by nebulae—we cannot see behind the Orion nebula for instance, and it is only about 400 parsecs away. The centre of our Galaxy is a more extreme example in that it is almost totally obscured by interstellar matter, but for which it would appear as a blaze of stars in a large telescope. Population surveys for stars are, therefore, fraught with biases (even more than population surveys for people!) and have led to long standing feuds between astronomers.

Any survey based on visible light gives only a partial, not to say distorted

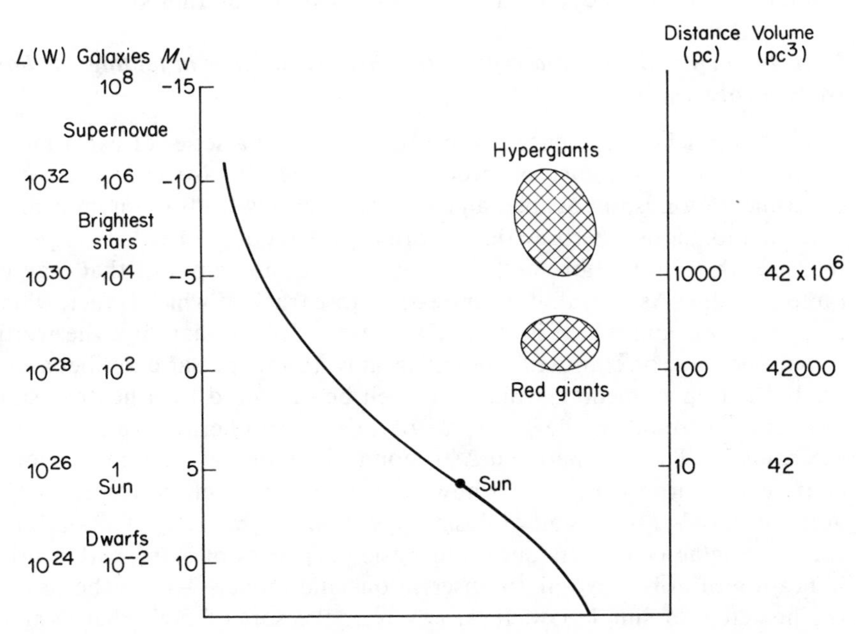

Figure 4.14 How many red giants should we see? Since highly luminous stars can be seen out to greater distances than weakly luminous stars, it is easy to get the false impression that bright stars are common. (100 times more luminous means ten times the distance, which is 1000 times the volume.) As the right-hand scale shows, the unaided eye can detect red giants in a volume a thousand times bigger than for stars like the Sun. But if we take a given volume and then check all the stars in it, then dim low mass stars will predominate. For instance, even among the 100 stars closest to the Earth, only 20 are visible to the unaided eye.

picture, biased towards objects with a temperature of a few thousand kelvin, since at that temperature the emitted radiation has a peak intensity in the narrow visible range. In terms of visual luminosity, Sirius and Canopus stand out as the brightest stars. But surveys at other wavelengths would look very different. For instance, in any *x-ray* picture of the sky, the hottest objects stand out; the brightest object is Sco X-1. An infrared picture is dominated by Solar System dust at 12 μm wavelength and by the Milky Way at 100 μm. In Orion, the nebula M42 and the Horsehead stand out more in infrared than any of the stars, even Rigel. For the radio astronomers, galaxies are brighter than the foreground stars. At 3 cm wavelength, the sky is still dominated by the radiation from the Big Bang! We should not, of course, think of these pictures as strange or unrepresentative. It is the tiny range of visible light, less than one octave, that is unrepresentative. False colour computer displays are extremely valuable in presenting a more complete picture rather as though we were in the enviable position of seeing the whole spectrum, with radio and x-ray as different 'colours' alongside the familiar set which make up the rainbow.

Note 4.1 How theory and observation help each other in finding out how stellar evolution works

In many branches of science it is possible to set up a series of experiments which sort out one fact or process at a time. In astrophysics, where experimentation is impossible, an observation rarely leads to an immediate and unambiguous interpretation. Normally, a model is needed simply to interpret the observation, and the model makes assumptions that may be quite complex. As a result it is very easy to lose track of which is fact, which is assumption and which is theory. For example one may find the words pulsar and neutron star used interchangeably. In fact pulsar describes a well established phenomenon, which may well be produced by a neutron star, but may be found in the future to be due to another, or various other mechanisms. The International Astronomical Union has held international conferences with the themes of how theory and observation interact in the areas of variable stars, and of cosmology, so it is a big subject. As a small start, the table below sets out a few basic properties of stars, with a brief indication of how they link to observations and models—or, in the case of the first item, a simple law. It is, however, the sort of table that is more instructive to make up yourself than to look at someone else's attempt.

Observation	Features	Theory needed
Absolute luminosity and temperature (from the spectrum)	There are distinct types of stars, of different sizes	The Boltzmann law that luminosity is proportional to r^2 and to T^4
Detailed spectrum	There are absorption and emission lines, with characteristic positions and shapes	Modelling of interactions of light in hot but often tenuous atmospheres
Mass (from well separated binary stars) plus luminosity	Mass–luminosity relation. Energy output depends strongly on (calculated) core temperature	Modelling of the structure of the star made mainly from hydrogen and helium
Lifetime (from HR plots for clusters) plus mass	There is a higher output of energy per kg than gravity alone can give	Calculation of energy release from thermonuclear reactions in stellar core
Population of the HR plot	Indications of evolutionary trajectories	Very detailed and lengthy modelling of complete structure, with a great deal of measured and extrapolated data
Variable stars	Period–luminosity relation	Calculation as above with details of energy flow in outer layers. Opacity model critical

Chapter 5

Radiation—a pause for reflection

The interesting thing about red giants in their later stages is their variability, and that usually arises from the damming up of radiation in the outer layers of the star and its periodic release. It is not too difficult to imagine light energy held in by an opaque layer, just as heat is held in by an insulating layer. The release mechanism, however, depends on particular properties of light in interaction with atoms, which are not shared by radiation of other wavelengths. These properties will also be important later on, in considering the finer points of how to detect light in photometers, so this seems a good time to gather together the basic facts about light. Although they seem simple nowadays, it has taken the greatest physicists of this century to put them together. If you or I had written this chapter 70 years ago we would have got a Nobel prize for it, as did Max Planck in 1918, Albert Einstein in 1921 and Franck and Hertz in 1925.

It is worth putting the story together from the beginning. As we know, all radiation is characterized by a frequency, higher for infrared radiation than for radio waves, and so on all the way up the electromagnetic spectrum, as set out in figure 5.1.

Blue light has a higher frequency than red light and ultraviolet frequencies are higher still. Another basic fact is that radiation of all types travels, in a vacuum, at the same speed, nearly 300 000 $km\,s^{-1}$. In fact the metre is now defined in terms of the speed of light and the speed has been chosen to be exactly 299 796.458 $km\,s^{-1}$. This number is directly related to the electromagnetic character of radiation because, as Maxwell showed, it is precisely equal to a combination of basic electric and magnetic constants, in a form that he was able to predict. Thus the value of the speed of light in itself confirmed the theory. It came as a complete surprise, however, that the electromagnetic energy is divided up into separate photons, and that one can no more have half a photon than one can have half an electron. Every *photon* travels in vacuum at precisely the speed given above. The greater the intensity of the light, the greater the number of photons. The amount of energy carried by each photon is fixed by the frequency of the radiation; in fact it is directly proportional to the frequency. That realization, which Planck said really did come to him as a flash of inspiration while he was out

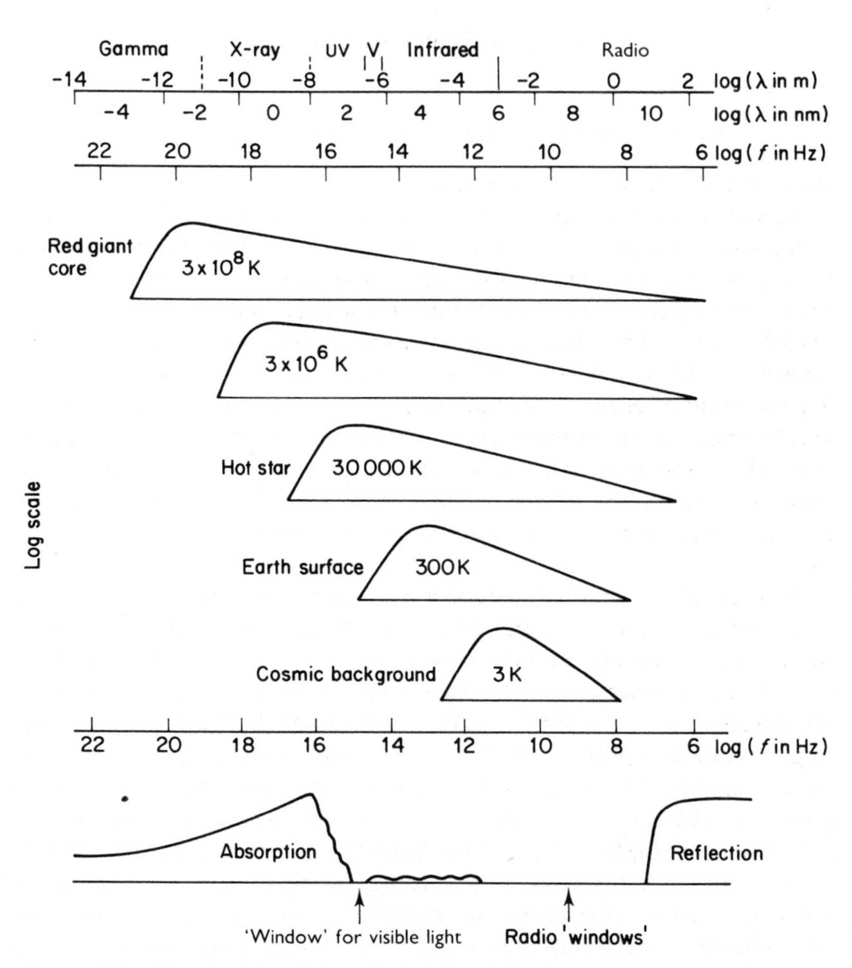

Figure 5.1 Wavelengths (λ) and frequencies (f) for thermal radiation. Electromagnetic radiation covers such a wide range of frequencies and intensities that logarithmic scales for both were unavoidable in making this diagram. The top two scales show that frequency is inversely proportional to wavelength. Try multiplying any corresponding values (or adding their logarithms) and the product will be the speed of radiation, 3×10^8 m s^{-1}. The blade-shaped diagrams are plots of the distribution of energy with frequency (running from right to left) at various representative temperatures. The peaks all lie in a straight line because the frequency there is proportional to temperature. (This is called Wien's displacement law.) Only visible radiation and radio waves traverse the atmosphere with only minimal loss by absorption or reflection—on good days.

walking, accounts for the 1918 Nobel prize. We can now write

$$\text{Photon energy} = \text{Planck's constant} \times \text{frequency}$$

or, equivalently

$$\text{Energy} = \text{Planck's constant} \times \text{speed of light/wavelength}$$

since speed = frequency × wavelength.

Nowadays any photomultiplier tube can be used to measure *Planck's fundamental constant of nature* since, in such a tube, light of a known wavelength releases electrons of measurable energy. The understanding of that process (plus a few other things he had been working on) got Einstein his 1921 prize. The value of the constant is 6.6×10^{-34} joule seconds. A photon of yellow light (5×10^{14} Hz) carries 3×10^{-19} joules, for example. That amount of energy is just about enough to activate a rod or cone in the eye, but that does not mean that every photon that enters the eye is detected. It is a debatable point whether we can claim to have firsthand experience of photons. Presumably physics would have unfolded very differently if the eye had been sensitive enough for the photonic nature of light to be obvious.

As soon as we fit up a telescope with any other detector the photonic nature of light becomes important in two different ways. First, any detector needs a certain minimum photon energy to record a hit. The easiest example is a grain of photographic film. There is a threshold energy which needs to be injected into a grain to make it developable. A single high-energy (high-frequency) photon can do this. A single low-energy (low-frequency) photon cannot. The same statement applies to a photomultiplier. Figure 5.2 shows the sensitivity of all detectors falling off towards low frequencies.

If a photomultiplier is used to count photon arrivals another limitation arises. First of all there is no such thing as half a photon, so it can only count in integers. A much worse problem however is that the photons do not arrive in a neat even stream but in a random way. Successive counts from a photomultiplier, each taken over, let us say, 5 seconds, might be 90, 103, 95, 93, 102, 100, 104, 99 and so on, purely as a consequence of the random arrival of the photons. This is a fundamental limitation in astronomy—and indeed in any communication over long distances—and we will have more to say about it in §10.3.

All the curves in figure 5.2 fall off towards higher frequencies, despite the fact that high-frequency photons have high energy. The reason is simply that the higher energy photons tend to get absorbed in the cover over the detector—the glass of the photomultiplier tube for instance. If such a high-energy photon does get to the detector itself it has a high probability of being detected.

Between the low-frequency fall-off and the high-frequency fall-off, each detector has a frequency of peak sensitivity. Figure 5.2 only shows *detectors* for visible light, so naturally these peaks are all in the visible range. It would make life simple if, at the peak, the detectors could all detect single photons. Unfortunately this is not the case. An individual detector in the eye (a 'rod' in the case of dim light) will certainly respond if at least two or three photons strike it within about 0.1 second. A grain in a photographic emulsion also needs a few photons if it is to become developable, but can wait longer. However the speed (sensitivity) of films is lower at very low light levels because a grain does tend to forget about the last photon after a few minutes. In a charge coupled device, however, the effect of single photons is stored for much longer, with very little loss. If we consider higher frequencies such as x rays or gamma rays, it is correspondingly easier to detect them singly. Indeed they can be dangerous, damaging plant and animal cells, and even destroying sensitive computer components in spacecraft.

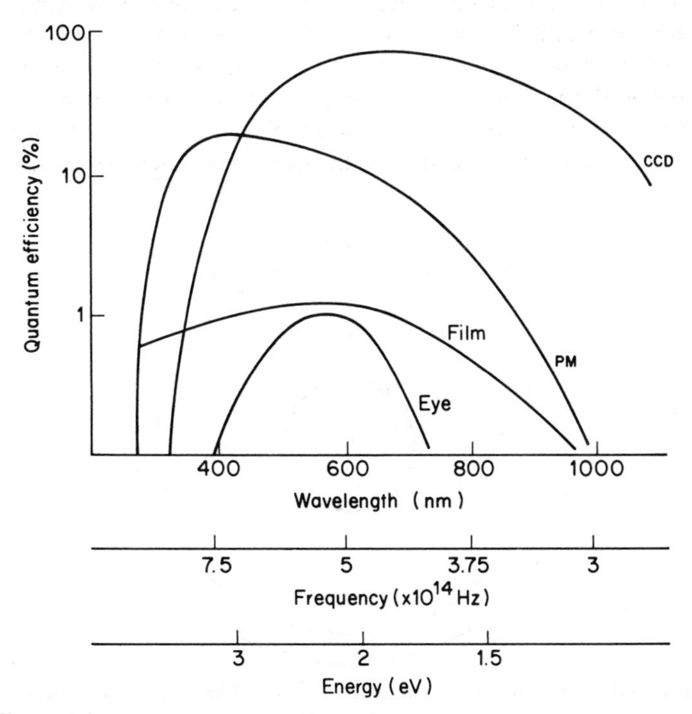

Figure 5.2 Is the eye the perfect detector? The eye does not excel in sensitivity or wavelength range (as shown here), or in resolution or speed of response, but it does excel in availability! More detailed curves are given in figure 10.2.

There is a useful difference in the meaning of the terms *x ray* and *gamma ray*. As we will use them, x rays are photons that arise from rearrangement of the inner electrons in an atom, and gamma rays arise from the re-arrangement of protons within a nucleus. Radiation is given out whenever charges are moved. So what about changes in the *outer* electrons of atoms? 150 years ago Sir William Huggins made a discovery ranking in importance with that of Planck. Indeed it is obvious nowadays that the two are very closely related. Huggins found that each element emits a set of colours of light—spectral lines—when heated, and that the set is quite distinct and different for the different chemical elements. *Spectroscopy* remains the most powerful tool for chemical analysis to this day, but covers the whole electromagnetic spectrum, not just the visible part which was accessible to Huggins.

White light is a mixture of all (visible) colours, so a beam of white light has photons with a continuous spread of energies with all colours equally represented. We have already mentioned another special case, the mixture corresponding to light inside a heated cavity, arising purely from the high temperature and called thermal radiation. Clearly different colours are not equally represented in this case—if the cavity is 'red hot', for instance, the colour red predominates. True thermal radiation corresponds to the case where the cavity walls are perfectly absorbing—black—so the term black body radiation is often used interchangeably with thermal radiation. It is

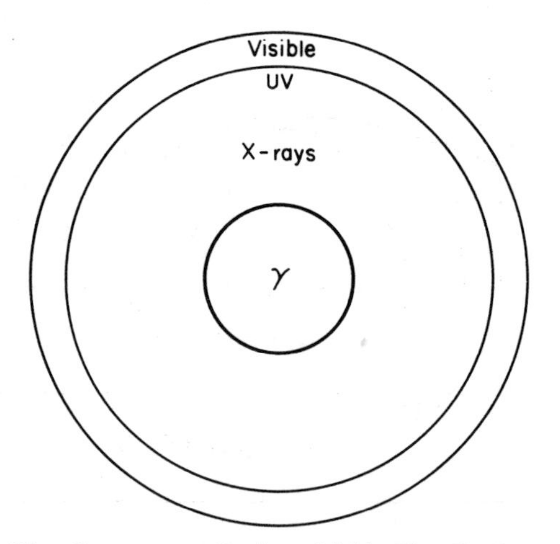

Figure 5.3 The Sun as a radiation shield. The Sun's core gives out enough gamma rays to give a lethal dose every microsecond at the distance of the Earth. In fact, each gamma ray only travels a few millimetres before scattering or absorption, so we are well shielded. As the radiation energy travels outwards it is shared among a progressively larger number of lower energy photons, as shown quantitatively in figure 5.5. Thus the radiation temperature reflects the matter temperature. The Earth's atmosphere is, of course, also a radiation shield. Without it we would all need parasols against the far ultraviolet radiation.

also often assumed that any incandescent body gives off radiation with approximately the same distribution of energies as a true cavity. Nature is kind in this case and one can often get away with this rather gross assumption.

In incandescent gases there are specific interactions between light and atoms which can either produce or absorb light at particular wavelengths. These are the interactions which produce spectral lines, either in *emission*, on top of the thermal spectrum, as Huggins first saw, or in *absorption* cutting into the thermal spectrum, as identified by Fraunhofer in the case of the Sun.

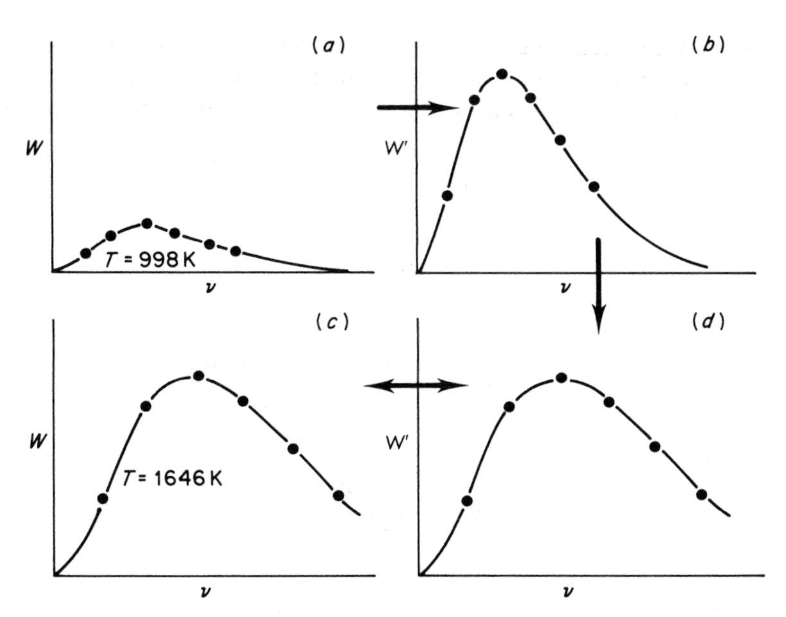

Figure 5.4 What does it mean to say that all thermal spectra are the same shape? All theoretical thermal spectra look the same shape, once you are used to logarithmic scales. The shapes are, indeed, identical, after the scale changes shown in this diagram. ($W' = W \times (1646/998)^3$, $\nu' = \nu \times (1646/998)$.) In practice spectra usually have some lines as well (one might, in analogy, say that all hands are the same shape but can be distinguished, when the occasion demands, by fingerprints).

We now have the vocabulary to describe the way radiation travels through a star. As depicted in figure 5.3, radiation is generated in the core of a star in the form of very energetic x rays or gamma rays from nuclear reactions. Clearly it is not these self-same photons which finally emerge from the surface of a star, otherwise stars would not emit light. In fact the original photon only travels a few millimetres before colliding with an electron or perhaps a nucleus. In using the word collision we are already using the photon idea—waves do not 'collide'. In any collision, energy is redistributed between the participants. The chances are that the higher energy participant will give up energy to the lower energy participant (using the word participant to cover both photons and particles—some people just use the word particle for everything). Since the nuclear gamma rays have high energy compared to the particles, the radiation will, in the early collisions, lose energy. This means that photons of lower frequency and correspondingly longer wavelength are produced. Only the basic shape of the spectrum remains unaltered, as set out in figure 5.4. All are of the thermal spectrum shape, but corresponding to successively lower temperatures.

The new photon can emerge in any direction from a collision, so the radiation gradually diffuses outwards. Gradually seems a strange word for photons travelling at the speed of light, but the collisions are so frequent

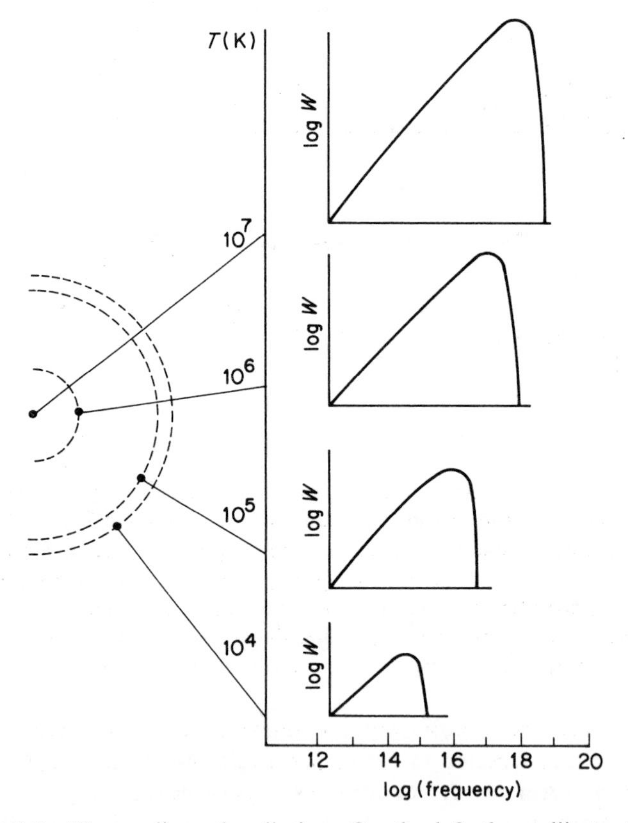

Figure 5.5 The cooling of radiation. On the left the radii at which particular temperatures are found are marked (cf figure 4.1). On the right the curves show the energy distributions corresponding to those temperatures (logarithmic on both axes).

that the radiation takes a very convoluted path and the energy is transported relatively slowly. We will take stock of the rate when we finally reach, in our imagination, the surface of the star. Clearly the degradation of the energy of the radiation can only go on for a limited time. Eventually the average energy of the photons will be reduced to the average energy of the particles and after that further collisions would not change the energy distribution. However, the radiation is gradually spreading outwards, and that means to cooler regions of the star. One of the simple basic laws of physics is that the average energy of particles in a gas is proportional to the temperature of the gas. Thus the average energy of the photons gradually decreases. Figure 5.5 represents this cooling.

In the inner regions of a star, atoms are fully ionized, and most of the *scattering* is off electrons. In the outer regions, there is a mixture of atoms

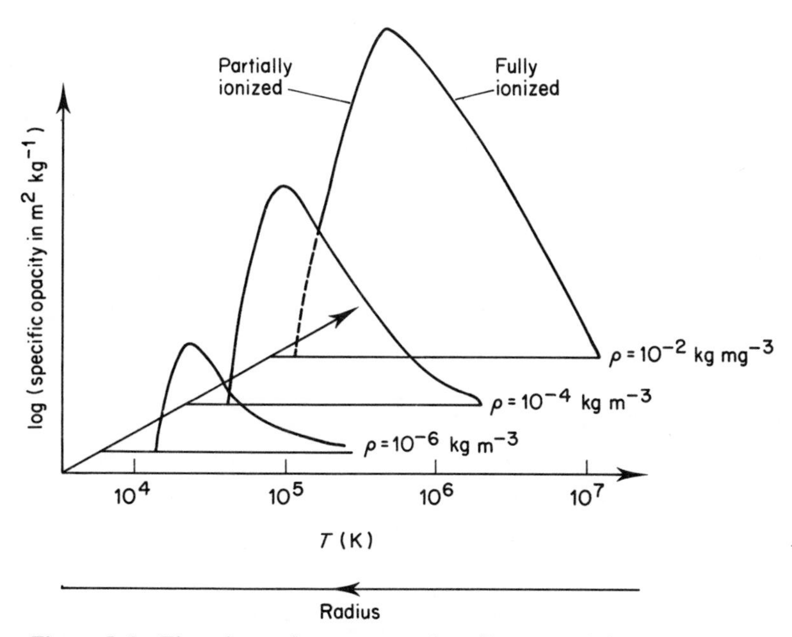

Figure 5.6 The absorption power of stellar material. The opacity (defined as the area blocked off by a kilogramme of material) increases with temperature at low temperatures near the surface, reaches a peak and then falls again at high temperatures deep inside the star (cf figure 4.13). This holds the key to the driving force of pulsations in stars. The three curves correspond to different densities. For instance for a point in the star where the density is 10^{-4} kg m^{-3}, one would look at the middle curve and read off an opacity corresponding to the temperature at the given point. In practice the data would be in the form of tables of numbers in a computer, making it easy to interpolate between the curves.

in various degrees of ionization. A photon may now scatter off an electron in an atom, either changing its energy level or removing the electron from the atom—that is adding a degree of ionization. The relative importance of these different processes depends not only on the density and temperature, but also on the chemical composition of the material, since this determines which energy states are available for the photons to hop in and out of during scattering. The *overall* rate of scattering depends primarily on the density. Figure 5.6, which should be compared with figure 4.13, shows how the scattering of material of a particular composition (chosen to be representative of the interior of the Sun as far as is known) varies with the density and the temperature.

Why go into all these details? Because this is where the variable star mechanism operates and that is the subject of the next chapter.

Chapter 6

Pulsating stars

6.1 WHY AND HOW STARS PULSATE

We will be concerned first with rhythmic variations, such as were first observed (by Goodricke, in 1784) in what we now know as Cepheids, but have since been identified in a wide variety of stars. The same pulsation mechanism operates at various stages in red giant and supergiant stars, but the oscillations of their huge distended atmospheres can be rather irregular. Isolated outbursts such as novae arise from different processes and of course eclipsing binaries, also called variables, are different again. The three types of process are readily distinguished by their different light curves as displayed in figure 6.1.

Cepheids pulsate in and out; they are radial oscillators. We cannot, of course, actually see the star getting bigger and smaller, but the varying velocity of the photosphere can be followed unmistakeably by Doppler measurements. Pulsation occurs among stars in particular well defined areas of the HR plot. Such stars must have very special internal structure to keep themselves surging in and out around their stable, equilibrium position. Variable stars are not unstable stars. If they were unstable, the oscillations would grow until the star was disrupted, as is thought to happen when a planetary nebula is ejected leaving a white dwarf. Normal variable stars can oscillate precisely because they do have a very definite stable configuration around which they can oscillate at their own natural frequency. There is a very handy rule about the natural frequency of stars, that it is proportional to the square root of the mean density of the star. Thus compact stars pulsate, in general, faster (i.e. with shorter period) than giant stars such as Cepheids. This feels sensible, intuitively, and, after describing how and why stars oscillate, we will be able to see how the particular square root form arises. First, why do they oscillate?

Any sudden perturbation could start an oscillation, like a tap on a bell. But, like a bell, it would soon die away. We must think of a mechanism which can feed a little energy into the oscillation to keep it going. That is how a clock works. It may be a pendulum or a balance wheel or a quartz crystal which is oscillating, but in all cases a small trickle of energy from a spring or battery is fed to it, in time with and enhancing the oscillation. In

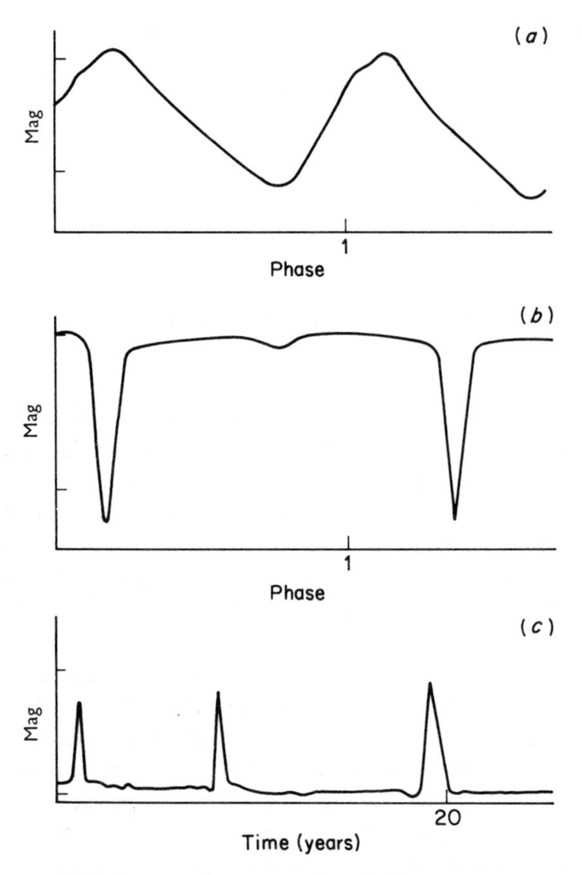

Figure 6.1 The three main types of variable star. (*a)* Waves in the stellar atmosphere cause roughly sinusoidal variations in luminosity. The cause is internal to the star and their study is important in revealing stellar structure. (*b*) Eclipses in binary or multiple systems cause sudden dips in the light output. Measurement of such systems can give very accurate information on the external properties of the star—radius, luminosity and total mass. In a few cases, a dust cloud rather than a star causes dips in the light curve. (*c*) Sudden outbursts, of light or any other type of radiation, have several causes. Small outbursts of, say, one magnitude may be flares in isolated stars. Larger ones, as here, are usually due to matter flowing spasmodically from one star to a more compact star in binaries. Type I supernovae are thought to represent an extreme case of this, where the matter transferred is extremely dense. There is a great deal to learn about both the matter itself, and the dynamics of the system, but it needs sophisticated modelling of the processes, as well as vigilant observation.

other words it is the oscillation itself which controls when the energy is fed in. This timing control is obvious in a pendulum or child's swing, but these involve separate pushes rather than an oscillatory mechanism as such, so the analogy should not be taken too literally. The basic mechanism in a star is a modulation of the energy which is flowing out through the star. There is such an enormous flow (10^{30} watts in the case of a Cepheid) that only a tiny fraction of it needs to be modulated. For instance, the total amount of energy tied up in Cepheid oscillations is about 10^{35} joules so even if only, say, 0.1% of the energy flux is fed into the oscillations (i.e. about 10^{27} J s^{-1}) it would only take a few years for them to build up. This corresponds to at most a few hundred oscillations. Several mechanisms by which this modulation can arise have been studied (eight were listed in a recent review), and no doubt others will be discovered. We start with one which, although only of secondary importance, does definitely act to foster oscillation. It will provide a simple introduction to the more subtle mechanism which is thought to be of primary importance in, for example, Cepheids and white dwarfs.

Imagine that by some magic a star—the Sun if you like—is compressed so as to reduce its radius by, let us say, 1%. If maintained, this will cause the star to readjust its whole structure in a complicated way. However, for the sake of this argument we need only consider the immediate effects. The compression reduces the area by 2%. (See note 6.1 for this piece of arithmetic.) We may assume that energy continues to be produced by the core at the same rate, for the time being at least. Thus the temperature must rise, so as to allow the radiation to escape, despite the reduced area. The increase in temperature raises the pressure so the star must *expand*. It overshoots its equilibrium position until it cools, pressure is reduced and it contracts. The contraction overshoots the equilibrium position, and the cycle starts again. The perturbation—the sudden compression—has started an oscillation, which can modulate the outflowing energy. The amplitude of the oscillations may well build up to a bigger amplitude than caused by the original 'tap', and that could never happen in the case of a bell. The reason, of course, is that a star is not an inert body and it can, in the right circumstances, react so that a small disturbance changes the structure in such a way that further disturbance becomes easier. In electronics terms, *positive feedback* can cause oscillation. In mechanical terms, oscillation arises when the flow of energy in the star is modulated by the oscillation in the right way to harness a little of the energy into maintaining the oscillation. So, why don't all stars oscillate because of this mechanism? Perhaps they do to a very small extent, but this mechanism (called the radius mechanism) is too weak by itself to lead to oscillations big enough to detect in distant stars. How big will the oscillations become? They will grow until the energy modulated by the mechanism is equal to the energy dissipated by processes such as turbulence, which are the equivalent of friction in a pendulum. To find a more powerful mechanism, which can lead to observable oscillations, we need to look at the role of varying opacity.

For a gas of given composition, the opacity per *unit mass* depends on density. (It is obvious that the *opacity* per unit *volume* depends on density; it is proportional to it since the more particles there are in a given volume the greater the chance of absorption. It is not obvious that there is still a dependence on density over and above this.) The opacity also varies with temperature, as figure 5.6 showed. For a given composition, the combination of the various effects outlined in the last chapter leads to a variation approximately as the inverse cube of the temperature. Thus an increase in temperature has an opposite effect to an increase in density, always assuming constant composition. However, if the temperature of a gas is raised, the composition may not remain the same. In particular if the heating causes ionization the number of particles per cubic metre will not remain the same. It will, of course, increase as electrons are liberated. Eddington was the first to realize that this has the makings of an excitation mechanism. Imagine radiation passing through a zone in the star containing partially ionized atoms. The star is in equilibrium and there is a steady passage of vast amounts of radiation. Now we imagine, as before, a magic compression. If there were no change in composition this would cause heating which would decrease the opacity and the radiation would pass through easily, more easily than before. But if the compression increases the amount of ionization, matters can be quite different. First and foremost, the act of ionization absorbs energy—13.6 eV per atom in the case of hydrogen—which in itself has the effect of increasing opacity. Second, the temperature does not rise so much, so the T^{-3} factor may no longer undo the effect of the increased number of particles. There can be conditions in which the opacity increases under compression, damming up the radiation. The rest of the argument is just as in the case of the radius mechanism. The temperature begins to rise and force an expansion, and so on. This is called the *ionization mechanism*. It may perhaps sound like a rather subtle and minor process, but it is able to disturb the very delicate balance of forces in a star, especially because it hits the star at just the right place to make it ring. Even with a bell, you have to hit it in the right place to make it ring properly, and this is true also in the case of stars. An excitation mechanism which works only out in the atmosphere of a star is ineffective because the heat capacity of the atmosphere is too small to be able to exercise much effect on the energy flow. It is too tenuous to catch hold of the energy and modulate it. Similarly, an excitation mechanism which operates deep in the star would, unless it is very powerful, be ineffective for several reasons. For instance, any oscillations which are produced get evened out during the slow diffusion of energy to the surface (i.e. slow compared with the periods of days or weeks which are typical of most variables). There is also more damping at the deeper levels. There may be convection and that kills pulsation. The two cannot coexist. However, at intermediate depths, any mechanism will work better. If, somewhere within those intermediate depths, the temperature is about right for the ionization mechanism, we would expect to find pulsation. And so we do.

It is easy to estimate roughly where this favourable region (called the *transition zone*) is and, as shown in note 6.2, it is usually not far below the surface. Thus we would expect the surface temperature of variable stars to be not much less than the temperature for the ionization temperature to operate. This feature stands out clearly in figure 6.2. Many variables, including the Cepheids, lie in a narrow strip which is not far off vertical, that is, at roughly constant temperature. That strip is usually called the *instability strip*. As so often in astronomy, that is a fine name provided you remember that it does not mean what it says. The stars in the strip are not unstable—indeed the technical name for such self-oscillating systems is overstable. The operating temperature for hydrogen and helium is around 10 000 K and all the stars in the main instability strip have a transition zone at about this temperature. For carbon and oxygen, which are abundant in highly evolved stars, the operating temperature is about 100 000 K. There should therefore be a second instability strip at the extreme left of figure 6.2. Such stars have indeed been found—very hot pulsating white dwarfs.

Figure 6.2 shows the relation, mentioned at the beginning of this chapter, between density and period in pulsating stars, and there is a simple way to see how this arises. Imagine that the outer layer of the star is in free fall when the star is in the contracting part of its cycle. The time taken in free fall is roughly the same as the orbital period at that radius—the Keplerian period. This is given by $(4\pi^2 r^3/GM)^{0.5}$ (remember Kepler's law: period^2 is proportional to radius^3). But, since in our case R is approximately the radius of the star, $4\pi r^3/M = 3/\rho$, where ρ is the density (remember volume $= \frac{4}{3}\pi R^3$ so period $= (3\pi/G\rho)^{0.5}$. Of course this is only plausible to within an order of magnitude. But it turns out to be true over a vast range of density from about 10^{-5} kg m^{-3} for red giants to 13 orders of magnitude higher for white dwarfs, showing that the basic idea is correct. Indeed it is an accepted rule of variable stars.

Observations of many Cepheid and RR Lyrae stars have established that the main instability strip is quite narrow—only about 1000 K. What defines the two edges of the strip? Points beyond the right-hand ('red') edge of the strip correspond to stars which have convection in the transition layer. The convection inhibits the pulsation which would otherwise occur there. Points beyond (i.e. to the left of) the left-hand ('blue') edge correspond to stars in which the transition layer (which needs a temperature of about 11 000 K) is near the surface of the star, where the density is low. The transition layer is, therefore, too tenuous—not beefy enough—to drive oscillations. However, other factors are also involved. We can see this by the complexity of the

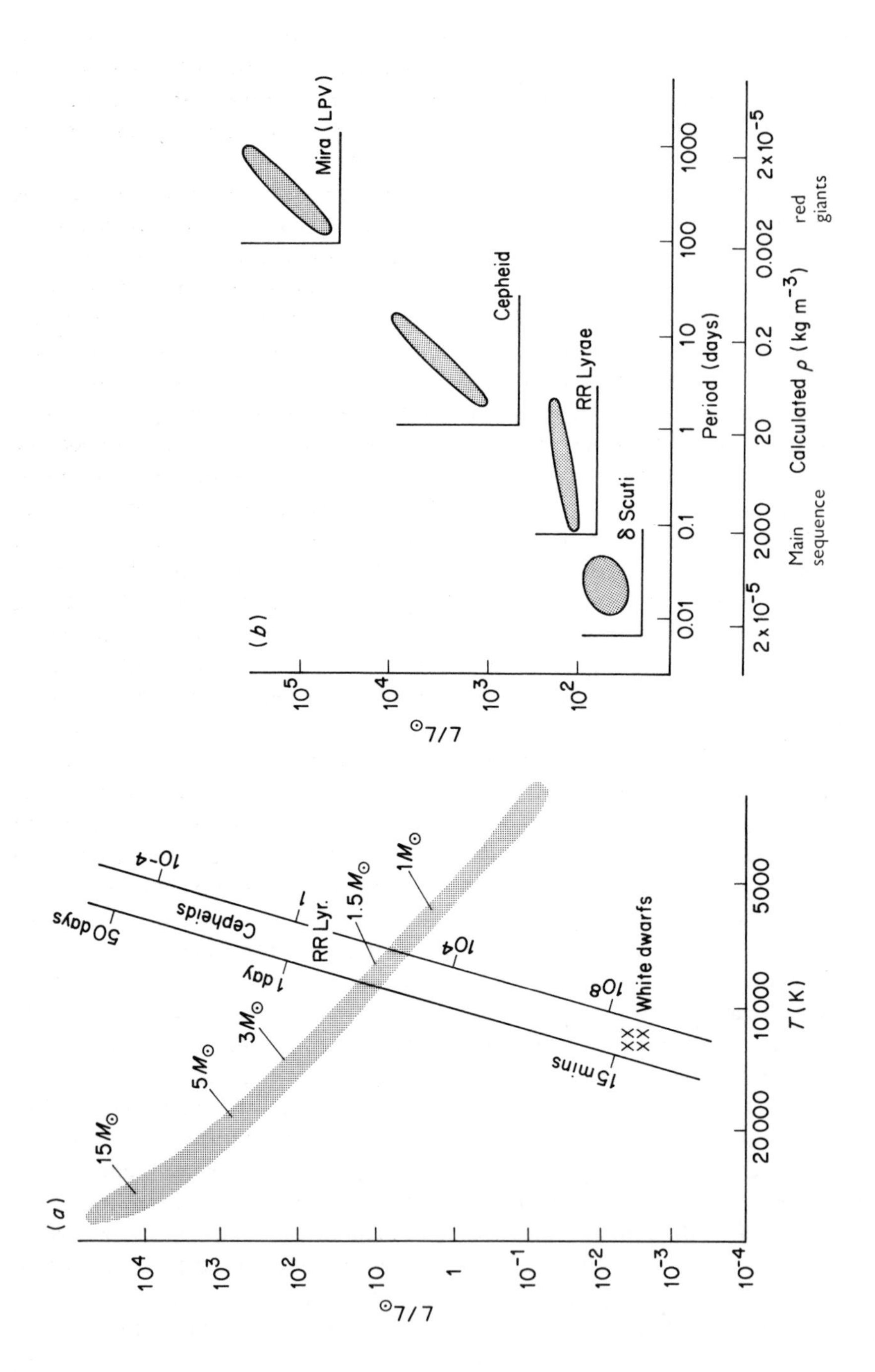
(a)
$L/L_\odot$
10^4
10^3
10^2
10
1
10^{-1}
10^{-2}
10^{-3}
10^{-4}
$15M_\odot$
$5M_\odot$
$3M_\odot$
$1.5M_\odot$
$1M_\odot$
50 days
Cepheids
1 day
RR Lyr.
10^{-4}
1
10^4
10^8
15 mins
White dwarfs
20 000
10 000
5000
T (K)
(b)
$L/L_\odot$
10^5
10^4
10^3
10^2
Mira (LPV)
Cepheid
RR Lyrae
δ Scuti
0.01
0.1
1
10
100
1000
Period (days)
2×10^{-5}
2000
20
0.2
0.002
2×10^{-5}
Calculated ρ (kg m^{-3})
Main sequence
red giants

models, but also by the very revealing observation that one finds stars which do *not* oscillate within the strip. About half the stars in the strip have no observable oscillation. Presumably they are slightly different in composition from the pulsating stars. Calculations confirm that the position of the edges is sensitive to the helium content. Although we have referred to the blue edge and the red edge there is in fact a series of each, corresponding to different *harmonics*. Conditions are more restrictive for the first overtone than for the fundamental, for instance, so the strip is narrower for that overtone.

Since the transition zone cannot be observed directly, being hidden inside the star, a good model is needed to link observations of the exterior of the star to conditions in the transition zone. Although such a model can afford to ignore the complicated nuclear reaction theory needed in stellar evolution modelling, it must take account of the detailed variation of the opacity with wavelength, which is different at each radius. Vast numerical computations are again needed. There are still too many discrepancies to claim that this modelling is satisfactory and the next decade may well see some interesting progress. Such calculations are inevitably the province of a few specialists. There is however one graph which can be used to take us a little beyond the vagueness of mere words. Equilibrium implies a balance of forces at all points in a star and two of the principal forces arise from gravity and thermal pressure. Numerical solution of the equilibrium equations therefore carries with it a value for the thermal pressure at each point (see note 3.1). Suppose we mentally focus on a particular group of atoms making up a cubic metre of material in the transition zone (see the insert in figure 6.3). In a non-pulsating star it would remain at fixed radius, but in a pulsating star we must follow the group of atoms as they bounce up and down. As they go up they will expand to more than a cubic metre and as they go down they will be compressed to less than a cubic metre. We can, in the computer, keep track of the pressure and volume of the group as they move about.

Figure 6.2 (*a*) The variability strip. Pulsating variables (figure 6.1(*a*)) of many types are concentrated in a strip rising steeply through the Hertzsprung–Russell diagram (i.e. the temperature does not rise greatly from top to bottom). The strip should perhaps be thought of as skipping over the main sequence stars since they are not usually variable. Indeed even inside the Cepheid region of the strip about half the stars do not show measurable variation. The period of the variations increases up the strip in a characteristic way shown in more detail in figure 6.2(*b*). Long period is linked to low density, as shown by the adjacent scales, for reasons as straightforward as Kepler's laws. There is, in fact, a second strip, at much higher temperatures, populated by only a few hot young white dwarfs. It is important, however, in confirming that the pulsation mechanism can operate with carbon and oxygen at high temperature, as it is believed to work in the main strip with helium and hydrogen.

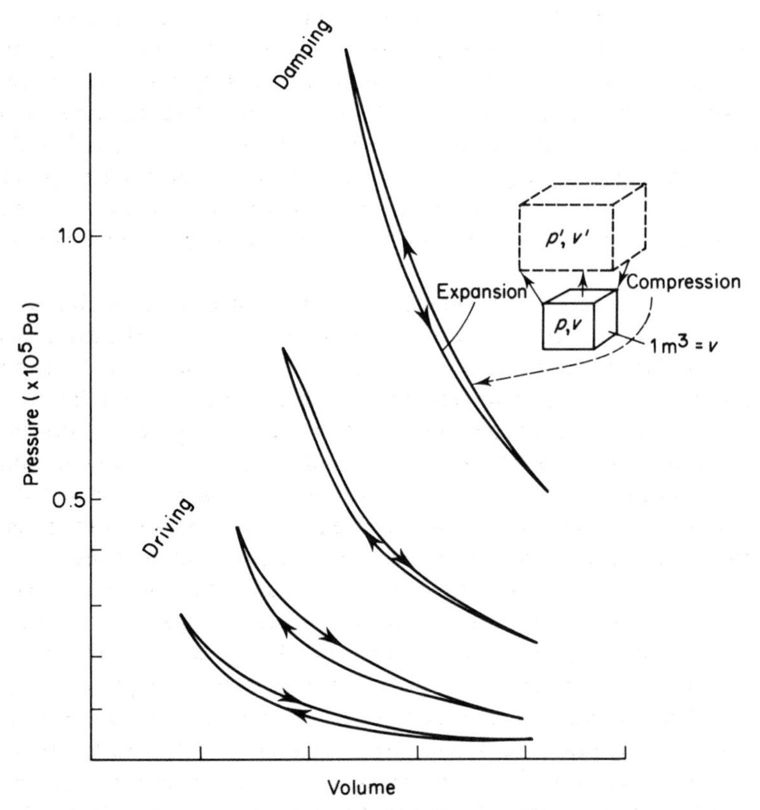

Figure 6.3 The driving mechanism. As the outer layers of a pulsating variable swing up and down, the pressure and volume of a particular cell of material (depicted in the inset) will vary, tracing out a loop on a pressure–volume diagram. If the loop is traced out clockwise, a driving force is created. A car engine provides an analogy for this. If the loop is traced out anticlockwise, the cell absorbs energy. This is more like a refrigerator. If the driving predominates, the star will pulsate.

That allows us to make use of a plot of pressure against volume used by engineers to understand and predict the efficiency of engines. The point representing the pressure and volume at a given instant traces out a line as they change. Imagine that the line corresponding to a compression stroke has just been traced out. Now the gas will begin to expand—either in the engine or in the star. The point will start moving in the opposite direction. It will not, however, simply retrace the same line, because conditions have changed. The composition of the gas will certainly be different in the case of the engine because of the explosion which produces the expansion. In the case of the star, the temperature and degree of ionization are likely to be different. The second line will be somewhere alongside the first. The two lines forming the graph must however join up to form a closed loop because the gas repeatedly follows a cycle of variations, coming back each time to the same starting conditions at the beginning of an engine cycle, whether the engine be in a car or a star.

Figure 6.3 shows a typical driving loop. Notice that the changes in pressure and volume draw out the loop in a *clockwise* sense. This is vital. If it were not so the engine would not work, or, more exactly, it would not do any work. But that is exactly what we need to know about the cube of gas in the transition zone. Will it do work on the surroundings, thereby driving oscillations? Or will it absorb work (i.e. energy) and so damp out any existing oscillations? It will be a *driving region* if the loop is clockwise, but a damping region if the loop is counterclockwise. The calculations split the transition zone into a series of thin shells, and find the work done or absorbed in each one. If the net amount of work done, added up through the zone, is positive then the prediction is that the star will pulsate.

So much for *why* a star can pulsate, now to the question of *how* it will pulsate. We have used the term 'natural frequency', but in fact most oscillators have not only a lowest fundamental frequency but also a series of harmonics. Any musical instrument can provide an example. We hear the fundamental frequency, which is the one which corresponds, all being well, to the note written in the score. But the quality of note we hear depends on the mixture of harmonics which are also present. The violin and the oboe have many harmonics but the flute has very few. By bowing in different parts of the string, a violinist can elicit different mixtures of harmonics, and produce a different quality of note. A pizzicato note is different again. Hearing a violin, we can tell immediately whether the excitation mechanism is the violinist's fingernail or his bow. With skill we could distinguish where the excitation is being applied—close to the bridge or further away.

Similarly, we can examine a plot of the variation of the luminosity of a star. (Such a plot is often called a light curve even when radiation outside the visible range is observed). It gives the basic period and hence the mean density of the star so there is a close analogy between the length of a violin string and the size of a star in determining the basic frequency. The details of the shape of the light curve give harmonic content and reflect, as with a violin, the nature and location of the excitation. To unravel this connection effectively needs, as mentioned above, rather better theoretical models than are currently available. However, it also needs better measurements. It is only in the last few years that harmonic analysis has become common in studies of variable stars, and there is much to be done. It needs a long series of observations, so it is appropriate for amateur observations. In the past, hand calculation of harmonic content was extremely tedious, but this is, of course, no longer such a problem, being good work for a computer. (Section 11.5 gives harmonic analysis programs in BASIC.)

The violin string has only one mode of vibration—sideways—but a star has two. It can pulsate in and out in a *radial mode*, as already described. However, it can also have waves running round the outer layers, as some earthquake waves do on the Earth. Solar oscillations are of this type. They are called *non-radial modes*. Both modes can break up into a series of harmonics. There are more possibilities in the case of the non-radial modes, since one can have different harmonics running round the equator, and round the poles. To make it more interesting, there are also two different ways in which the non-radial waves can travel. One type is, in effect, a sound wave, but at very low frequency. They are called *p waves*, meaning pressure waves. Sound waves are also p waves, because it is air pressure which acts, first one way and then the opposite way, to restore any given group of atoms to the equilibrium position. (In radial waves, pressure is also the restoring force but since it is the only possibility it is not usually specified. 'Radial' is adequate, there is no need to say 'radial p wave'.) The second possibility for non-radial waves comes from the fact that a combination of gravity and buoyancy provides a net force which can act either up or down, as any bobbing ship testifies. Non-radial waves in which this is the restoring force causing the oscillation are called *g waves*. They are in general of longer period than p waves. Incidentally, g waves can occur in the Earth's atmosphere but they are small in amplitude and, in the absence of a driving mechanism, soon die away. p and g waves have different speeds and so a different set of frequencies. There is a connection between the radial and non-radial modes, which will appear in figure 6.5. However, we will regard them as separate modes. Some writers in fact use the word mode to refer to different harmonics, but this does not seem necessary since the word harmonic is quite explicit in itself.

A particular excitation mechanism can give rise to either radial or non-radial modes, or both. For instance, both Cepheids and pulsating white dwarfs are believed to be excited by the ionization mechanism. However, Cepheids pulsate in the radial mode—the surface movement is clearly indicated by Doppler shifts—and white dwarfs pulsate non-radially. The surface movement cannot be seen but the typical periods of white dwarfs are too high to correspond to radial modes (see below).

If we now consider how to make harmonic analyses in these two cases it will bring us to the heart of the question of how stars pulsate. The violin string again gives a lead on how to proceed. The string can vibrate in any of the ways shown in figure 6.4, in which the string is an integral number of half wavelengths. Notice the meaning of '*node*' shown in the figure. Because of the relation between wavelength and frequency it follows that the frequencies of the harmonics are simply an integer times the fundamental frequency, the integer being the number of half wavelengths, and is the number given to the harmonic. When a violinist plays an A (440 Hz) the note we hear is a mixture of 440 Hz, 880 Hz, 1320 Hz and so on. Many harmonics may be involved. In the oboe, for instance, the seventh and

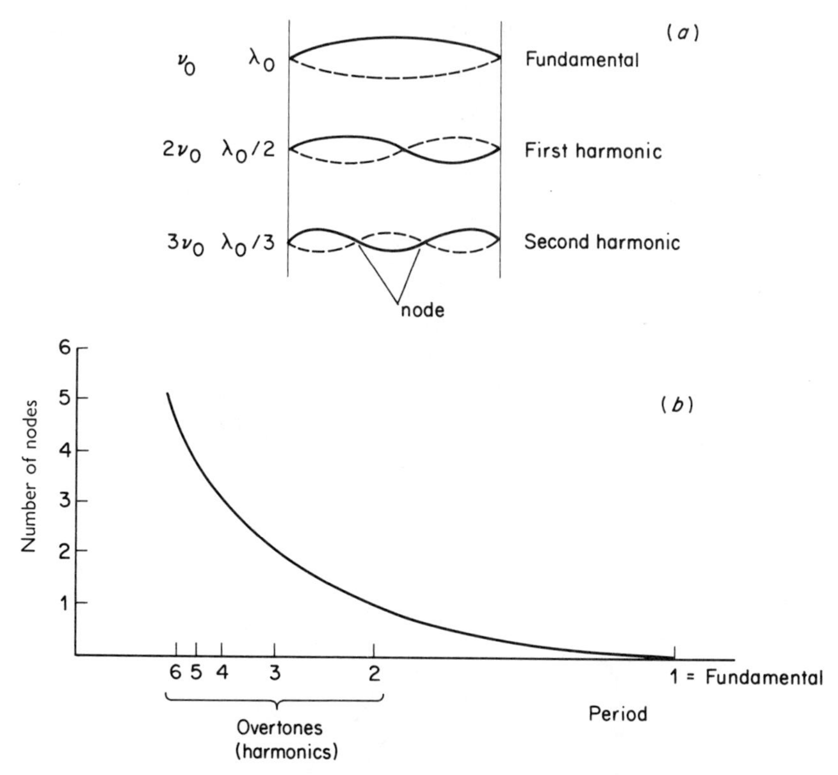

Figure 6.4 Harmonics on a violin string. (*a*) This is a representation of the string itself, with the displacement greatly exaggerated, for a pure fundamental and two pure harmonics. The points with zero displacement are called nodes. A plucked string can vibrate in nearly such an idealized way. A bowed string cannot, because of the friction of the bow, so it exhibits a complicated mixture of vibrations. In a broadly similar way, a star, which is a good deal more complicated than a violin string, would be expected to show a mixture of harmonics rather than a pure note. (*b*) The graph shows the periods of the harmonics along the horizontal axis, and the number of nodes along the vertical axis. For each harmonic there is a single point. The line is merely to make the pattern easier to see. The string cannot oscillate with, say, 3.27 nodes, so fractional values on the vertical axis have no physical meaning. Don't look at figure 6.5 before it is clear what this one means!

eighth harmonics may actually be stronger than the fundamental. The resulting waveform may be quite complicated, but it is made up from a simple series of frequencies.

In the case of a star we still have a series of separate frequencies for the harmonics but the relation between them is not so simple. Moreover, and this is perhaps the most important point, the series are different for p and g

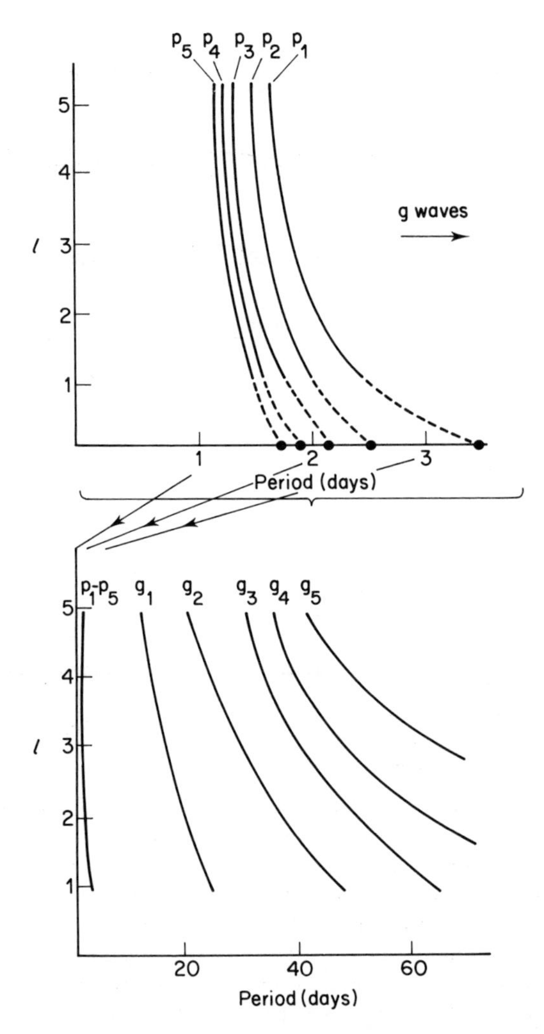

Figure 6.5 Can a *mode* of oscillation be identified from a *period*? This is a key question. The mode reflects the structure and the period can be measured. It is the central aim of astrophysics to deduce structure from measurements. This figure shows the theory, figure 6.7 makes the link to measurements. The two main modes of oscillation are p waves (pressure waves like sound) and g waves due to gravity and buoyancy like a bobbing ship or water waves. The group of lines giving (calculated) p-wave frequencies is well separated from the group of lines giving g-wave frequencies, so that is hopeful, provided the calculations are reliable. The lines join points plotted at the separate frequencies for the different harmonics of oscillation, just like the different harmonics in figure 6.4. With a violin string, we only have to consider one set of nodes, spaced out along the string. In a star, there can be two sets of nodes, one set

waves. Figure 6.5 summarizes the relations between the series. There is quite a lot of information in it, so we will go through it step by step. Don't hurry, it's important!

For the purely radial modes there is a set of periods corresponding to the series of harmonics, and these are plotted *along the horizontal axis*. Look only at the axis for the moment. The longest period is for the fundamental, so that is at the right-hand end. The graph is made for a system in which the fundamental period is 3.5 days. The harmonics are at shorter periods. For the violin string, they would be at (3.5/2)days, (3.5/3)days, (3.5/4) days and so on. (That would be a big violin, of course.)

For the non-radial modes, we have to remember that it is not good enough to just count the number of half wavelengths round the equator to determine the harmonic. The period depends also on the number of half wavelengths in the radial direction. It needs two numbers to specify a harmonic. The second number is on the vertical axis. Looking first at the p modes—the sound waves—there is a curve labelled p_1, meaning one radial node. At the height of 1 on the vertical axis, meaning one half wave round the equator, the period can be read off as 2.5 days. The way the surface breaks up into waves for this harmonic is shown in figure 6.6 in the line labelled $l = 1$, and named *dipole*. As you see there, there can also be nodes along lines round the poles. These also have a number to specify them (usually called m) but we do not have to bother about it because it does not affect the frequency, except in complicated cases where the star is aspherical or rotating. Now, still remaining on curve p_1 for one radial mode, move up to $l = 2$, for which the period is a little shorter, about 2 days. Figure 6.6 shows the surface patterns for this case, named *quadrupole*. One could also have a quadrupole case with two radial nodes, and you can read that period off the p_2 curve at the height $l = 2$. If a star is thought to be pulsating non-radially with p waves then any or all of these periods could appear in the light curve. The astronomer's job is to analyse the shape of the light curve in terms of a sum of these harmonics. We will come to an example of this soon. First we have to take account of the completely separate possibility of g waves. Fortunately this can be handled in exactly the same way. The periods are all longer, and so there is a separate set of curves for g

---

spaced out along the radius and a quite separate set spaced out round the surface. By convention, the number in the radial set is given as a subscript (e.g. the 2 in g_2) and the number in the surface is given by the number l. The different ways that a surface can be broken up by nodes is sketched in figure 6.6. If $l = 0$, i.e. there are no nodes on the surface, which means the whole surface moves in and out together, the oscillation is a particular radial mode called, less technically but more graphically, a breathing mode.

waves, shown in the lower part of figure 6.5. Read them in the same way as the p-wave curves.

The case with $l = 0$, the horizontal axis in figure 6.5, is the case with no radial nodes, i.e. the whole star moves in and out in unison. This has a name which is easy to remember—the breathing mode.

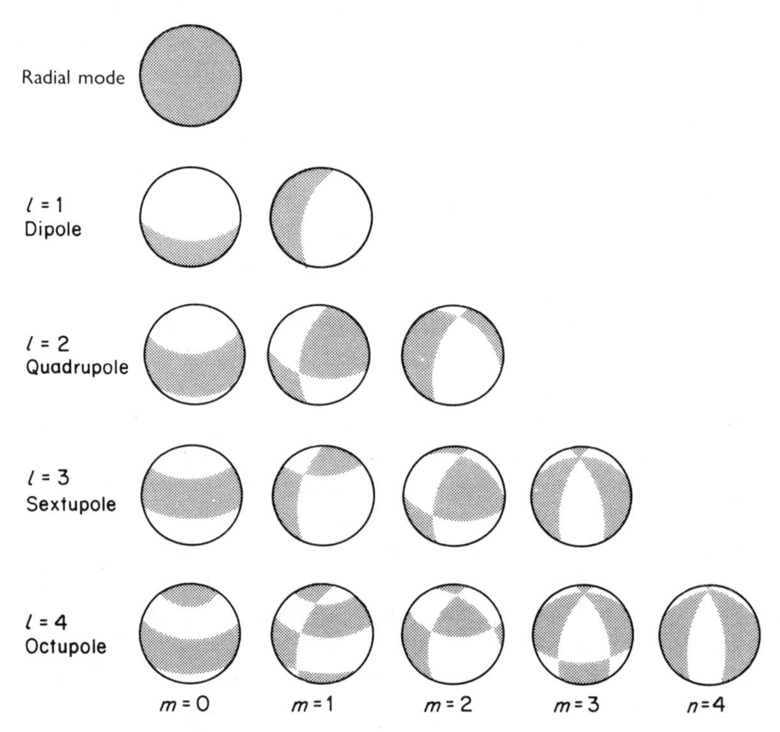

Figure 6.6 A stellar surface broken up by waves. Imagine the cross-hatched areas expanding while the clear areas contract, until maximum expansion, when the roles reverse. The l value is the number of nodes around the equator, as used in figure 6.5. However there can be nodes at other latitudes. These are labelled by m. The m value has no effect on the period of oscillation, provided the star is spherical, so it was not necessary to introduce them in figure 6.5. Stellar research is not yet at a stage where m values can be identified.

If you wish to carry these ideas further, look for a mathematical book with a chapter on Legendre functions or spherical/surface harmonics, which will explain the following statement. 'All the waves are described by a set of spherical harmonics, usually denoted by $Y(l, m)$, multiplied by a function of radius with k nodes (i.e. zero values). The radial oscillations are those with $l = 0$. The frequency depends on k and l, but not on m.' Many books on quantum mechanics explain this piece of mathematics at length, because precisely the same formalism is used to describe atomic excitations.

6.2 WHICH MODE AND HOW MANY HARMONICS?

Thanks to home computers the analysis of light curves to reveal intimate details of the interior oscillations of stars is well within amateur possibilities. Moreover there is a substantial subjective element involved, and re-analysis of already published data is by no means an empty exercise, especially if there are several papers on the same object. There is also a vast amount of unpublished data available through various societies (see appendix 6) or of course from colleagues. We will take two examples to illustrate the principles involved and further examples will arise in relation to particular types of variable. The first example will be a white dwarf, and the analysis quoted will in fact be a re-analysis, disputing results from an earlier paper. The second example will be concerned with RR Lyrae stars, but looking for evolutionary effects rather than a single specific light curve.

6.2.1 Analysis of a white dwarf

The star catalogued as GD66 was first measured and analysed in 1983 by Dolez *et al*. Their analysis was disputed in 1985 by Fontaine *et al*. The light curve is shown in figure 6.7(a) and was accepted as the basis for both analyses. It is clear just looking at the curves that the amplitude is not constant. Usually this arises from '*beats*' between two similar frequencies. As an example, figure 6.7(b) shows the graphs of $\sin 4t$ and $\sin 3t$ and the sum $(\sin 3t + \sin 4t)$ which shows the beats. Every periodic curve can be made up from, or alternatively broken down into, a sum of sine curves in this way. If there were no experimental errors, the breakdown would be unique. Of course there always are experimental errors, so the result of the breakdown depends to some extent on the way the errors are handled. The amplitude (strength) of each sine *component* can be given as a set of strength factors (*coefficients*) but it is easier, and usual, to display them as a graph or histogram (figure 6.7(c)). Usually the squares of the amplitudes are given corresponding to the power (energy per second) of each component. This is called *power spectral analysis* (a term to look for in books on statistics). The background noise at the bottom reflects the effects of errors, and shows the corresponding limit of sensitivity. It is impossible to know whether the small peaks represent real components or not. For several components, however, there is very little doubt. Let us take first the peaks at 3.3 and 3.65 mHz (1 mHz is a period of 1000 s). These must surely represent real components of the curve. We now want to know which harmonics they represent, but we cannot get any further without a specific

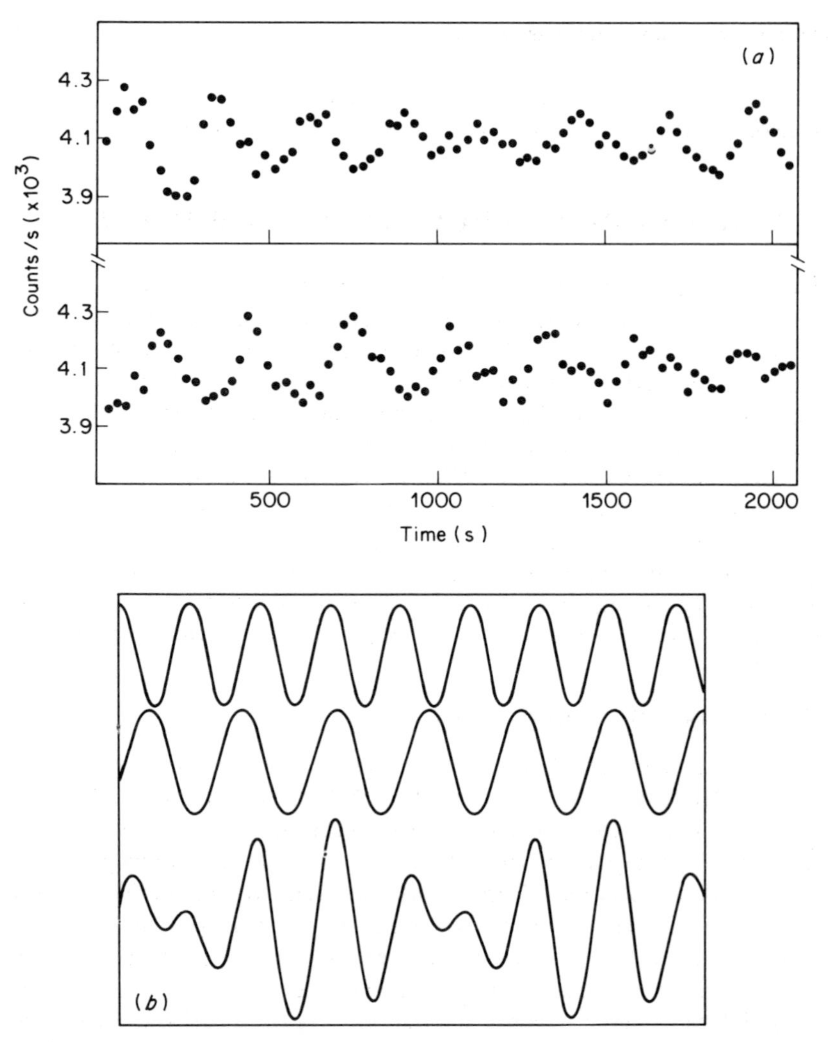

Figure 6.7 See a harmonic, pick a mode. (*a*) The light curve of a white dwarf star (catalogued as GD66) measured using a veteran telescope, the 100 inch at Mount Wilson (Fontaine G *et al* 1985 *Astrophys. J.* **294** 339). The period is too long to be due to p waves (see figure 6.5 for this crucial point) so they are taken to be g waves. But which harmonic(s)? (*b*) Even by eye one can see the beating of two frequencies, giving a periodic change in amplitude. (*c*) Analysed by mathematical filtering, called Fourier analysis, two main frequencies are indeed found (see Dolez N *et al* 1983 *Astron. Astrophys.* **121** L23). But what harmonic do they correspond to? Here they are interpreted as third and fourth harmonics and are so labelled on the graph. The reasoning is that by making this interpretation, other frequencies more weakly present can be understood with the interpretations marked on the plot. If the predicted frequencies were reliably and accurately known, this would be a strong argument. In practice, the predictions arise from a very sophisticated modelling process, and must be regarded as uncertain. This is the state of the art in variable star analysis.

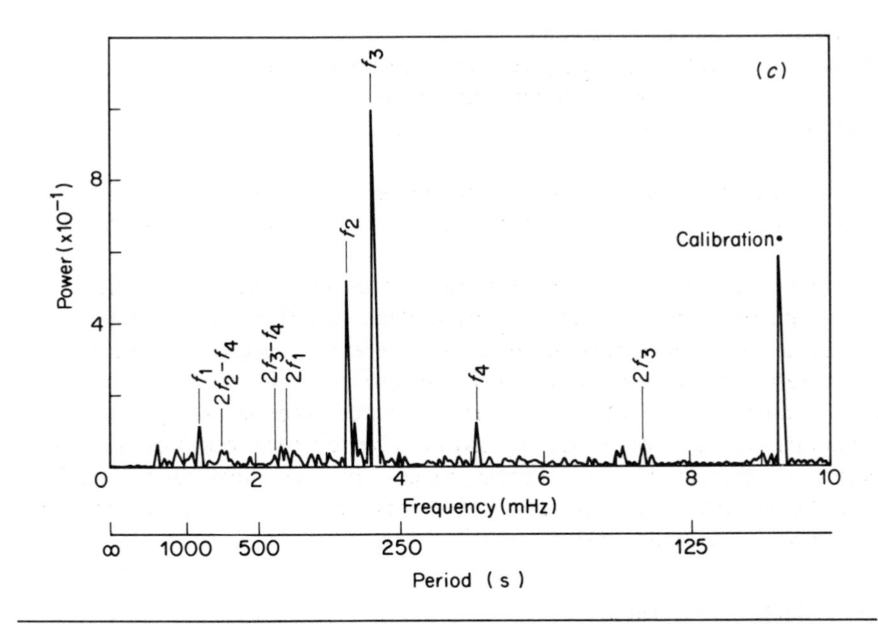

model, such as the one used to calculate the curves in figure 6.5 (which was a model known as an $n = 3$ polytrope). Fontaine *et al* had a different model from Dolez *et al*, hence they disagreed about the assignment of the two frequencies to harmonics. Dolez *et al* interpreted them as some high harmonics, Fontaine *et al* as the second and third harmonics, as marked on the figure. Suppose we accept Fontaine *et al* for the sake of argument. Having made that identification they now want to read off the predicted frequencies for all other harmonics off their equivalent of figure 6.5. To do that they need to know whether to read off the p-wave curves or the g-wave curves. If we knew the fundamental frequency it would be easy to decide, because the predicted p_1 frequency for white dwarfs is about 10 mHz whereas the predicted g_1 frequency is much lower, about 1 mHz (i.e. period about 15 mins). Since there are frequencies much less than 10 mHz present the non-radial g-wave interpretation is to be preferred. Having made such a choice, there is no further flexibility. One has to read off all the frequencies and fit them, or their sums or differences, to all the peaks which are regarded as real in the 'power' spectrum. Models do not currently give reliable predictions of strength so it is unfortunately only the positions and not the heights of the peaks which need to be matched to the theory, so this may leave a lot of flexibility, and room for judgement. If you wish to follow the detailed arguments which led to the rest of the identifications of the peaks, read the original papers. For our present purpose let us summarize the steps in the analysis.

Compute a frequency spectrum
Decide which peaks are real, given the experimental errors
Decide on a model for the stellar atmosphere
Compute (or look up) the p-wave and g-wave curves
Choose p wave or g wave
Read frequencies from the curves and *assign harmonics* to the observed peaks in such a way as to interpret as many of the peaks as possible.

It is clear even from this brief example that uncertainties in the measurements, and probably still more in the stellar modelling, leave room for a good deal of difference of opinion on interpretation. Conversely, a clear cut interpretation, which points unambiguously to a precise model of the white dwarf atmosphere, would be a very valuable piece of astrophysics. In particular, the different modes are favoured by different conditions, and so operate in different, but overlapping zones. Thus the study of the oscillations allows, ideally, a peeling off of successive layers, to reveal interior conditions. This is any stellar modeller's dream!

6.2.2 RR Lyrae stars

The pulsations of *RR Lyrae* stars (see §14.3) are interpreted more simply than in the above case, with only the fundamental and first harmonic radial waves being considered. That is the assumption in the following case, and it will allow us to concentrate on a different aspect of the topic. RR Lyrae stars are on the horizontal branch and so have similar luminosities but differing temperatures, straddling the instability strip and forming a band across it. Figure 6.8 shows the small section of the Hertzsprung–Russell diagram in the RR Lyrae region. 35 well measured stars are plotted on it, using three different symbols to represent the stars interpreted as showing only the fundamental, or only the first harmonic, or a mixture. (This is a simpler situation than in the first example.) Fundamental mode pulsations appear close to the 'red' (i.e. right-hand) edge, first harmonic near the blue edge and stars with a mixture appear somewhere near the middle. Could it be that the star starts pulsating in the fundamental mode, later does a '*mode switch*' (harmonic switch in our nomenclature) during which both periods are present, and ends up with the harmonic only? It would be nice to follow a single star to see first of all whether it does cross the strip, and to look for changes in harmonic content. The predicted time to cross the strip is around a thousand years. If this is correct one could only hope for about 1% change of harmonic content in a span of, say, 10 years, which is too small to detect. What is feasible is to look for changes in period (as distinct from harmonic content), because period can be measured very accurately (1 PPM or better is reasonable) and changes might be seen in a few years. Indeed some authors report detecting such changes. This leaves the problem of relating changes of frequency to physical changes such as a slight drift of the driving region, and hence to changes in harmonic content. This is the type of development to expect in the near future.

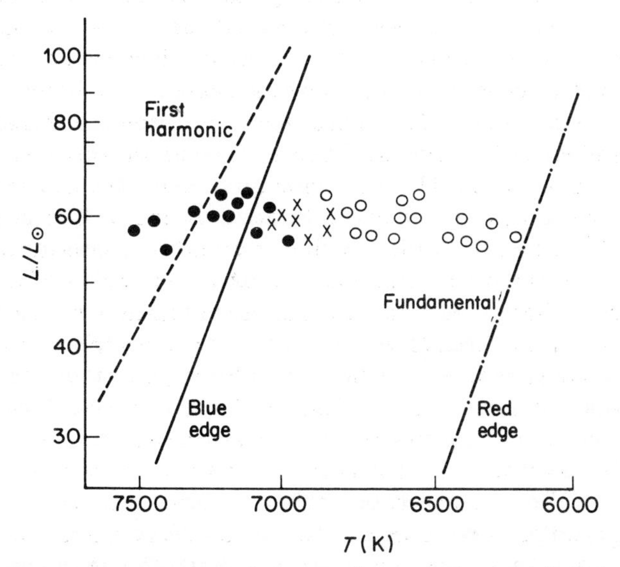

Figure 6.8 Does a star switch harmonics as it heats up? This is a section of a Hertzsprung–Russell diagram showing some of the RR Lyrae stars in the cluster M15. They are grouped into a band because they are horizontal branch stars with little spread in luminosity (see figure 6.2). The symbols indicate the harmonic which they exhibit. There is a strong suggestion that if a given star entered from the (cool) right-hand edge and heated up it would first oscillate in the fundamental mode, then begin to show an increasing amount of first harmonic until that dominates at the (hot) left-hand edge. It would be nice to watch a single star go through this act. It may take a thousand years to complete the trip, so that is too long to follow. But it is not out of the question to see some change. Harmonic analysis is quite sensitive and might pick up changes of a few per cent. The frequency can be measured much more accurately, so a better bet is to look for changes in the fundamental frequency itself with time, leaving theory to connect that with a temperature change and, hopefully, with a predicted change in harmonic content ○, fundamental; ×, two frequencies mixed together; •, first harmonic.

6.3 FREQUENCY ANALYSIS IN GENERAL

There are (at least) three different ways of defining the component frequencies in a given light curve, i.e. a curve giving the variation with time of whatever part of the electromagnetic spectrum has been observed. The white dwarf example used a frequency (power) spectrum of sinusoidal waves. This is straightforward in application but, for reasons we will come back to, the sum and difference frequencies may not be easy to interpret, and it also uses quite a lot of computing time. To match the analysis to the physical situation more directly, one would ideally like to have a model which predicts an unequivocal set of waves of given shape and frequency. They will not in general be sine waves because a star is too complicated ('non-linear') for that. (The bigger the amplitude the bigger the non-linearity is likely to be.) One would then make up the observed curve using *only* those basic waves. The RR Lyrae curve used this method, but in an extremely simple way with only two frequencies and without addressing the question of the non-linearities. A third method is to decompose the wave into a series of sine-wave harmonics (i.e. sin ωt, sin $2\omega t$, sin $3\omega t$ and so on, with different phases). At first sight this seems ridiculous because we know that the oscillation frequencies are not in those simple ratios. There is, however, a considerable mathematical advantage in that the calculation is fast thanks to some nice mathematical relationships. Moreover, such a decomposition will always work, as was proved by the French mathematician J-J Fourier, who has given his name to the process. (Sine waves form what is called a complete set. This is the phrase to look for if you want to take this further.) With such a *Fourier decomposition*, the comparison with theory becomes a two-stage process. Both the observations and the predicted light curve from a given model are Fourier analysed, and the strengths of the various components compared. Section 11.5 gives some details of this technique. Cepheid variable stars have been analysed in this way in a series of papers by Simon. Their plot of the way the relative strengths and phases vary with period certainly seems to reveal some change in the physical mechanism at about ten days, showing up as sharp dips in the curves. No such structure is shown in a simple period–luminosity plot, which passes smoothly through ten days. So far so good. But further progress is hampered by the fact that the Fourier components do not have a direct physical significance, so one can only guess at the cause (Simon says that the first overtone has just half the period of the fundamental when the fundamental period is ten days). The problem of interpretation is linked with the question of non-linearities, which introduce sum and difference terms into the decomposition. If the basic frequencies used are in simple arithmetical progression (ω, 2ω, 3ω, 4ω, ..., etc) the sums and differences of frequencies do not introduce any new frequencies, so the effect is not evident, being hidden in changes in the amplitudes of the basic terms. If the basic frequencies are anything other than an arithmetical sequence, then new frequencies are introduced, as we saw in the white dwarf example. In one paper, Simon used a Fourier decomposition which had these sum and difference terms built in. However that does not solve the problem of how to relate them to physical processes.

Note 6.1 Dealing with small changes

There is a short cut to deal with small decreases (and increases) in the area or volume of a sphere. If the radius decreases from 1 m to 0.99 m, a decrease of 1 per cent, then the surface area decreases from $4\pi \times 1^2 = 4\pi$ to $4\pi \times (0.99^2) = 4\pi \times 0.9801$. The fractional decrease is

$$4\pi(1 - 0.9801)/(4\pi) = 0.0199$$

or, near enough, 2 per cent. The volume, which is proportional to the cube of the radius, decreases by just over 3 per cent. The smaller the change, the more accurate the rule.

The algebraic basis of this rule, which is quite general, is

$$(1 + x)^2 = 1 + 2x + x^2$$
$$\sim 1 + 2x \text{ if } x \text{ is small } (\ll 1)$$

and

$$(1 + x)^3 = 1 + 3x + 3x^2 + x^3$$
$$\sim 1 + 3x \text{ if } x \text{ is small } (\ll 1).$$

Note 6.2 Transition layer

The rule of thumb for finding the depth of the transition layer is that the energy in the material outside the layer be about the same as the total amount of energy radiated by the star in one period. For Cepheids, which typically have a luminosity of about 10^{30} W ($= 10\,000\ L_\odot$) and a period of, say, 10^6 s (12 days) this energy is about 10^{36} J.

Now the energy in a kilogramme of the material is the sum of the products of its specific heat (in J $\text{kg}^{-1}\text{K}^{-1}$) multiplied by a step up in temperature, added up all the way from 0 K to the actual temperature. (This is a numerical integration.) The specific heat of hydrogen gas at high temperature is about 20 000 J K^{-1} ($\frac{7}{2}k$, not $\frac{3}{2}k$, if you were wondering) and the temperature in the outer zones of a Cepheid is about 10 000 K. Thus we can get an (over)estimate of the internal energy by multiplying the two to give 2×10^8 J kg^{-1}. Hence an (under)estimate of the mass above the transition layer is $\sim 10^{36}/(2 \times 10^8) = 5 \times 10^{27}$ kg. For a Cepheid of mass $10\ M_\odot$ this means that only $\sim 10^{-4}$ of the mass is above the transition layer. Of course, the density is very low in these outer regions, so the thickness of the layer as a fraction of the diameter is $\gg 10^{-4}$. If we take the density as $\sim 0.05\ \text{kg m}^{-3}$ with a radius of 10^{10} m, i.e. a surface area of $\sim 10^{21}\ \text{m}^2$, the depth comes out at $(5 \times 10^{27})/(10^{21})\ \text{kg m}^{-2}/(0.05\ \text{kg m}^{-3}) = 10^8$ m, which is about 1 per cent of the radius. Needless to say these are only order of magnitude estimates.

Chapter 7

Binaries

More than half of the well observed stars have been found to have one or more companions in mutual orbits. Since the masses of the members are normally similar we should picture the orbits as concentric circles, or interlocked ellipses round each other, as in figure 7.1, and not as one star in the centre of a circle. (One can of course always choose a coordinate system with one star at the origin, but such a system would itself be in an orbit relative to the Earth.) More than 65 000 stars with companions have been analysed and separated, with difficulty, into those with two, three, four and more members. Figure 7.2 shows, in the form of a histogram, the result of several investigations. Clearly the first question to ask when trying to understand the behaviour of a star is 'are you disturbed by neighbours'. As with people, the answer will rarely be 'not at all'. Fearing complications it is tempting, as with people, to try to ignore the problem unless it is really serious. In trying to use observations of binaries to understand stars better, this approach must be turned on its head. Every 'complication' is a reflection of some facet of stellar behaviour. The most obvious example is the mass of a star whose measurement is confined, with few exceptions, to binaries with well known orbits. This brings us immediately to the question of what information binaries can yield.

First, some nomenclature. The term *visual binary* (or triple or quadruple) refers to two (or more) stars which do not occult each other as they orbit, but which are sufficiently well separated to resolve the two stars. In other words the orbits are more or less face-on rather than edge on and the stars are well separated. Unfortunately this implies long periods, since the separation must be large, so their study is a very slow business. What if the two images are not resolved? If they are in a face-on orbit, the binary nature would probably pass unnoticed. The combined spectra might look rather odd if interpreted as if from a single star, but otherwise there is not much to give away their secret. This classification is therefore rather arbitrary, depending on the telescope used and the observations made. If their orbital velocities have a sufficient component along the line of sight to the Earth to produce a measurable Doppler shift, alternating in sense of direction, then they are classified as a *spectroscopic binary*. This can only happen, however, if the system has some nice sharp spectral lines and someone has taken the trouble to record them on a number of occasions well spaced out

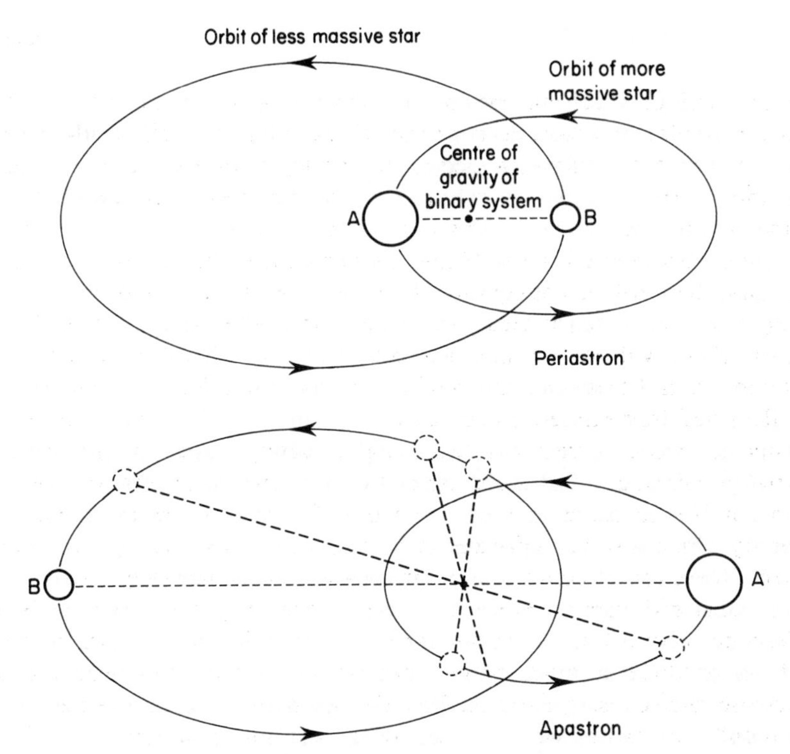

Figure 7.1 Mutual orbits. This is simply a reminder that stars, being of comparable mass, revolve around the centre of mass. A mental picture of one body revolving around a parent body, which is adequate for the Solar System, should not be carried over to binaries.

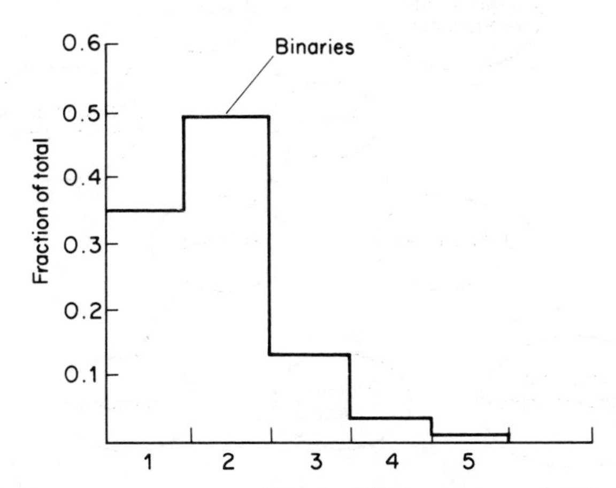

Figure 7.2 How common are binaries? Very common, and if an object is not a binary it may well be a more complicated system. It is a fair bet that some stars regarded as single do in fact have a faint companion or a planetary system.

within a period, which may be anything from a day to several years. Clearly many examples may have been missed. If the orbits are sufficiently close to edge on for occultations to occur, the binary nature is revealed unmistakeably by the light curves, whether or not the spectra are recorded. The shape of the light curve in this case depends first and foremost on the separation between the stars. If this is several times the sum of the radii of the stars, there will be flat regions where the two stars are not occulting each other, cut into by sharp dips when one or the other star comes in front. Clearly there will be two dips per rotation period. Such curves are quite different from the sinuous curves of an intrinsic variable, and the object can be identified immediately as an 'eclipsing binary'. (The term occulting is technically more correct but 'eclipsing' is always used, in the sense of partially eclipsing.) The great majority of known multiples are spectroscopic and/or eclipsing. It needs too much of a coincidence for the plane of a binary which is so far separated to be resolved to pass exactly through the Earth. There are of course many binaries which are not only resolved but have measured orbits. None of these is eclipsing. The Hubble Space Telescope, with 0.1 arcsec resolution or better, will resolve closer binaries, and so produce a much bigger overlap of 'visual' (astrometric) and spectroscopic/eclipsing binaries. Perhaps new names will be invented. Even the number of names we have used so far warrants a summary:

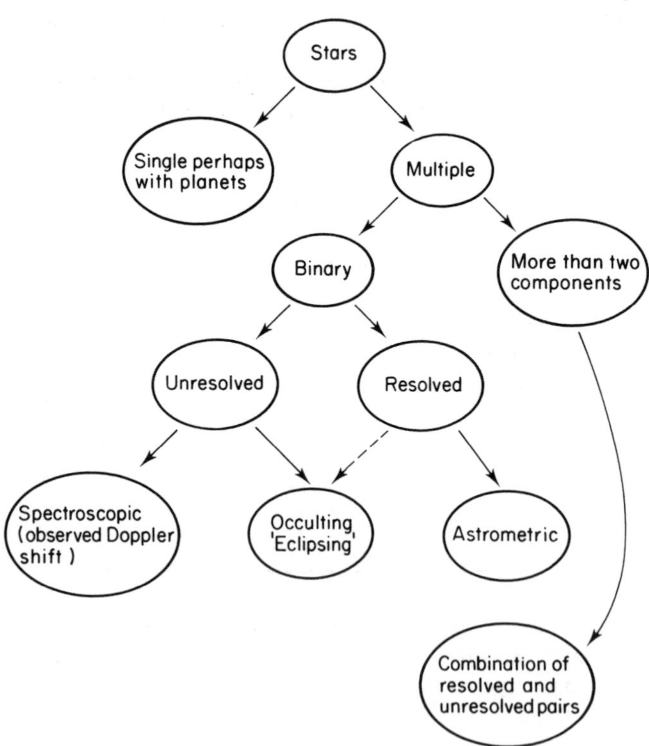

Measurement of spectroscopic binaries needs spectroscopes of high resolution to measure Doppler shifts so it is normally a job for the professional. Systems with more than two stars are difficult to model, so most studies are, so far, restricted to binaries. Let us concentrate, therefore, on some relatively straightforward cases, sketched in figure 7.3, and see how their light curves can be deciphered. The ratio of the depths of the two dips is equal to the ratios of the brightness of the two stars (i.e. luminosity per unit area). The relative duration of the two dips is related to the ratio of the radii of the two stars, but again not necessarily directly, as a further glance at figure 7.3(*c*) shows. Such ratios can in any case only be converted into absolute values if the distance of the system is known. This may not be too difficult if the system is within a cluster. But if not one must be very wary of judging the absolute luminosity (and hence distance) of a star in a

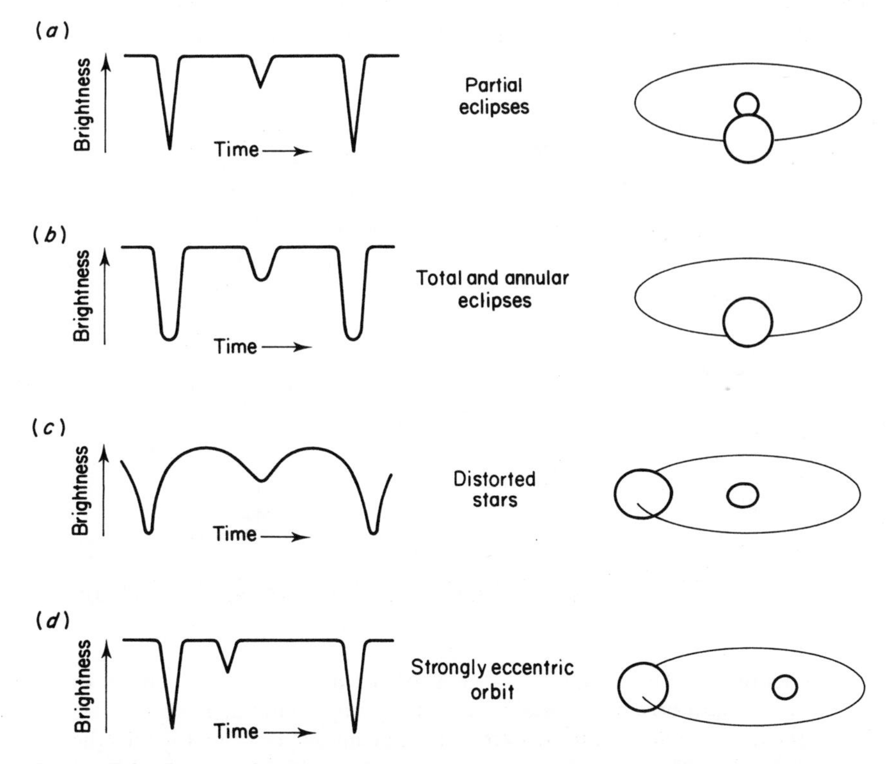

Figure 7.3 Types of eclipse. The shape of the light curve will depend on the chance degree of alignment of the orbital plane with the Earth and the size and luminosity of the components. L = light flux reaching the Earth. From *Astronomy: the Cosmic Journey* 3rd edition, by William K Hartmann © 1985 Wadsworth, Inc. Reprinted by permission of the publisher.

binary (multiple) system by trying to place its spectral class on the conventional main sequence. Mass transfer between the members drastically affects the evolution of close binaries, and they will never, thereafter, lie on a main sequence line as calculated for single stars. Figure 7.4 shows the type of wandering path, on a HR diagram, which calculations reveal. The eccentricity can be judged by any skewness of the timing of one dip between its neighbours, as shown in figure 7.3(*d*). Since eccentricity is itself a ratio, this does not need a known distance. The precise shape of the dips reflects a mixture of asphericity of the stars (which can be considerable in close binaries) and the effects of limb darkening. Again these effects are parametrized by ratios, rather than absolute values, so the distance problem is side-stepped. The detective work of matching up measured light curves to models such as in figure 7.3 was pioneered by Shapley and H N Russell among others. It demanded arithmetical work of, at the time, heroic proportions. At that time the word computer meant a person, not a machine. Nowadays, of course, a home computer could repeat a lifetime of calculations in perhaps a few hours. By the same token, modelling can try to

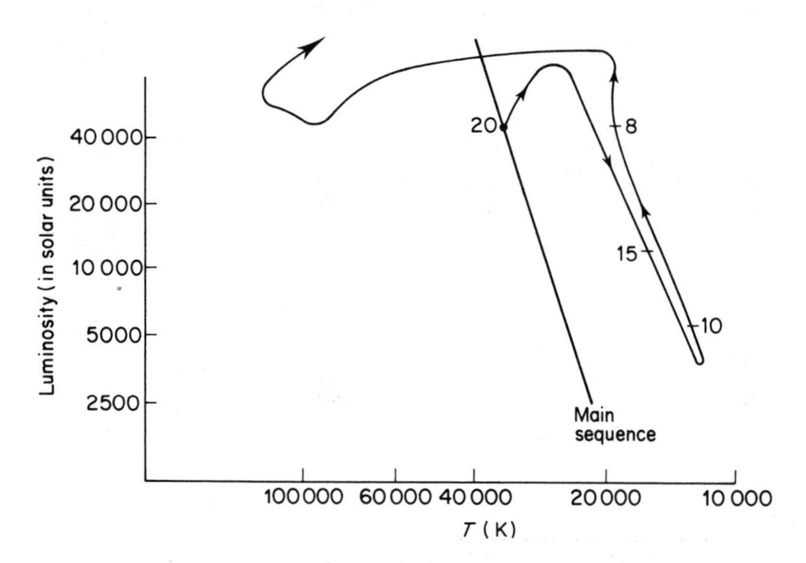

Figure 7.4 The evolutionary track of the donor star in a binary in which both members initially have high mass. Without mass transfer the track would be a short spur from the main sequence up towards the top right. The other star gains mass and so undergoes a very rapid evolution towards the top of the main sequence. After calculation by various authors, principally de Loore, de Greve, Kippenhahn and Paczynski.

avoid being misled by undue simplification. Let us come back, therefore to some of the complexities which are there to be unravelled.

Figure 7.2 shows that, judging by this sample of 66 multiple stars, about 15% of *all* stars may be expected to be more complex than binary. (The Alpha Centauri system and the Mizar system are two examples which can be picked out by the unaided eye.) In addition, there may be non-stellar components such as rings of gas or dust. This is an extremely important topic in astrophysics, with a direct bearing on planetary formation, star formation and various stages of stellar evolution. For instance, most protostars, or very young stars, many of which are presumably binary, have a cocoon of dust and gas. (Dust may well be necessary to initiate the condensation and fragmentation of the parent cloud.) Protostars also commonly display back to back ('*bipolar*') jets of outflowing material. As far as late stages of evolution are concerned, R CrB variables are interpreted as stars which periodically throw out, into the line of sight to the Earth, obscuring clouds of dust. Thus if a binary (multiple) system includes such objects, the light curves of the system are liable to be affected in a significant way. Binaries with mass transfer exhibit marked variations due to the heating of the matter in transit. They are called, as a group, cataclysmic variables. The following chapter is devoted to their structure, and Chapter 15 is devoted to measurements on such systems. Such systems have received too little attention in the past. One reason no doubt has been that they are complicated to model but large computers are changing that. Another reason is that, as described in the next chapter, x-ray measurements are needed to give a guide to the temperatures involved, and this of course requires x-ray telescopes on satellites. Developments in these very expensive areas of astronomy are good news for amateurs, because they create a need for much more associated data on the variations in the visible output.

Kopal has eloquently described the interrelation of measurement and modelling of binaries in his book '*The Language of the Stars*':

> 'eclipsing variables have always been favourites for pioneers of accurate photometry of any kind—visual, photographic or photoelectric—and the total number of observations made in this field run into millions. Observations alone are, however, insufficient to disclose to inspection a wealth of information which they contain. To develop this information calls for introduction of systematic methods rooted in physically sound models of the phenomena we observe to decipher what observations have to say ... The problem at issue is indeed one of astronomical cryptography: the messages these stars send out on waves of light are encoded by processes responsible for them; while the task of the analyst is to decode the photometric evidence to yield the information which it contains. To do so

requires some knowledge of the code; and to provide it is the task of the theoretician. ... The decoding process constitutes an essentially mathematical problem; but the identification of the code is primarily one for the astrophysicist.'

Chapter 8

Cataclysmic and eruptive variables

If we took the above adjectives literally they could refer to any outburst of radiation from a star, from a solar flare to an isolated massive star which after millions of years of evolution suddenly collapses in a few seconds, forming a type II supernova. In a loose sense all such events might be called *novae*, literally meaning new stars, and originally applied to all stars showing surges of luminosity such as that shown in figure 8.1. There are however many ways that such brightenings can occur, and the term nova is now too vague. The title of this chapter is coming to mean those objects where the outburst arises from matter transfer in a binary. Flares and type II supernovae, being single stars, are excluded from this category. In some cases only one outburst has been seen within the brief span of human astronomy and these are called classical novae. The great majority, however, exhibit one or more well defined periodic features which provide detailed clues to the mechanisms involved. A cataclysmic variable is in fact a complex system, not a single star. There are at least three sources of light involved—the two stars of the binary, and the matter being transferred from one to the other. Often, as we shall see, this matter gathers in a more or less stable disc around one star, and the disc may well outshine both the other stars. Furthermore, the radiation output of the disc may be concentrated in a 'hot spot', emitting copious x rays. Of the several periodicities that may be involved one will be the orbital period of the two stars, and another may be the rotation period or oscillation of the disc. Either or both of the stars may pulsate. There is a lot of information to be gained from light curves. Often the periods are short and the stars bright so that novae are good objects for an amateur. They need repeated observations over a long period which the amateur can provide, and the harvest of astrophysics can be high. We will try to describe the processes in the triple light source, so as to suggest what to look for and measure. Perhaps the best place to start is the new feature, the disc. Incidentally, there is a summary table of the different types of cataclysmic and eruptive variables at the end of this chapter, and you may find it useful to refer to it and Appendix 8 as new types are introduced.

One of the commonest forms of cataclysmic variable, dwarf novae, consist of a white dwarf and a star which we may call 'main sequence', though the paths of evolution for a star in a close binary are always

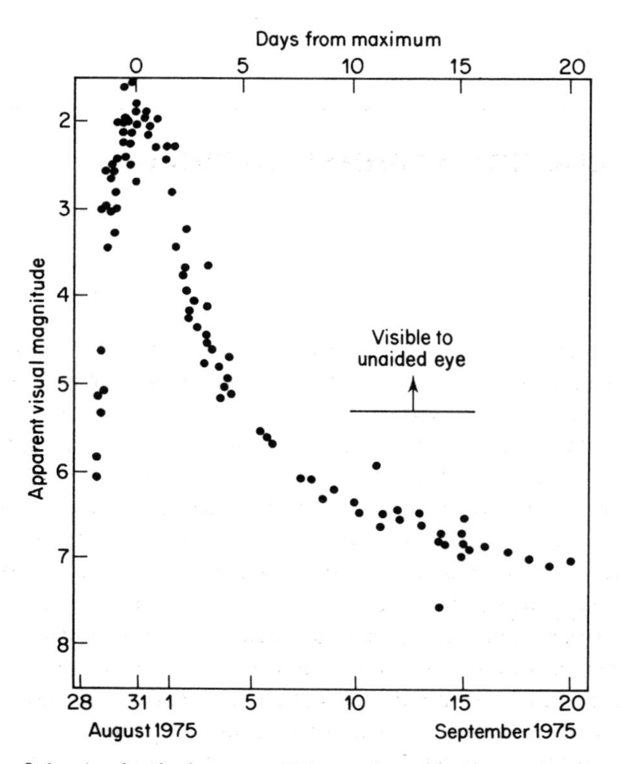

Figure 8.1 A classical nova. This outburst was recorded in 1975, having actually occurred about 4000 years before. It was first named Nova 1975 Cyg, and this has become a standard notation for a variable, V 1500 Cyg. No outburst has been seen since, but this probably simply means that such outbursts only occur every few hundreds or thousands of years.

modified to a greater or lesser extent by mass loss. A useful term for this star is the 'donor star' since the compact white dwarf attracts and captures matter from its much more diffuse partner. The technical term for this process is quite expressive: the white dwarf is said to produce a deep gravitational potential well, and once matter is over the edge of the well it falls into it, as indicated in figure 8.2.

In falling it speeds up in the normal way, gaining kinetic energy. Because the white dwarf is compact and dense beyond anything in our direct experience the kinetic energy available is very high. Sooner or later most of the kinetic energy of bulk movement will in turn be converted into the random motion which appears as heat. A waterfall is a good and often quoted example. As water drops down Niagara Falls its temperature does not change, but when it strikes the bottom and becomes a white foaming turbulent mass its temperature increases. One only has to convert the

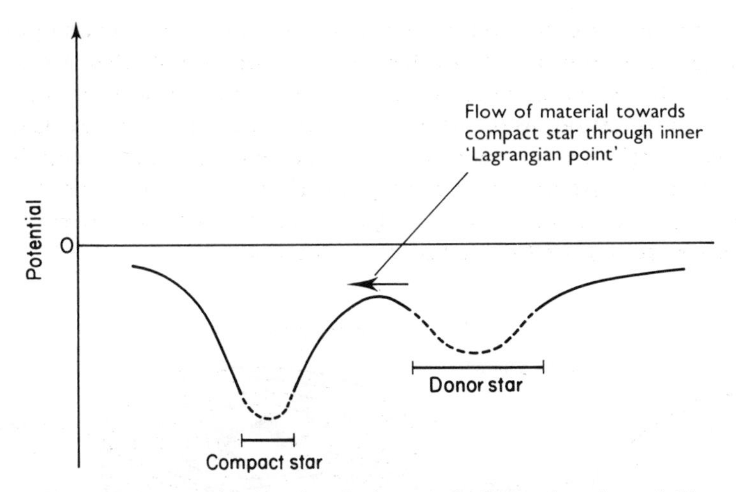

Figure 8.2 Gravitational wells in a typical cataclysmic variable.

gravitational energy loss into calories to find the temperature rise which this process will produce—just one degree in the case of Niagara Falls. A jet of water directed from, say 10^9 m away, and scoring a bull's eye on a white dwarf would lose perhaps a billion times as much gravitational potential energy per kilogramme and the temperature rise would be a billion degrees (10^{13} J kg^{-1}). Of course, it would not by then be water, and stars do not 'direct' the stellar wind, so we have to consider a more realistic case. A white dwarf is a small target from 10^9 m ('bull's eye' is about right, if you work out the angle) and most of the matter will miss and swing round behind the white dwarf. It would follow a very long thin elliptical orbit if it was travelling in essentially empty space, like a comet in the Solar System. In fact, there is enough gas and plasma round the white dwarf (as evidenced by the shape of the emission lines) to cause drag, so the matter slows down, like a Shuttle beginning to re-enter the atmosphere. Eventually, indeed, the matter will land on the white dwarf. Before that, there will be a range of altitude in which the drag gives rise to a path spiralling down towards the white dwarf. The same thing happens to artificial satellites freely and slowly drifting down into the atmosphere. The Shuttle short circuits this process by deliberately losing energy by firing retro rockets. Apollo spacecraft used a carefully adjusted burn of retro rockets to enter an orbit round the Moon. The orbit can in that case be circular because the Moon has no atmosphere, and so there is no drag. Where there is drag, the orbit will be a spiral. The range of altitudes at which matter is 'parked', in a slowly descending spiral, constitutes the disc of accreting matter in a cataclysmic binary. The total mass held in the *accretion disc* depends on the length of time in 'parking' orbit, just as the total mass of aircraft above a crowded airport depends on how long they are being delayed. Assuming that all the matter which escapes from the potential well of the donor star eventually ends up on the white dwarf, the overall rate of heat production is not affected by the existence of the accretion ring, and can be found by a simple gravitational energy loss calculation, as for the waterfall. But since the disc may be giving off as much light as the two stars, the appearance, and certainly the variability of the system, depends very much on the form of the disc. To calculate the shape realistically enough to account for the observed light curves other factors need to be taken into account.

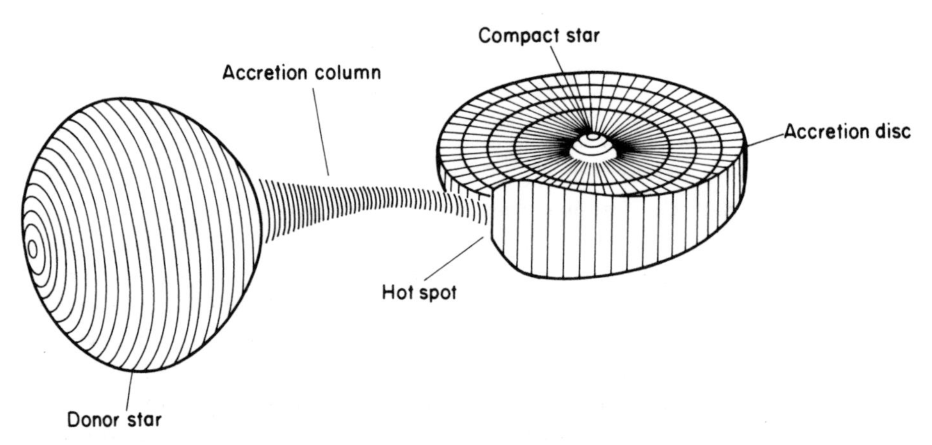

Figure 8.3 A computer artists' impression. This shows a perspective view of a computer simulation of the matter transfer into an accretion disc round a white dwarf, by P Eggleton and J MacDonald, Institute of Astronomy, Cambridge.

Among those factors, spin is of prime importance. Both stars have their own spin, and the angular rotation of the white dwarf is commonly very fast, with periods of the order of minutes. The stars have angular momenta about each other, as will the matter in transit between them. Although the total angular momentum of the whole system will remain constant, this is not (in this case) a very helpful law because the drag will transfer angular momentum between the components. A great deal of skill and computer time has been expended in the last ten years in calculating steadily more refined and realistic cases. Figure 8.3 shows the result of an accurate computer simulation of a particular case. The whole structure is contained within ~ 1 AU. The characteristic feature, besides the disc itself, is the long 'elephant's trunk' funnelling the matter onto the white dwarf. This concentration of energy can cause a visible hot spot on the accretion disc. It is tempting to think of the matter crashing onto the accretion disc. Whilst this conveys the idea of heating it is certainly not good enough to calculate a model. Both the funnel and the disc are very tenuous in terms of our everyday experience, and are probably in the form of plasma rather than gas. They may be strongly influenced by the magnetic field of the white dwarf. In fact the relative sizes of the funnel and the disc, and also the shape of the funnel, are governed by the magnetic field, so the magnetic field is an important physical factor and it is appropriate to distinguish classes of cataclysmic variable by the strength of the magnetic field of the white dwarf (as is done in the summary table at the end of the section). It would be even more helpful if only it were easier to measure the field! To model heating in a magnetized plasma an extra idea is needed, that of a *shock wave*. The shock wave from an explosion is a transient which passes by and is dissipated. The shock wave from Concorde, or any other supersonic plane, is a stable feature moving along with the plane. Extending this idea slightly gives the idea of a permanent shock where two winds—the funnel and the disc—meet. If the whole system settled down to a steady equilibrium we would see the combined radiation of the components but from our standpoint it would be modulated by the effects of rotation. The *hot spot* swings round the star, for instance, with the orbital period, which is a few hours for these stars. In fact, the flow of matter is not steady, so there can be on the one hand rapid small irregular variations ('flares' and 'flickering') or, on the other hand, the major outbursts of energy which gave rise to the name of nova in the first place. Periodic variations with low amplitude (< 0.1 mag) and short periods (< 75 s) have been interpreted as either the white dwarf rotation period, or as oscillations of the disc. The mere fact that these periods are so short is very significant, because they can only occur in connection with white dwarfs. It is a simple application of Newton's laws to show that this is the case (see p 65).

The three components in a cataclysmic variable—the two stars and a stream of heated plasma—can between them set up a baffling range of behaviour. Moreover, the observed features depend not only on the structure of the binary system but also on the angle at which it is viewed from the Earth. If the disc is seen face on, its flickering can be well observed. At oblique angles it is less apparent. Different light curves do not, therefore, necessarily mean different types of object. The changing spectra help to indicate which processes are being observed—for instance, emission spectra seen during quiescent phases are masked in outburst by a powerful thermal spectrum, cut into by absorption lines.

Since a comprehensive review would be beyond the scope of this book, we attempt only to indicate the main types. When you have chosen a particular type to concentrate on, look for further description in, for instance, one of the books listed under '*Astrophysics*' in Appendix 5. There you will find a detailed but unfinished detective story of the search for a plausible model, to which you may well wish to add your evidence.

If we had good models (or even maps from very high resolution interferometry) of the systems which appear as cataclysmic variables, we would probably classify them by their different structures. Since we do not,

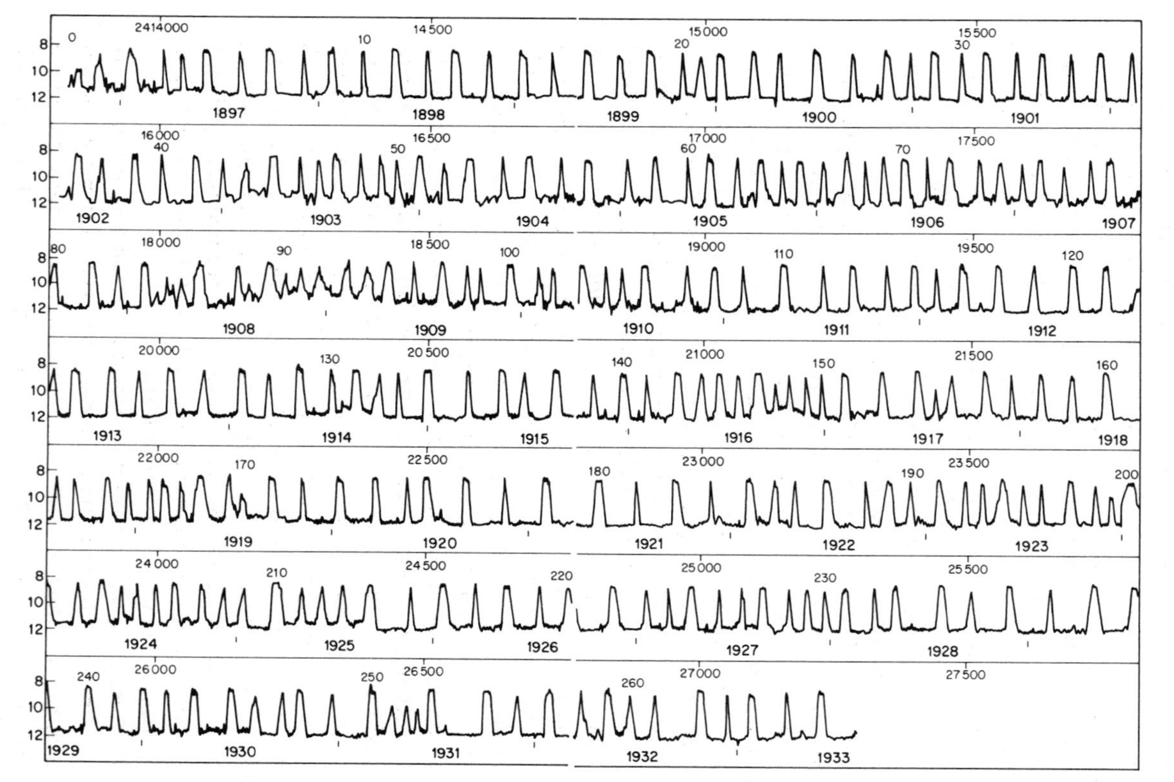

Figure 8.4 A beautiful recurrent nova, SS Cygni. In each outburst, about 5×10^{19} kg of material must be transferred to give such outbursts. The white dwarf on the receiving end of this material must have gained about 10^{22} kg (1% of the mass of the Earth) between 1896 and 1933, because it is certain that once the material is on the white dwarf surface, it will never escape again.

they must unfortunately be categorized by their observed characteristics. This means that the notation changes from time to time as more features are discovered. A large group of cataclysmic variables come under the observational classification of 'dwarf novae'. They are also called after the archetypes, U Gem and SS Cygni (shown in figure 8.4). The outburst amplitudes are 2–4 magnitudes (in the visual), superimposed on the system luminosity of about +3 visual magnitudes. They last a few days. The outbursts are repeated every few tens or hundreds of days. The binary periods (i.e. the variations interpreted as due to the mutual orbital rotation) seem to split into two groups, less than two hours (in the SU UMa type) and greater than three hours—remember these are close binaries. The disc speed corresponding to these periods is high—typically 500 km s^{-1}, which serves to broaden the Balmer lines appreciably, to about 2 nm. Such line broadening is commonly used to estimate orbital speeds. The puzzle to be solved with SU UMa stars is that they show 'superhumps' and 'supermaxima', presumably due to 'hot spots'. Their appearances seem to be related to some sort of beat frequency. Various patterns have been discovered, which narrow down the range of possible models. It is thought that the magnetic field of the white dwarf is important, leading to what is known as the 'intermediate polar' model. This is a key word to follow up.

There are various other subgroups of dwarf novae. The Z Cam stars show a 'standstill' on the light curve (figure 8.5). The reason for this is not understood so these are well worth measuring. The WZ Sag stars show long periods between outbursts. An example is shown in figure 8.6. If you decide to observe such stars, you will need to look up the most up-to-date ideas on their structure, so as to decide which features to concentrate on.

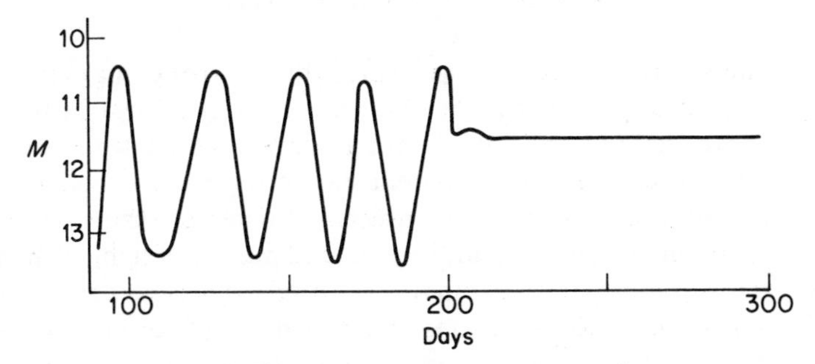

Figure 8.5 A Z Cam light curve.

The group known as recurrent novae are rather easier to describe. The companion of the white dwarf is a giant rather than a 'main sequence' star, and the orbit radius is bigger. The times between outbursts are typically tens of years rather than days. About a dozen or so are known. In their spectra, absorption lines of the star are mixed with the emission lines from the disc. *Recurrent novae* tend to throw off a nebula in the process so although they are rare they are interesting objects. 'Recurrent' is not a particularly good name, because all dwarf novae show repetitive behaviour.

There are interesting correlations between the outburst size, the time between outbursts and the decay time, which apply, roughly at least, across all these types. For instance, one such relation is

$$\text{amplitude (in mag)} = 1.5 \log(\text{mean period in min}) + 2.$$

Is this a good enough fit to be significant, and does it reflect a real, identifiable, feature of the processes? Modelling still needs to make too many simplifications to know, but one day we will know whether such regularities arise from the basic physics of the system.

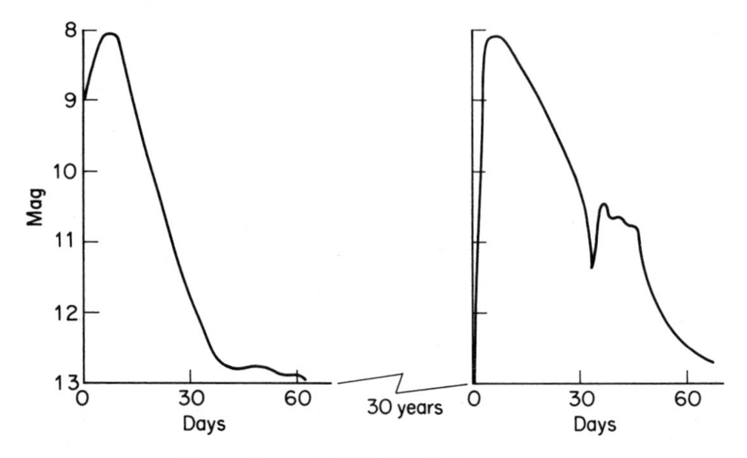

Figure 8.6 A WZ Sag light curve.

Light emitted from a region pervaded by a strong magnetic field is generally polarized, that is, the direction of the electric field in the electromagnetic wave making the light is not random. When white dwarfs are formed, the magnetic field of the parent star is trapped and concentrated in the contraction, and many white dwarfs do indeed show evidence of possessing a strong magnetic field, by emitting polarized light. When such a white dwarf is in a cataclysmic binary we would expect to see variations in the polarization for various reasons—as the angle between the line of sight and the magnetic field changes, for instance, and as various parts of the system come into view. Objects where these effects are very marked are called polars, and the archetype is AM Her. It shows polarizations (linear and circular) which are observed to vary by about 10% with a 3.1 hour period, presumably corresponding to the orbital period. Such stars are suitable for study by polarimeters, even on small telescopes (see §10.5.2). Any observations linking changes in polarization with other changes in the light curve would be strong clues to the processes involved.

Novae generally contain a white dwarf, but what if its companion is also a white dwarf? Such a binary system can be extremely compact, with a correspondingly very short orbital period. AM CVn is such a case, with an orbital period of 18 min. Another diagnostic feature of such systems is that one does not expect to see hydrogen lines, because white dwarfs have very little hydrogen left. There are a handful of cases with both these characteristics (see figure 8.7). AM CVn also shows a 2 min period, thought to be the pulsations of one of the stars as in the ZZ Ceti variables (Chapter 6). Normally there is steady loss of orbital energy in binaries because of the large, and usually increasing, effects of atmospheric tides. Such effects are minute in white dwarfs, which are for practical purposes solid bodies. Will they continue orbiting for ever? It is thought that the emission of gravitational energy will make them very slowly sink towards each other. This is an interesting point, because if they ever come close enough to collapse into a single body there will be a huge and sudden release of gravitational energy. This process is the currently favoured hypothesis for the mechanism responsible for type I supernovae, because the predicted rate (about one per 100 years per galaxy) and energy release fit well.

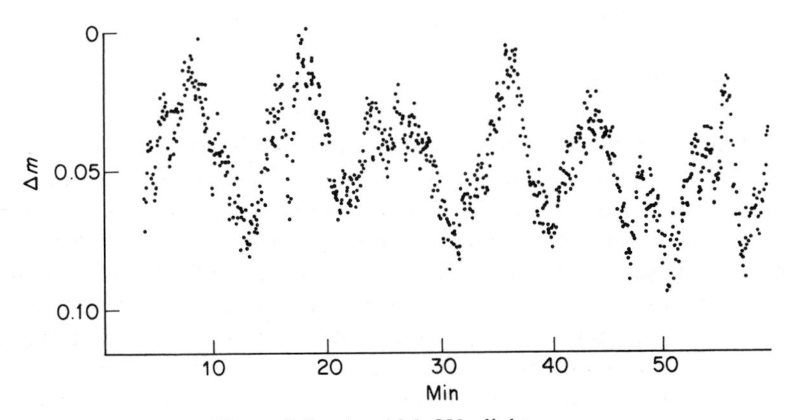

Figure 8.7 An AM CVn light curve.

Type II supernovae on the other hand are not thought to be due to binaries. They are in a class of their own, so do not fit easily in any chapter. They are briefly described in note 8.3 below.

We have saved till last the novae which show the largest outbursts and have only been observed once, probably because outbursts only occur every few hundred or few thousand years. They are called *classical novae*, because they were the first to be recognized, and have a special feature of ejecting a shell of matter, and this corresponds to being much brighter (absolute mag rising to -7), so that radiation pressure becomes important in the process. During the decay of the outburst, fast oscillations are sometimes seen—the period is 71 s in the case of DQ Her, for instance. Clearly this is again evidence of behaviour near or on the white dwarf. The power surge in classical novae is very large (see figure 8.1). Indeed, some classical novae have initially been mistaken for supernovae. Novae and supernovae are very exciting when they occur in our Galaxy, but one needs a big telescope to observe them in other galaxies. The best ones are also well observed by professionals, photometrically and spectroscopically, in a range of wavelengths, so it is difficult for amateurs to contribute. The exception to this, as mentioned in note 8.3, is at the very start of a nova or supernova. At that stage, any record, even with a fixed ordinary camera, can be of great value.

Visual observations alone can never give a complete picture, and this is especially true in the case of cataclysmic variables. Because of the high temperatures which can be produced, many cataclysmic variables (for example AM Her) are very bright in x rays, indeed the whole subject expanded greatly as soon as x-ray detectors were lifted above the atmosphere, first in rockets and later in satellites. Figure 8.8 shows a dramatic example of a type of behaviour unknown until it was revealed by an x-ray telescope in a satellite. It is a trace of the x-ray intensity of the star later identified as 2S 1636-536 and observed in July 1983 by the x-ray telescope on board the satellite EXOSAT. It reveals brief but powerful surges of x rays, repeated at almost regular intervals of less than two hours. EXOSAT also carried a series of x-ray filters which allowed the intensity of different parts of the x-ray spectrum to be compared (broad band x-ray spectroscopy, analogous to UBV photometry in the visual). This showed that the spikes contain x rays with a wide spread of energies, with relative intensities consistent with a thermal source. The temperature needed to match the distribution is very high—about 100 million degrees. We are used to such temperatures in the core of a star, but no x ray produced there can escape from the star. To find such temperatures in plasma so thinly veiled that the x rays escape was something entirely new. A temperature of 10^8 K is not only enough for the thermal spectrum to peak in the x-ray region but also enough, as we have seen, to allow thermonuclear reactions to occur at a rate which gives a significant yield of energy, as occurs steadily in stellar cores. In this case the resulting explosion would expand the material and so cut off the reactions. Perhaps that is the reason the energy is released in pulses.

The fact that we observe radiation in pulses does not necessarily mean that it is produced in pulses. The familiar example is the rotating light of a lighthouse which is often used as an analogy for a radio pulsar. In such pulsars, such as the famous Crab pulsar, radiation is generated by electrons spiralling in the rotating magnetic field of the neutron star. This radiation is mostly in the radio region—only a handful of the 300 or so known pulsars have been observed in the visible range of wavelengths. There are a few stars which exhibit pulses with the same machine gun like rapidity, but in the

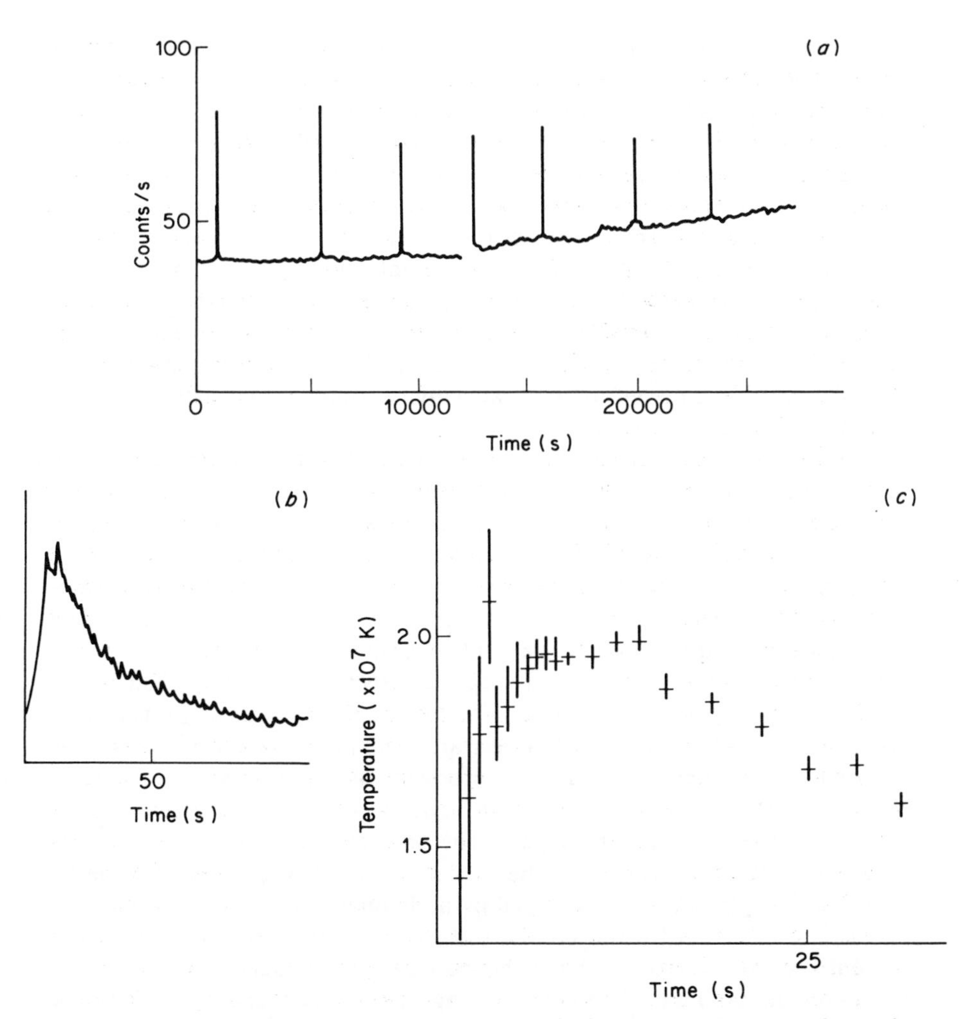

Figure 8.8 A very dramatic variable. (*a*) and (*b*) X-ray telescopes above the atmosphere have revealed one of the most violent forms of x-ray variability. This one shows, every hour or so, a burst of quite hard x rays lasting only about a minute. (*c*) X-ray filters (absorbing plates) split the energy range into two, yielding an estimate of temperature. The temperature rises to over 10^7 K in a few seconds. This is high enough for thermonuclear reactions, but is the density in the emitting region high enough to rise above critical conditions (the same sum that has to be done for thermonuclear reactors on Earth)? Matter falling on to a neutron star could rise to the same temperature thanks to purely gravitational energy. So does the few seconds represent the free fall time on to a star or the time for thermonuclear reactions to get up to power? In this case the answer is probably the latter. In other novae this is an open question, with an intriguing link to unsolved problems in laboratory physics. After Turner M J L and Breedon L M 1984 *Mon. Not. Royal Astron. Soc.* **208** 29. Reproduced by permission of Blackwell Scientific Publications Limited.

x-ray region. These must have a different explanation. They can be understood if the thermonuclear source of x rays is confined to the polar regions, which flash into sight as the neutron star spins. Since the infalling matter is guided by the huge magnetic fields to the polar regions, this is very plausible. Such a model has been worked out in detail for HZ Her (Her X-1). The x-ray pulse rate, interpreted as the rotational period, is 1.24 s, and is subject to a strong modulation with a period of 1.7 days. This modulation is almost certainly Doppler shift due to orbital motion, since eclipses within the cycle are also seen. Unlike 'ordinary' pulsars, x-ray pulsars emit visible light, copiously produced by the high-temperature plasma, so amateurs are not excluded from their study. In fact, Her X-1 was discovered as an optical variable in 1936, long before its x-ray activity could be revealed by satellite or rocket borne detectors.

We will take one more case involving a white dwarf or a neutron star in a close binary. Suppose now the donor star is a giant or supergiant with an envelope big enough to engulf the compact star. Because the envelope is so tenuous it is quite possible for the compact star to continue orbiting inside the giant for thousands or even millions of years. Gravitational or thermonuclear energy release produces x rays in such cases. An accretion rate of 10^{-8} or more solar masses per year will be enough for the luminosity due to the accretion to be comparable to the luminosity of the stars themselves, and so stand a good chance of being detected. Such binaries, in which the compact star orbits within the large star, are called 'symbiotic' stars. The thermonuclear energy may again be produced as explosions. The technical term is 'flashes'. Flashes occur inside single stars, but are not seen because they are hidden by the outer layers. The duration of a flash is in this case much less than the time for the radiation to diffuse out. The theory developed for these cases has helped in modelling flashes in symbiotic stars.

It is not necessary for the smaller star to be as compact as a white dwarf or neutron star. It may still be at the main sequence stage of evolution if it is of low mass, and therefore evolving slowly. The energy release per kilogramme for matter falling on to a main sequence star is much less (about a hundred times less) than that for falling on to a white dwarf. Thus for the luminosity so generated to be detectable, mass transfer rates of 10^{-6} or more solar masses per year are needed. Naturally this can not be sustained for more than a few million years, so we would expect such symbiotic stars to be fairly rare. This is indeed the case, and only a few dozen are known. They are however of great interest and, because of their complexity, more observations would always help unravel their structure. The envelope of the giant is tenuous enough for much of the radiation to escape, especially if the transfer is by a thin stellar wind rather than tidal distortion.

The expected behaviour of stars in different stages of development in a symbiotic binary, with different types of accretion disc, is well reviewed by

S J Kenyon in '*Interacting Binaries*', ed Eggleton and Pringle. However, it would be as well to look also for the most up-to-date review available, or seek professional advice, since this is a rapidly advancing field.

Note 8.1 How to calculate temperatures

The loss in potential energy of a kilogramme of matter in falling on to a star of mass M and radius R is not quite GM/R—it would be equal to that value if it fell literally from infinity. This energy is converted automatically to an equal amount of kinetic energy in the fall. On impact a fraction of this energy will be converted into heat (as with Niagara Falls), which is random motion of molecules. This conversion will never be complete. Thus for this reason too the maximum value of GM/R will never be reached. Nevertheless it is useful to calculate this value as an upper limit. This needs an assumption about the composition of the matter. Suppose the kilogramme of matter is made up of n particles of mass m. Each particle has gained an energy of, at the upper limit, GM/Rn. But the average energy of particles in a gas at temperature T is $3.5kT$ at high temperature (it is $1.5kT$ at low temperature). k is a physical constant, named after Boltzmann, whose value is 1.38×10^{-23} J K^{-1}. Equating the two gives $T = GM/3.5kRn$. But $nm = 1$ so this is equal to $GMm/3.5kR$. It is easy now to find the maximum temperature rise of, say, hydrogen falling on to stars of different mass and radius. Table 8.1 gives some typical values, starting with the x-ray burster case.

Table 8.1

Star	Mass (kg)	Radius (m)	Maximum temperature rise (K)	Energy (eV per atom)
Neutron star	3×10^{30}	10^4	10^{11}	10^7
White dwarf	10^{30}	6×10^6	10^8	10^4
Sun	2×10^{30}	7×10^8	10^6	10^2

Although these numbers are only rough upper limits, they are adequate to highlight some of the features of matter transfer between stars. First, it is clear that x rays can easily be produced by matter falling on to a neutron star—the x-ray burster and x-ray pulsar cases. Figure 8.9 shows the temperature needed to produce thermal x rays. However, it is important to remember that x rays are stopped by only a thin veil of matter, so if they were indeed produced at the surface of a star they would be unlikely to escape. The x rays we see come from further out, and the gravitational energies available are less than those shown above.

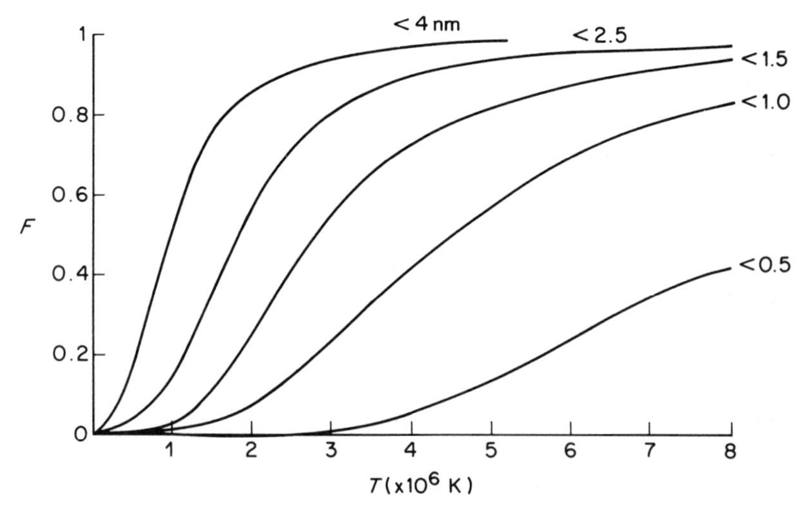

Figure 8.9 How hot does a body have to be to emit x rays? It depends what you mean by x rays. Here, the fraction F of number of photons in the x-ray region, defined by the upper limit labelled on the curve, is shown as a function of temperature (since x rays are detected one by one it makes more sense to plot numbers rather than total energy). Thus for a body at 4×10^6 K, 44% of its photons have a wavelength shorter than 1 nm (corresponding to 1.24 keV). However, only 5% of its photons have a wavelength shorter than 0.5 nm (2.48 keV).

It is revealing to work out the rate at which matter must fall in to a star like the Sun to sustain an increase in luminosity—let us say by a solar luminosity which is 3.8×10^{26} watts. We merely have to solve the equation $(GM/R) \times$ mass per second $= 3.8 \times 10^{26}$ watts, to give a value of order 10^{14} kg s^{-1}. This is much bigger than the rate at which the Sun sweeps up typical clouds of interstellar matter. The thin rain of ISM does not make novae. However, 10^{14} kg s^{-1} is only 10^{-9} solar masses per year. This is a typical rate of mass transfer in the evolution of close binaries containing, say, a main sequence star and a red giant. The red giant partner may, for instance, lose half its mass in a few hundred million years to the other partner (which then rapidly becomes a red giant itself). Given that most stars are in binaries, this is apparently a normal process which proceeds smoothly, without the outbursts that characterize novae.

Note 8.2 Summary

Since the cataclysmic variables form such a complicated group, a summary may be useful. For details of particular types, see one of the books listed under '*Astrophysics*' in the further reading list.

Cataclysmic variables are binaries (this excludes flare stars and type II supernovae).

One member, the 'donor', is sufficiently large that matter overflows towards the other, compact, member.

Matter builds up into an accretion disc, often with a 'hot spot', and is parked there for a time which is long compared with the orbital time of the binary. It is often difficult to decide which if any of the periods seen in a cataclysmic variable is that of the orbital rotation.

Some mechanism causes instability in the accretion disc. Ionization instability and gravitational instability may be important.

Variations of the rate of matter transfer onto the compact member cause irregular variations on a wide range of amplitudes and timescales. Large outbursts may be enhanced by thermonuclear reactions.

Table 8.2 attempts to summarize the classification, but the notation is not universally agreed, and in any case changes as further discoveries are made. It might be a good idea to amend or remake the table as you read further into the subject.

Table 8.2

First component	Magnetic field	Second component		
		Main sequence	*Giant*	*White dwarf*
Main sequence		Eclipsing interacting binaries	Symbiotic stars	
White dwarf	low	Dwarf novae U Gem SS Cygni Z Cam–UX UMa WZ Sag SU UMa	Recurrent novae Symbiotic stars	Type I supernovae
	high	Polars AM Her		AM CVn
Neutron stars	low	X-ray bursters		
	high	X-ray pulsars		

Note 8.3 Supernova 1987A

Supernovae are distinctly rarer than recurrent or classical novae, there being on average one per galaxy per 30 years. By scanning the nearest 10 000 galaxies, one could presumably be seen almost every night, but it is only the nearby ones, which can be studied in detail, that are really of interest. A supernova occurred in our own Galaxy in 1604, now called Kepler's supernova after his descriptions of the event. This was before the invention of telescopes, and in recent years, as a repeat performance for the benefit of modern instruments seemed long overdue, it has been suggested that perhaps even a supernova might be hidden by the dust and gas which obscures the centre of our Galaxy. In February 1987 a supernova was observed in the Large Magellanic Cloud, a close companion of our Galaxy prominent in the southern sky. There must scarcely be a telescope in the southern hemisphere that has not been turned to it. Amateurs cannot compete with the detector and spectrometer power of all the professionals in such a case. Notably missing from all their data, however, are observations from the first few hours. Any measurements or photographs taken during that crucial risetime of the light curve would have been extremely valuable—and could have come from a sleepless amateur.

SN 1987A has turned out to be a rather strange supernova, not fitting the common classifications of type I (thought to be a collapsing white dwarf pair) or type II (the final stages of a massive single star). The debate is in full swing as this book is being prepared.

It would not be quite correct to say that SN 1987A was not observed in the northern hemisphere. A great pulse of neutrinos, created in literally the first minutes of the explosion before the rise in visual luminosity, passed right through the Earth. A handful of them were detected in particle physics equipment in Japan and Italy. They provide the first direct confirmation of the neutrino flux predictions of models of supernovae, and a zero time against which to set the visual observations, which as mentioned above are unfortunately not available for the explosion itself. If only the neutrinos could have rung a bell in an observatory!

SN 1987A is 170 kpc away, and reached an apparent visual magnitude of about 4. A supernova well within our Galaxy might be only 10 kpc away, and, if of the same type, would then be magnitude -2, brighter than Sirius. Clearly that event is being reserved for the northern hemisphere!

Chapter 9

Minor planets

9.1 INTRODUCTION

The minor planets presumably had their origin in the same accretion process that made the major planets in the Solar System, though not necessarily at exactly the same time. In fact, in one sense they have never finished the formation process because collisions must still be taking place. That is their main interest: they are the closest we can get, in time and in place, to a planetary formation process. That is reason enough for plans for minor planet flybys to be considered attractive. Indeed there is just a chance that an asteroid will one day provide a cheap base for an interplanetary space station. What we can see and do much more cheaply from the ground is not nearly so dramatic but it can also shed light on the outstanding problem of the origin of planetary systems. We do not even know whether there is one or hundreds of millions of planetary systems in our Galaxy. The modelling problem is so difficult that it is even debatable which mechanisms contribute most to the formation process, without even getting to the details of how they operate. For instance, did magnetic forces play an important role? We will start with one fairly straightforward link between observations of minor planets (or 'asteroids') and the formation process, and then go back to try to fill in a more general picture.

9.2 ARE COLLISIONS DETECTABLE?

About 3000 *asteroids* have been found, nearly all with orbits between 2 and 4 astronomical units, so that their orbital speeds are between about 20 and 15 $km\,s^{-1}$. Most are to some degree irregular in form, so that their period and possibly direction of rotation can be found. The fifteen biggest have diameters greater than 250 km, and there can scarcely be any more like that to be found. The smallest seen are, judging from the amount of light they reflect, only about ten kilometres in diameter. Doubtless there are many more of this size or smaller, but if we take the observed number as a lower limit, and use the so-called particles-in-a-box model which assumes that they move randomly around in their allotted space between 2 and 4 AU, we can easily make a rough estimate of the length of time a given minor planet will travel between collisions. It is about 100 000 000 years. Thus one or other of the 3000 will collide every 30 000 years. If it was necessary to see a collision to extract any useful physics out of minor planet work, it would therefore be a waste of time. Fortunately a collision is likely to leave a lingering record of the event. To see why involves the following rather nice piece of the mechanics of rigid (well, nearly rigid) bodies. Minor planets are small enough to be of irregular shape, as are the smaller 'moons' in the Solar System: Phobos, for example. This is crucial to the argument, which would not work for perfect spheres. At a collision, we may imagine two protruding bits of the minor planets catching each other and swinging the planets round, and pushing each other into new orbits with different inclinations and eccentricities and perhaps different semimajor axis, as they separate again. (We will come back to the case of catastrophic collisions in which one or both of the bodies are smashed into pieces.) The spinning bodies will *precess*. This is a word which has to be used very carefully. It is most often used to refer to a gyroscope. A gyroscope spins on a fixed axle which exerts a *couple* (torque) on the body. There is no axle in a minor planet so it is not that sort of precession which is involved. The word is also used of the precession of the Earth (and hence of the equinoxes). That is a precession caused by a couple (force $\times$ radius) produced by the gravitational force of the Moon (and Sun) on the bulging equator of the Earth. That effect (called a forced precession) must exist for a minor planet, but it is small because it is further from the Sun and has no major disturbance from a nearby neighbour such as the Moon provides for the Earth. The Earth also has a *free precession*, which can occur for any freely rotating *aspherical* body. It is small for the Earth, only a few seconds of arc, with a period which happens to be about a year but has nothing whatsoever to do with the motion of the Earth around the Sun. It would happen in a force-free space which, to a first approximation, is the case for minor planets. This, then, is the type of precession we are concerned with. It is small in the case of the Earth because it is nearly a sphere. Minor planets are very far from being spheres and the free precession can be very large. In fact it is wrong to think of them rotating about a given axis. Their axis of rotation itself rotates so that they rotate about, in general, three axes and, moreover, the angular velocities need not be constant. Figure 9.1 shows the various axes involved. This is a strange idea when set up alongside the idea of the conservation of angular momentum in the absence of torques. Torques are nearly zero for minor planets between collisions, and angular momentum is indeed very nearly conserved. What do change are quantities called the products of inertia. These are similar to moments of inertia but they happen to be zero for spherical bodies, so we do not normally come across them. It is these effects of asphericity which make a minor planet

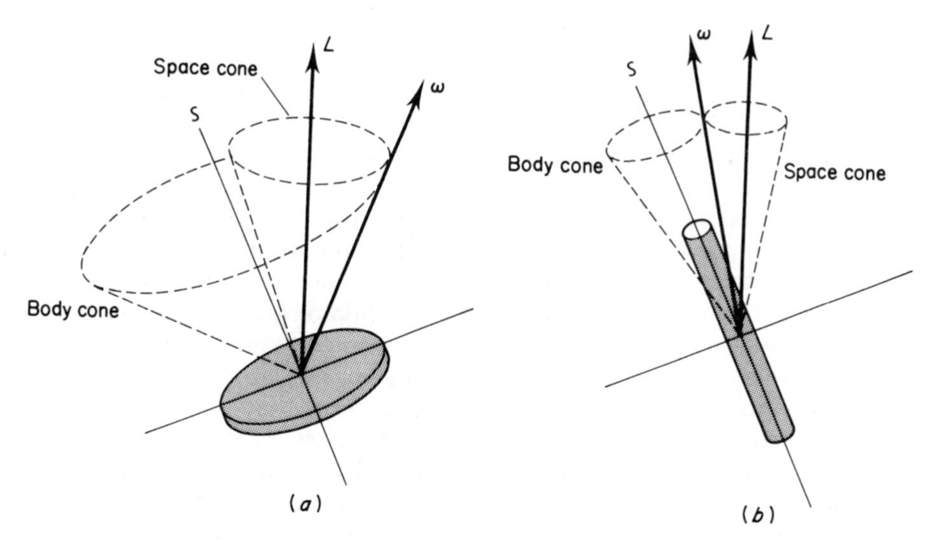

Figure 9.1 How can there be more than one axis of rotation? A disc may rotate about an axis perpendicular to it and through the centre, like any well behaved gramophone record. The axis of rotation ω and the angular momentum L are then along the same line. But, if accidentally dropped, the axis of the spinning disc may itself rotate about a symmetry axis S, drawing out an imaginary cone called the body cone. (It is easier to try it—perhaps with a plywood disc!—than to explain it.) It also rotates about the angular momentum, drawing out a 'space cone'. If you want to go into this more deeply, it is the properties of these two cones which are the key. In the case of an irregular asteroid, three axes and three periods may be involved.

wobble and twist even in the absence of torques. Any oddly shaped body thrown into the air will show the effect. If a minor planet were perfectly rigid (or perfectly elastic) these wobbles would persist forever, and every minor planet which had ever had a collision would exhibit them. In fact the effect of all its collisions would be inextricably superimposed. Fortunately for us minor planets are almost certainly inelastic. Judging from meteorites many are stony and they may even be piles of rubble, rather loosely held together by gravity. They must grate and rumble as they rotate, and the energy so lost reduces the wobble. An early estimate put the time for the oscillations to die away as about a million years. Let us hope that is nearly correct. If it is, we have only to divide a million by 30 000 to see that about 30 minor planets should exhibit, through multiperiod light curves, the wobbling left over from the last collision. If, however, the relaxation time were much shorter, or even if it were much longer, it would be difficult to detect or disentangle the evidence.

9.3 COLLISIONAL EVOLUTION

These ideas give some hope that it will one day be possible to learn about the formation of the Solar System from observations of minor planets, so let us take a more systematic look at the whole process, to see where the greatest difficulties in interpretation lie. We need, ideally, to follow the changes in diameter, period and orbit at each stage. Cosmogonists long ago gave up hope of working *back* from a particular observation to a particular feature of the initial state of the Solar System. Rather one must set up a model of the initial state, and follow it through using all known physics to see if it would produce the system we see today. Many simplifications have to be made, but how serious an effect do they have, and, at the end of the calculation, how close a fit to the present day pattern should be considered acceptable? With, at present, only one planetary system to look at, these are unanswerable questions, and cosmogony is a difficult subject. However, it is common to many theories that there was a first stage in which gas played an important part. It would damp the motion of dust, and drag both gas and dust down from a roughly spherical region to a disc where the dust began to gather into planetesimals—bodies of typically a few metres in diameter. The important point is that, thanks to the gas, this can happen very quickly—in thousands or tens of thousands of years. A striking feature of accretion in the Solar System is that the *spin period* of the accreting body is predicted to remain constant. Is this the reason most bodies in the Solar System have a period in the region of 8 h? Even though there is initially more gas than dust (in terms of average density) it is steadily depleted, and a turning point comes when the gas no longer has a significant effect on the motion of the planetesimals. Subsequent evolution can then only take place through the process of actual collisions. They are rare, so the evolution becomes very gradual. This sequence is shown schematically in figure 9.2. We need to classify the effects of different types of collisions and refine the estimate of the probability of occurrence of the types.

To estimate the collision time we can improve on the random-motions-in-a-box picture mentioned above by taking random orbits within the asteroid belt, or better, orbits clumped into families as they are observed to do, as shown in figure 9.3. The families are concentrated into small ranges of the values of the semimajor axis and eccentricity, with a fairly wide range of orbital inclinations. There is also a definite avoidance of some values of semimajor axis, called the *Kirkwood gaps*. This is equivalent to the avoidance of some particular values of the period which turn out to be simply related to the period of Jupiter. Since this allows more chances for Jupiter's gravitation to disturb the orbit of the minor planet, it looks at first sight as though there may be an easy explanation of the gaps. However, periods with equally simple relations are not avoided, or even preferred, so the effect is not so straightforward, and has yet to be modelled in a

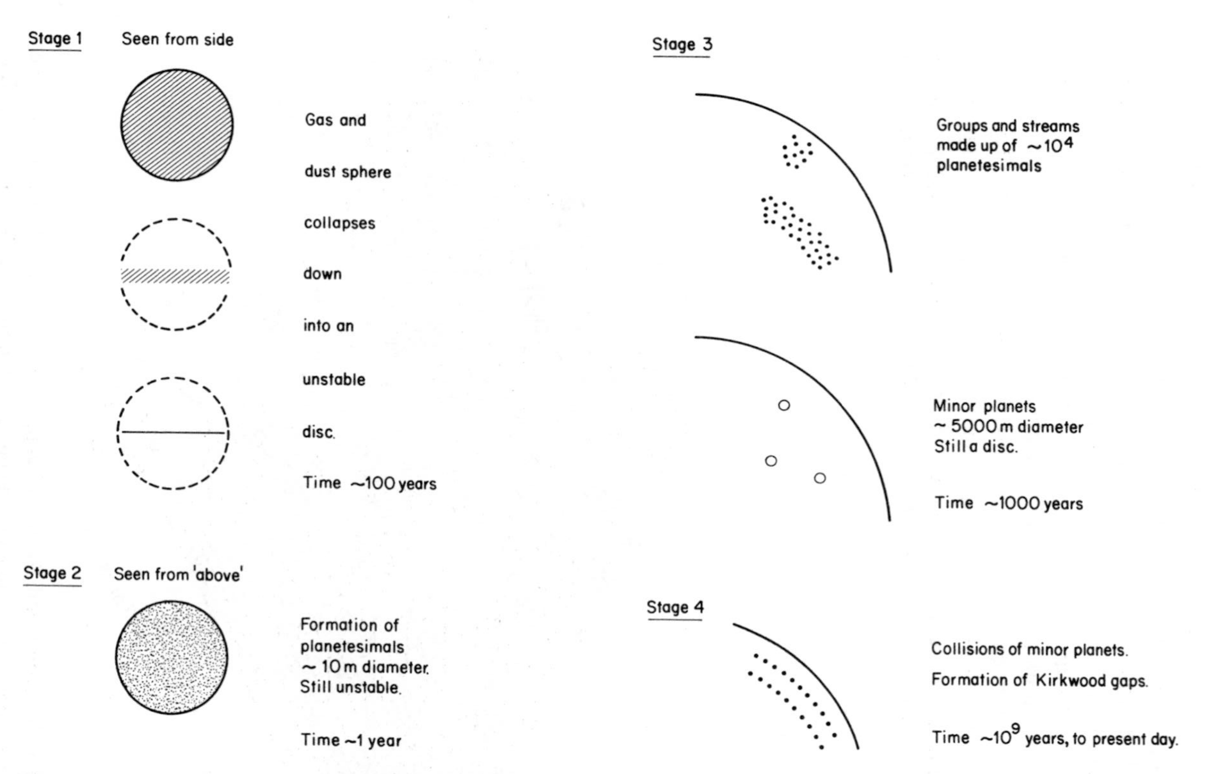

Figure 9.2 Stages in the formation of planets, including minor planets. The early stages proceed quickly, under the influence of frictional drag from gas. When the gas is used up, evolution proceeds through infrequent collisions, so the timescale is much longer. All models would agree on this overall feature. These diagrams are typical of low mass nebula theories, currently favoured. However, some small minor planets may derive from the Oort belt of comets, which does not figure specifically in this picture. After Farinella P 1982 *Icarus* **52** 409.

satisfactory way. The kinematics of the orbits will yield a distribution of the relative velocities at collision, which is what governs the violence of the encounter and hence whether it is glancing or catastrophic. Head-on collisions will of course be very rare since the orbits of most minor planets are prograde. Collisions will arise from the intersection of orbits of different size and shape. The present distributions are shown in figure 9.3, but they were presumably different in the past. To find the rate of collisions, however, we still need to know the distribution of radii of the minor planets. The differential distribution observed today is shown in figure 9.4. The flattening in going to low values of R is no doubt largely due to observational losses. At greater R, the distribution is approximately an

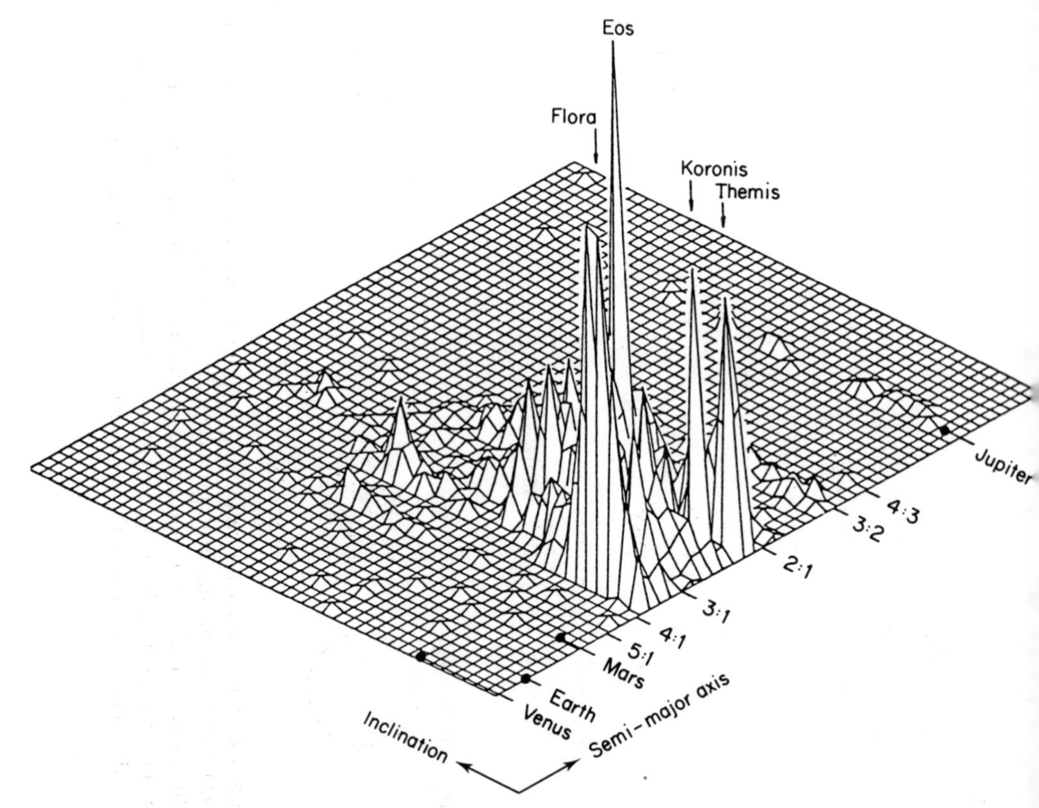

Figure 9.3 Family groups and Kirkwood gaps. Minor planets clearly avoid some orbits whose period is simply related to the period of Jupiter—called resonance. But other resonances are slightly favoured. Strong concentrations of minor planets are evident at some combinations of orbit size and eccentricity. Such concentrations and gaps are also seen in the rings round planets, but precisely how gravity acts to produce these effects is not yet clear. After Eaton N 1986 *Phys. Rep.* **132** 262. Reproduced by permission of Elsevier Science Publishers.

inverse cube, and there may perhaps be a break at about 8×10^4 m. However, one cannot simply assume that an inverse power law covers the whole range. It would mean an infinite number of bodies if the power were greater than -4 and, even worse, an infinite mass if the power were less than -4.

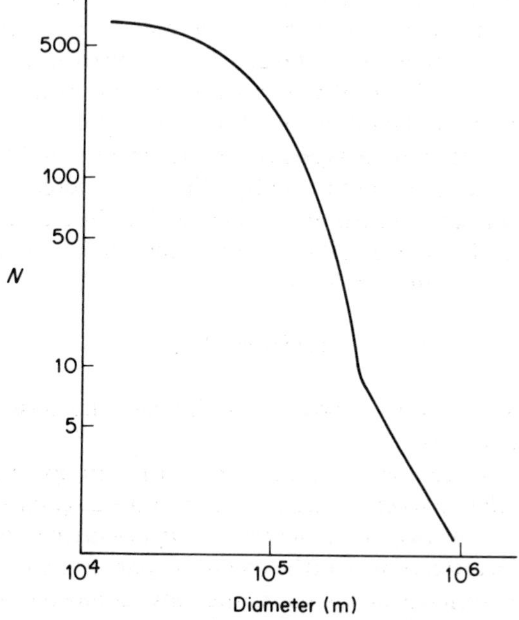

Figure 9.4 Are big asteroids primeval? The distribution of diameters of asteroids shows severe observational losses at small diameters, but probably most of the larger asteroids have been picked up. Some theories suggest that the largest asteroids should have survived the collisional processes. Is there a break in this distribution between the primeval bodies and those which show the results of collisional evolution? After Hughes D 1982 *Mon. Not. Royal. Astron. Soc.* **199** 1149. Reproduced by permission of Blackwell Scientific Publications Limited.

The cratering of the Moon also carries information about the distribution of radii. Most of the craters are due to quite small bodies, a few kilometres or less in radius, and radioactive dating shows the material to be very old—more than 3.5×10^9 years. There has been very little weathering of the surface because the Moon lost any atmosphere at a very early stage. This therefore is consistent with the idea that most small bodies were 'mopped up' in the early stages. Other moons and planets with negligible atmosphere show similar cratering, but they cannot (yet) be radioactively dated. There is evidence hidden in the number of big craters which have smaller, subsequent, craters within them, but it is not easy to disentangle.

Different types of collision will have different effects. A useful parameter in classifying collisions is the kinetic energy per kilogramme of the target body, that is $0.5mV_{\mathrm{rel}}^2/M$, considering each in turn as the target. Based on laboratory experiments, which can duplicate most of the speed range, but for which a vast extrapolation in mass range is necessary, it seems that if the energy is greater than about $10^5\ \mathrm{J\,kg^{-1}}$, the outcome will be catastrophic. For an encounter speed of $5\ \mathrm{km\,s^{-1}}$ this means a projectile mass greater than 1% that of the target. For comparison, burning petrol gives about $10^8\ \mathrm{J\,kg^{-1}}$. Figure 9.5 shows predictions of how various types of catastrophe vary with the radius of the bodies involved. A notable effect of break-up is that the angular speed of the fragments will be in general be higher than that of the parent body. This is a direct result of the conservation of angular momentum. In an oversimplified case of a split into two equal spheres with no loss of angular momentum, it is easy to calculate that the factor is $2^{2/3}$, since we have

$$\tfrac{2}{5}Mr^2\omega_0 = 2(2/5)(M/2)(r/2^{1/3})^2\omega$$

where $r/2^{1/3}$ is the radius of a body containing half the mass of the parent body, at the same density.

This is important because anything to do with periods may perhaps be observable in the light curves. There is a limit to the angular speed possible without break-up. It is easy to find when the *surface acceleration* is equal to the gravitational acceleration for the case of a sphere; the results are given in figure 9.6 for a given density, which is the only parameter involved in this simple case. In practice, however, a rapidly spinning body will bulge at the equator, so that the matter there will be less strongly held. The limiting period is therefore longer than in the simplified case. Careful estimates give about 4 h. Many minor planets have periods not much longer than this. This begins to be interesting especially when we go to consider that the two other types of collision will tend to slow rotations.

If the energy is between $10^3\ \mathrm{J\,kg^{-1}}$ and $10^5\ \mathrm{J\,kg^{-1}}$, there will be cratering, with ejecta at speeds of up to perhaps $1000\ \mathrm{m\,s^{-1}}$. This is greater than the typical escape speed, so some will escape. The ones ejected in the same sense as the direction of rotation will be more likely to have a speed greater than the escape speed so there will tend to be a net loss of angular momentum and a consequent increase in period. Figure 9.7 shows this effect.

If the energy is less than $10^3\ \mathrm{J\,kg^{-1}}$, the collision is what was called glancing in an earlier paragraph. Such collisions are difficult to analyse or simulate. They are not like a billiard ball collision on three counts: because the surfaces are rough, because a significant amount of kinetic energy is lost in the collision, and because the bodies are aspherical. All these properties are important as we have already seen. As far as angular speeds are concerned, a glance at figure 9.8 will show that the effect of friction will be

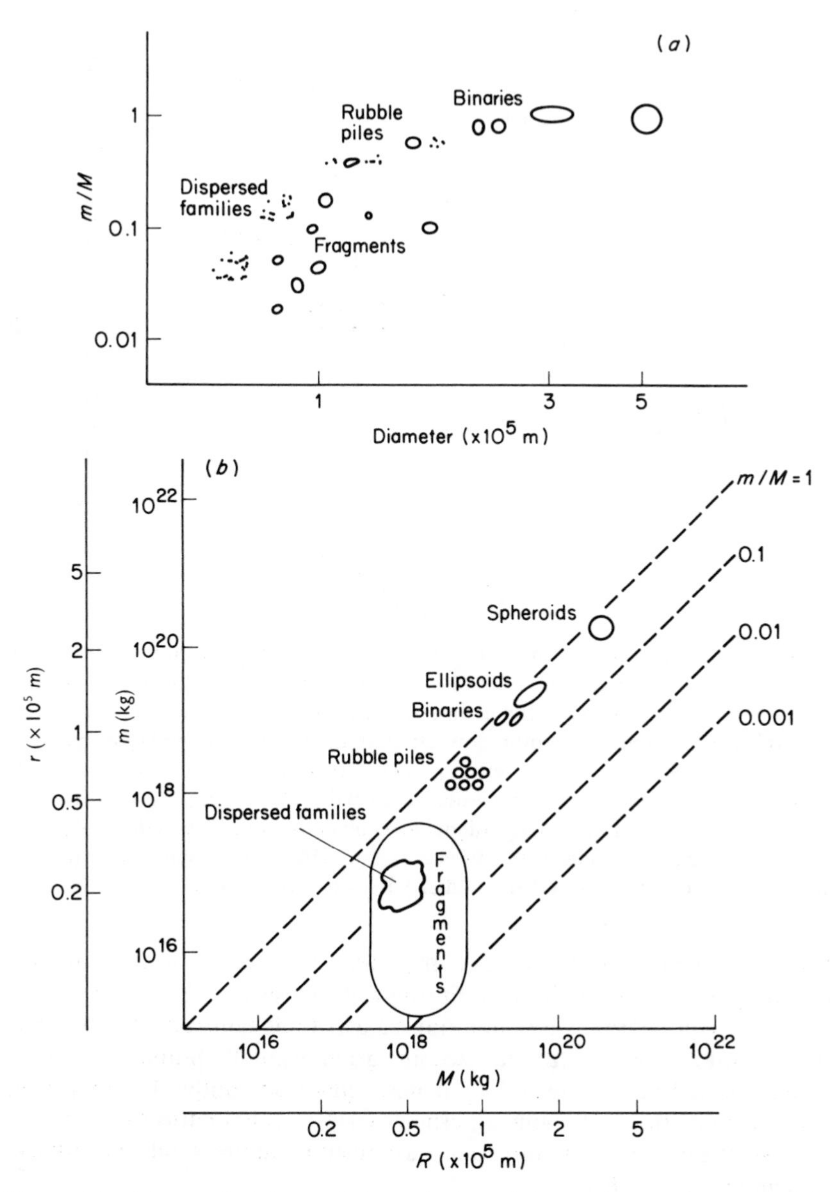

Figure 9.5 Effects of violent collisions. Only collisions with energy greater than 10^5 J kg^{-1} are considered in this diagram, which is based on calculation and experiments in which rocks are fired at each other in the laboratory, since no asteroid collision has ever been seen. Notice the disruption of smaller bodies, whilst larger bodies remain intact, and compare with figure 9.4. One possible outcome is a binary asteroid, but no clear evidence for a binary has yet been found. (*a*) and (*b*) shows the same results, plotted in different ways. (*a*) after Farinella P 1982 *Icarus* **52** 409.

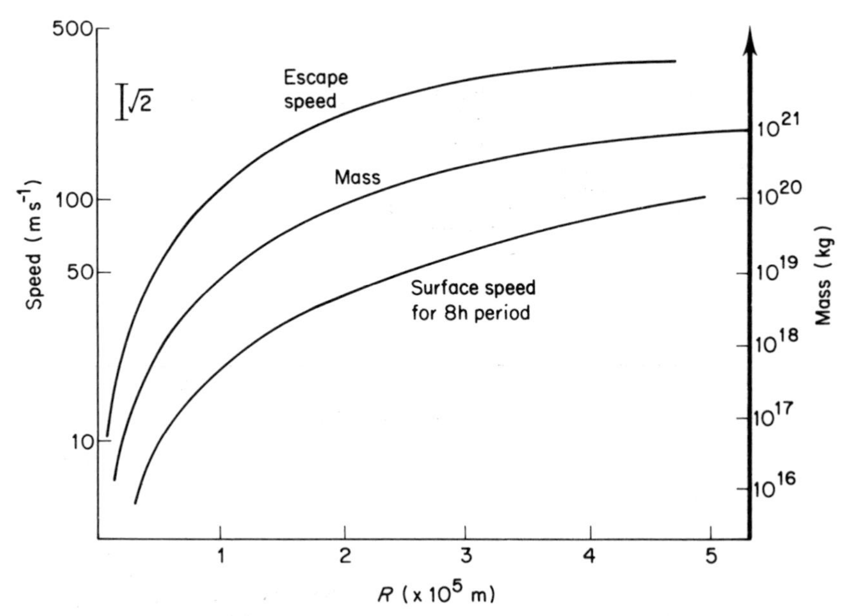

Figure 9.6 Vital statistics for a minor planet. Assuming a period of 8 h, which is typical for minor planets, and a density of 2500 kg m^{-3} (2.5 times that of water) it is a simple matter to calculate the escape speed ($(2GM/R)^{1/2}$) and the surface speed ($2\pi R/8$ h)). Both speeds are quite low, compared with those for a larger planet such as the Earth. The escape speed given is that perpendicular to the surface. At other angles the surface speed must be added vectorially. This is important in figure 9.7. The mass corresponding to each diameter is given as well. The density of a minor planet is thought to range from 2500 kg m^{-3} to 3500 kg m^{-3} with the lower values applying to stony bodies and the higher values to iron bodies.

to reduce them for both bodies if they are both rotating in the same sense, and to average them if they are in opposite senses.

If sufficient energy is lost in a non-catastrophic collision, the two bodies (give or take a few ejecta) may become gravitationally bound to each other producing a binary asteroid or a new, coalesced body. The binary may coalesce in its turn following a perhaps very slow dissipation of energy, e.g. by tidal effects. Let us assume that the collisions can be modelled, otherwise we can get no further.

The final question is how long to let the collision process continue in the model. Meteorites again help to answer this question. Radioactive nuclei found in them are consistent with a picture in which the Solar System has been evolving undisturbed for the last 4.5×10^9 years. In this time, a given minor planet, if it stays intact, will have been involved in a few dozen collisions. This is barely enough to say the minor planets have come to an equilibrium, and is the justification for saying at the beginning of the section that we are witnessing a planetary system in the process of formation.

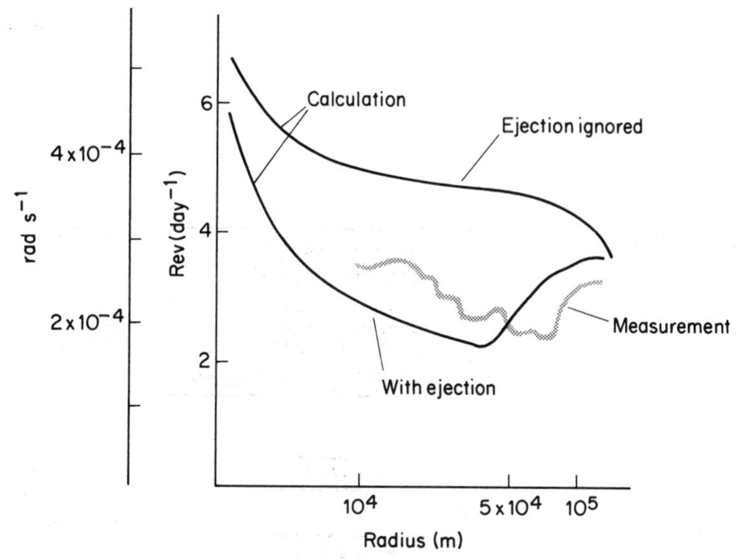

Figure 9.7 A way of slowing down rotation. Over an intermediate range of diameters, material ejected from cratering at collision can escape in some directions (with the rotation) more easily than others (against the rotation). The result is to slow down the rotation of the body itself, because the ejecta carry away angular momentum. Similar considerations lead to spacecraft being launched eastwards rather than westwards. After Dobrovolskis A R and Burns J A 1984 *Icarus* **57** 464.

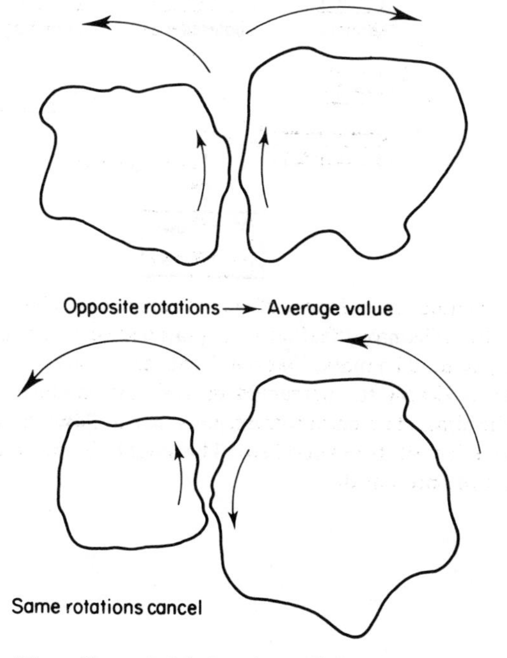

Figure 9.8 The effect of friction in collisions. If all asteroids were spinning in the same sense, collisions would slow them down.

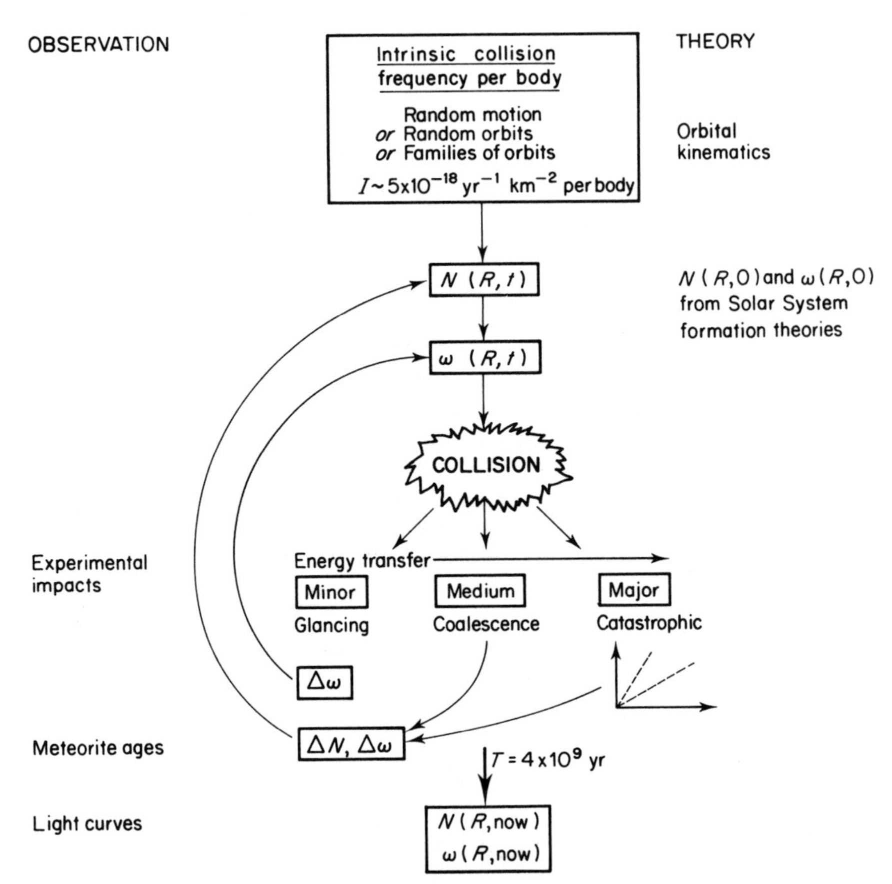

Figure 9.9 A simulation of the effect of collisions. The starting point is a calculation of the collision probability per square kilometre of minor planet. This is easy if they are assumed to move randomly but they clearly do not (see figure 9.3). Next one needs to know the current distribution of diameters and periods—it is changing all the time. The computation must then follow an asteroid and assign, statistically, the effect of its next collision. This might be continued from the time of formation until the present day.

9.4 SUMMARY

There are some definite patterns in the characteristics of minor planets which may well bear witness to the collisional evolution. They include:

The small range in periods
The fact that periods are quite close to minimum
The Kirkwood gaps
The fact that multiple periods are rare but not absent
The distribution of radii.

Figure 9.9 summarizes the steps to be considered in stimulating an evolutionary process to compare with present observations.

Part II

Photometers and Photometry

Chapter 10

Photometers and photometry

If you have no immediate intention of doing photometry, you may wish to skip from the *end* of §10.1 to Chapter 11, or even to Chapter 12, where the projects start. We hope, of course, that you will find some of the projects sufficiently interesting that you want to come back to the photometry.

10.1 INTRODUCTION

The earliest attempts to quantify the light received from stars were made by the early Greeks who divided the range of stellar brightness into six 'magnitudes', the apparently brightest stars being first magnitude and the faintest visible stars being sixth. This system was not refined until the nineteenth century when it became possible to use laboratory instruments to measure the luminosity. It was then found that there was approximately a factor of 100 in luminosity between the first magnitude and sixth magnitude stars and, after some discussion, the system proposed by Norman Pogson, an English astronomer, was adopted. This *defined* one magnitude as the fifth root of 100, i.e. 2.512 and thus placed the ancient Greek system into a quantitative, modern system. The magnitude scale is therefore logarithmic, which is an advantage because the response of the eye is also subjectively logarithmic. Pogson chose the base of the logarithm as 2.512, rather than the more familiar base 10. Thus a first magnitude star is ~ 2.5 times brighter than a second magnitude star, 100 times brighter than a sixth magnitude star and 10^4 ($= 100 \times 100$) times brighter than an eleventh magnitude star. (We refer to 'star', but the same applies to a minor planet or indeed any object with a point image.) The *apparent magnitude* is proportional to the logarithm of the *received* intensity.

The *absolute magnitude* is found by calculating, using the inverse square law, the magnitude which would be observed if the star were at a standard distance of 10 parsecs. The absolute magnitude is proportional to the logarithm of the *emitted* intensity. In practice, an observer rarely deals with the absolute luminosity, in watts, of a star. It is much more usual to work with the ratio of brightnesses of two stars. The logarithm of this ratio is simply the *difference* between the magnitudes. A difference between two numbers is traditionally represented by the symbol delta (Δ), so a common piece of astronomer's jargon is 'delta mag' to mean the difference in the magnitude of two stars. Although the logarithmic scale is appropriate for visual estimates, it is not particularly useful for photometers, which are linear, that is, record intensity itself, not its logarithm. However, tradition dies slowly and photometric results often end up plotted as 'delta mags' (Δm).

With the advent of laboratory measurement and of the photographic plate it became clear that a pair of stars with a given magnitude difference as judged by the eye did not necessarily have the same magnitude difference as measured off the photographic plate. This is not surprising since the eye and the photographic plate cover different spectral ranges, and have different variations of sensitivity with wavelength within their range. We have evolved on a planet with an atmosphere of a certain density and chemical composition, circling a G2V star, and our eyes have presumably evolved to be most sensitive to those colours most useful to us under those conditions. The sensitivity of the photographic emulsion on the other hand depends on photon energy, and it is a basic physical law, true everywhere, that blue photons have more energy than red photons. It soon became clear therefore that it was not sufficient to state that a certain pair of stars had apparent magnitude difference +1, for example, but it was also necessary to specify how the magnitude difference is being measured—by visual, photographic, or other means. The apparent visual magnitude difference is usually written Δm_{vis} or Δm_{v}. Thus, we can write

$$\Delta m_{\mathrm{vis}} \text{ or } \Delta m_{\mathrm{v}} = 1.$$

Suppose the range of sensitivity is restricted by a filter, as one often needs to do. Measured through a blue filter, it might have a higher apparent magnitude difference written as, for example,

$$\Delta m_{\mathrm{b}} = 1.5$$

whereas through an ultraviolet filter it might have

$$\Delta m_{\mathrm{u}} = 0.8.$$

As long as we deal strictly with magnitude differences (i.e. brightness ratios) there is no need to set a zero on the magnitude scale. For some purposes however—listing magnitudes in a catalogue for instance—it is useful to choose a zero. The technicalities of this choice are of no particular interest at the moment. When applied, they give the bright star Vega an apparent magnitude very close to zero. This is an adequate reference point.

Many systems of filters have been developed over the years, some for very specific purposes. Some of these filters have very narrow transmission bands so as to isolate the light from a particular element. This may be very useful, but also means that they pass much less light than a broad band filter, which may make them difficult to use with small aperture telescopes. There is a standard broad band system, called Johnson UBVRI, which virtually any telescope can use, and, because it is so standard, provides very useful classifications. There are many other systems, the Stromgren system and the Geneva system being two of them, which have their particular uses but are relatively narrow band systems not only letting through less light but also being intrinsically much more expensive than broad band filters—perhaps twenty times more expensive. We will therefore concentrate our future discussions on broad band UBVR and I filters. For some special projects one may want to use narrow band filters to cut out or include narrow spectral features (for instance see §10.5.1). Unfortunately, the Earth's atmosphere is also a filter and since it is variable it is always a possible source of difficulty or error. It is obvious from everyday experience that the Sun is apparently much brighter when it is high in the sky than when it is rising or setting; apparently brighter on clear days than on foggy or hazy days and so on. This apparent variation (i.e. not arising from the body itself) affects all the stars, and consequently all measurements of stellar brightness must be made against a reference of other stars. This is discussed in more detail in §10.3.4. The reference stars must be nearby, and their measurements must alternate with (or ideally be simultaneous with) measurement of the star being studied. Brightness can be measured by eye, by looking through a telescope and estimating whether, and by how much, the star of interest was brighter or fainter than reference stars visible in the field of view at the same time. The eye has to be trained by looking at stars with given apparent visual luminosity and imagining the difference between them split into the appropriate number of tenths of magnitudes. Provided that there are comparison stars near the object of interest, with similar magnitude, consistent and accurate results can be obtained. Addresses of some sources for standard comparison stars are given in note 10.1. More sophisticated methods used variable filters to allow progressive dimming of one star's image against the other until the two images had equal apparent brightness. This can be remarkably accurate. By such techniques accuracies of 0.1 or even 0.05 magnitude can be obtained and much useful work is carried out by amateur astronomers using this method.

Technology, however, has moved on and with the invention of the *photomultiplier tube* in the 1930s it became possible to turn even very faint light into a measurable electric current. This gave birth to photoelectric photometry, and brought at tenfold increase in accuracy, that is to 0.01 or even 0.005 magnitudes. Early equipment displayed the photomultiplier current directly with moving coil galvanometers, and later with amplifiers connected to chart recorders. However, a digital display is easier to read, and can be input to a microcomputer. Currently it is possible to buy amplifiers of sufficient sensitivity and stability which, combined with analogue to digital converters, allow fast and accurate measurement of photomultiplier electrical currents, all for a few tens of pounds. The photomultiplier assembly itself is more expensive but even so can be bought for as little as £150. We have thus moved through a full circle from when an amateur astronomer with a practised eye could produce results of just the same accuracy as a professional, through an era when cost and technological complexity worked against the amateur, to now, when an amateur can afford equipment superior to that used by most professionals even ten years ago. Indeed, amateur recording equipment need be no less accurate than that of a professional. Of course an amateur will have a smaller telescope and so cannot measure very faint stars. Paradoxically, this can confer an advantage over professional astronomers, who find it increasingly difficult to get large amounts of telescope time because only the largest telescopes are now maintained for professional use and, whilst they are splendid instruments, they are heavily overbooked. The amateur can concentrate on one object, or one type of object, and spend years doing a thorough job investigating it. The professional is under pressure to publish papers and may tend to, or indeed be forced to, avoid long-term projects. Long-term astronomical projects seem likely to depend in large measure on small telescopes in the future.

10.2 DETECTORS

Although the most common way to measure stellar brightness nowadays is to use a photomultiplier tube, solid state devices such as PIN photodiodes are being used increasingly. We will therefore describe both types of detector.

10.2.1 Photomultiplier tubes

The photomultiplier is contained within a glass cylinder from which air has been evacuated. Light to be measured shines through the glass onto a plate known as the photocathode, with an area of light-sensitive compound, which has the property that when light shines upon it, electrons are liberated from within its molecular structure. This photoelectric effect was discovered by Hertz in 1887, and was explained in 1905 by Einstein as arising from the quantum mechanical nature of light. The released electrons are accelerated in turn by a series of charged metal plates, sometimes ten or more in number, each of which has an electrical potential of about 100 volts greater than the preceding one in the chain. These metal plates are known as *dynodes* and the whole assembly as the dynode chain. Each dynode has a coating which releases electrons when struck by an energetic electron (whereas the photocathode releases electrons when struck by a photon).

Each electron gains energy in travelling to the next dynode, and on colliding with it is usually able to release more than one electron. This results in an avalanche effect, whereby for each electron released from the photocathode perhaps 10^6 or 10^7 arrive at the final collecting plate, called the anode. This forms a pulse of electric current at the output. If the pulses are allowed to merge together they form a current which is proportional to the amount of light falling on the tube. This is called *DC photometry*. The pulses may merge together either because they arrive in such rapid succession or because they are deliberately smoothed out with a capacitor. Alternatively, one may count the pulses, which is equivalent to counting the photons. This is called *photon counting* photometry. However, not every photon which arrives at the photocathode succeeds in releasing an electron, and not quite every released electron manages to initiate a cascade down the dynode chain. Typically, the photocathode will be 20% to 25% efficient at its most sensitive wavelength but at much of its wavelength coverage it will be 10% efficient or less. Of those electrons which are emitted from the photocathode, perhaps only 90% will succeed in initiating a cascade. Hence the number of pulses is proportional to the number of incident photons, but not equal to the number. Notice that intensity is photon count times photon energy. Energy depends on wavelength, so intensity is *not* proportional to photon count, except for a fixed wavelength.

Many early photomultiplier tubes had the photocathode positioned so that light was shone onto it through the *side wall* of the tube and the dynodes arranged in a zigzag form around the walls of the tube. All the early standard photometry was done using such tubes and some observers feel it necessary to continue to use this form of tube in order to try to reproduce the same photometric system as the early observers. Modern tubes are different in that they typically have the photocathode on the *end window* of the tube and have the dynode chain arranged linearly down the length of the tube. Figure 10.1 below shows the layout of such a tube. We feel that there is much more to be gained from using modern tubes than by continuing to use the older tube types. The most serious defect of the older tube is the relatively small size of its photocathode and the large variation in sensitivity over its area. This renders the photometer sensitive to small movements of the image on the photocathode such as can easily arise from imperfect guiding.

Apart from not detecting all the photons which arrive, photomultiplier tubes also have other faults. Not all electrons which are emitted by the photocathode are due to the arrival of a photon. A significant number are spontaneously emitted from the cathode or one of the dynodes due to statistical fluctuations of electron energies within the structure of the photocathode. The larger the photocathode area the larger the number of these '*noise*' electrons. These in turn cascade down the remainder of the dynode chain. The spurious pulses due to emission from the dynodes are smaller than true pulses, because they have been amplified over a smaller voltage range, so they can be suppressed by a discriminator. However, no such discrimination is possible in the case of electrons spontaneously emitted from the photocathode itself. Cooling the photocathode can help to reduce this *dark noise*, and a good professional system will often only produce one, or a few, such counts per second, called the 'dark count'. As against this, cooling decreases sensitivity so there is a limit to how much improvement in performance can be obtained by cooling. Bialkali photocathodes, for example, lose as much as 50% of their sensitivity if they are cooled from 0 to $-25\,^{\circ}$C. In general, the more red sensitivity a particular photocathode possesses, the greater the amount of 'noise' it generates. Approximate temperatures at which photocathodes are normally used are

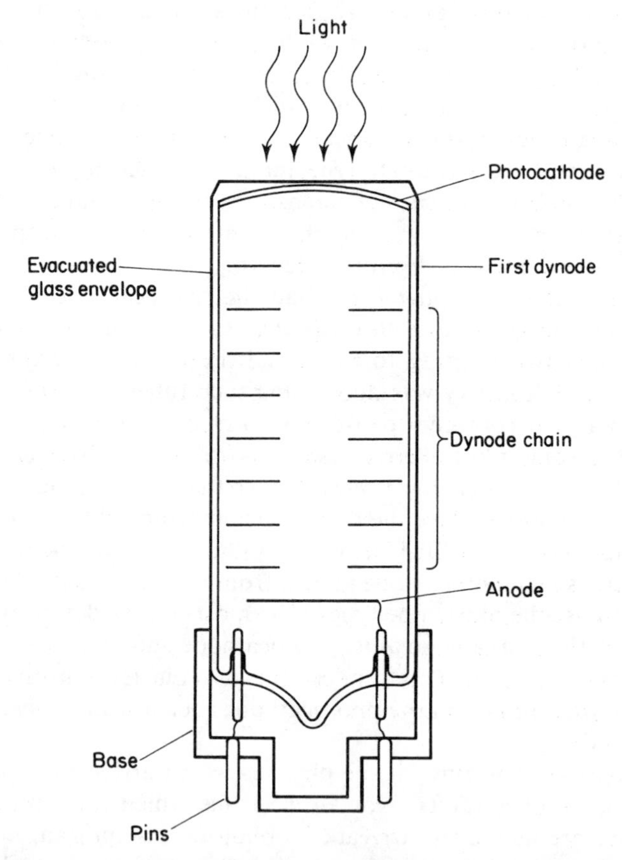

Figure 10.1 Light enters the photomultiplier through a flat window, which ideally should be transparent at all frequencies, but in practice cuts off most ultraviolet and infrared light, and electrons liberated from the photocathode by the impact of the photons are accelerated down the set of dynodes, which are at successively higher voltage.

~ − 10 °C for S11 and − 25 °C for S20. Figure 10.2 shows the sensitivity against colour for several different types of photocathode and window material.

In summary, one can never achieve a situation in which there is a one-to-one relation between incoming photons and output pulses, though one can get fairly close to it by careful monitoring of the size and shape of the pulses. In DC photometry this information is lost, since the individual pulses are not seen. DC photometry is simpler, but inherently more susceptible to systematic error.

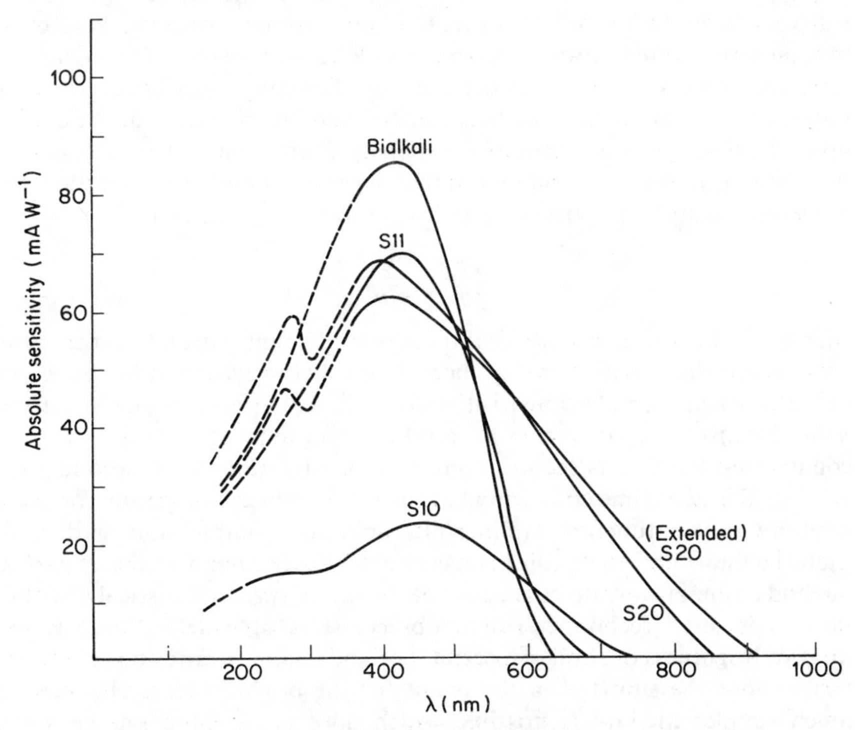

Figure 10.2 Photomultiplier sensitivity drops to zero because long wavelength photons have too little energy to release an electron. Ultraviolet photons are absorbed in the window, so sensitivity also drops off there. The curves show the variation of sensitivity in between, which depends on the detailed structure and composition of the device (courtesy Thorn EMI).

10.2.2 The dead time correction

With a photomultiplier tube there is a limit to the rate at which photons can be counted. This is due to the finite length of time which it takes for an electron to traverse the length of the photomultiplier tube, and to the minimum time in which the amplifier can respond. The electrons reach speeds of $\sim 1\%$ that of light. Fast as this is, it means that they take several nanoseconds to pass the length of a typical tube. (A convenient unit for the speed of light is 1 ft (30 cm) per ns!) Thus photon rates of greater than say $10^9\,s^{-1}$ can never be recorded. In practice, the maximum rate is much lower than this. This is due to the fact that the photons which arrive are not spaced at regular and equal intervals of time. Indeed they are not even random, but arrive (irregularly) clumped together. (This is due to a deep principle of physics—they are massless bosons and follow Bose–Einstein population statistics.) If two photons arrive in a shorter time than the response time of the tube and amplifier then the pulses merge together and will very likely be recorded as single photon, rather than two. Hence the photon arrival rate is always underestimated to some extent. The higher the rate, the more serious the underestimate. Typically such effects become important at photon rates of between 10^5 and 10^6 per second, depending upon the tube type (and amplifier response). Somewhat higher rates can be handled, provided the undercounting is compensated by correcting the observed counts to the true counts by the following formula:

$$N = \frac{n}{1 - n\delta}$$

where N is the true count rate, n is the observed count rate and δ is the '*dead time*' correction. The term has been handed down from other types of detector which are indeed insensitive for a time after each count. Clearly if $n\delta$ is close to 1 the correction is very large and unreliable. Units for both count rates are counts/second and for δ are fractions of a second (e.g. 10^{-7} s). The dead time correction has to be determined empirically for every photomultiplier tube/preamplifier/discriminator combination and it is useful to know how to do this in an easy and efficient manner. The empirical method automatically takes account of the rather special statistical distribution. It is often recommended to observe stars of widely different, but known, brightness of similar spectral type and colour near to the zenith and hence judge the shortfall in the count for the brightest star. However, a much simpler method is possible which does not require known stars, accurate guiding, or even for the telescope to be moved. It depends on the steadily brightening sky before dawn.

The telescope is pointed towards the sky in the area above where the Sun will rise. It can be moving, or even left fixed, provided that bright stars are avoided. Two apertures whose areas differ by a factor of up to, say, ten are chosen, but the true ratio of their areas need not be known beforehand as this will be determined. Some time before dawn the sky background counts are determined, first through the small aperture and then through the larger. This will be repeated continuously until the sky is too bright for observations to continue, that is, when the count rate is, say, more than $10^6\,s^{-1}$. It is then necessary to plot a graph of the results. Along the horizontal axis plot the counts through the larger aperture, n_2. Along the vertical axis plot R, the ratio of counts through the small aperture, n_1, divided by the counts through the large aperture (i.e. $R = n_1/n_2$). Note that because the sky is continuously brightening it is useful to take the mean of two readings through the small aperture which bracket a large aperture reading, and to compare this mean with the large aperture reading. If the readings are changing rapidly then the geometric mean should be used, which is the square root of the first reading multiplied by the second, etc, in general

$$\bar{n}_1 = (n_{1(i)} n_{1(i+1)})^{1/2}.$$

(For slow changes, the arithmetic mean is almost the same as the geometric mean.) In figure 10.3 we show a typical plot of the results and show how it is used to obtain δ.

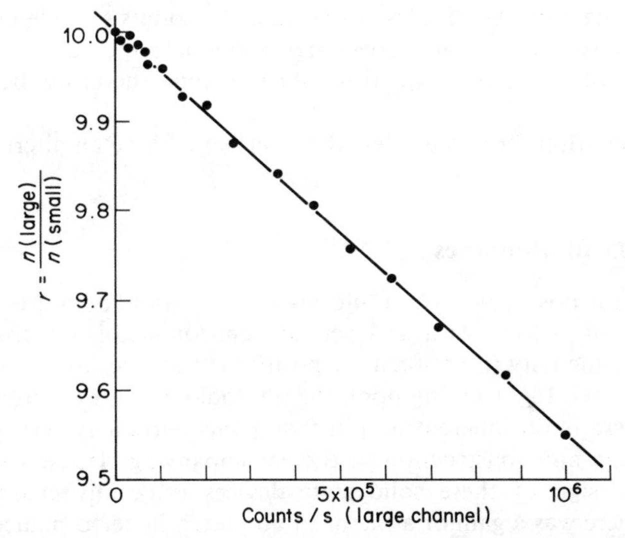

Figure 10.3 By comparing count rates with large and small apertures at different levels, the loss due to dead time can be found (δ = gradient/$(r-1)$). Neither the light levels nor the apertures need be known!

Note that there will be many observations at low count rates near to the vertical axis of the graph as the sky brightness will hardly be changing provided that one starts long enough before dawn. As the sky becomes brighter, and the counts become more affected by dead time, the ratio increases and this is the basis of the method. A straight line through the points intersects the vertical axis at exactly the ratio of the two aperture sizes. The gradient of the line equals $(r-1)\delta$ where r = ratio of apertures.

Therefore

$$\delta = \frac{\text{gradient}}{(r-1)}.$$

In the example shown the exact ratio is 10 : 1 and the gradient of the line is

$$\frac{0.45}{10^6} = 45 \times 10^{-8}\ \text{s}.$$

The dead time, $\delta = (45 \times 10^{-8})/(10-1) = 5 \times 10^{-8}$ s or 50 nanoseconds. This is typical of the value for a good professional tube and amplifier. Cheaper equipment could well have dead time corrections several times this.

50 nanoseconds sounds like a short time, but let us work out its effects. At a count rate of $10^4\ \text{s}^{-1}$ there will be undercounting of about 0.05% or half a millimagnitude, which one might safely ignore. But at 10^5 counts s^{-1} this will have risen to ~5 millimags while at 10^6 counts s^{-1} it will be ~5% or a staggering 0.05 magnitudes. At around 10^7 counts s^{-1} one may decide the uncertainty is too great, depending on the nature of the project. The importance of correcting for this effect cannot therefore be overemphasized.

If photomultiplier tubes are less than perfect is there an alternative?

10.2.3 PIN photodiodes

PIN stands for positive–intrinsic–negative. *PIN photodiodes* are solid state semiconductor diodes with a layer of semiconductor material having intrinsic conductivity (I) between the positive (P) and negative (N) semiconductor material. Light falling upon the photodiode causes current to flow. Because there is no inherent amplification this current is very small and careful design and construction of the accompanying electronics is necessary. The design of these solid state devices is rapidly changing. For example, there was a gain of a factor of about ten in performance between 1982 and 1984. It is therefore not possible to predict with any accuracy whether, eventually, these devices will seriously compete in overall performance with photomultiplier tubes. The intrinsic spectral response of PIN diodes differs markedly from that of photocathode based devices. They have effectively zero sensitivity at 300 nm, rising to a maximum near 800 nm and falling to zero again near 1.1 microns (1100 nm), see figure 10.4. For the U, B and V bands the solid state devices still have some disadvantages compared with photomultiplier tubes but even here their advantages of very low weight and absence of high voltage supplies make them very useful for some applications. For R and I bands in fact the photodiodes are already serious contenders. They have much higher intrinsic sensitivities than the more normal (i.e. low cost) photocathodes and their other advantages make them worthy of serious consideration for light-weight R and I photometers.

Optec Inc of Lowell, Michigan in America produce a commercial photometer based upon a PIN photodiode and their figures suggest that with a 12″ (30 cm) telescope useful results on stars at the eighth to ninth magnitude level can be obtained, whereas a photomultiplier tube on the same telescope could produce the same accuracy on a star 4 magnitudes fainter. Because of the small size of the photodiode, the image magnification must be low and this means that even very small particles of dirt on the detector may obscure all or part of the small image. On the other hand, one major advantage of its small size is its rapid response time, typically about one nanosecond (10^{-9} s) compared with perhaps a few to ten times that value for photomultiplier tubes. This means that it can be used on bright stars at intensities of illumination which would saturate (or even destroy) a normal photocathode.

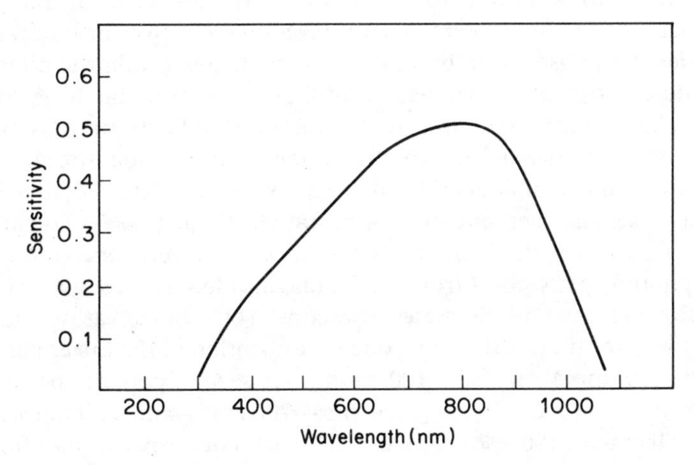

Figure 10.4 The variation of sensitivity with wavelength for a PIN diode is to be compared with those shown in figure 10.2 for different types of photomultiplier.

10.3 LIGHT AND ITS MEASUREMENT

10.3.1 The nature of light

The liberation of electrons in photocathodes and PIN diodes can be readily explained in terms of the impact of individual photons, and the separate pulses emerging from photomultipliers confirm this view. It is this view which is needed in §10.3.2. However, light is not so simple. We will need, in §10.3.3, to describe the scattering of light from surfaces, and diffraction of light from edges. These processes are easier to picture in terms of waves of light rather than photons. Physicists have long since become resigned to carrying along these two points of view, without trying too hard to reconcile them, and we shall do the same. Many physics textbooks go into this question, often under the heading 'wave–particle duality'.

10.3.2 Photon statistics

There are some interesting consequences of using photon counting photometry which bear on the limitations which face all observers, and the reasons why the small telescope has so much to offer.

Suppose we had a perfect detector above the atmosphere. Then we could use the following rule of thumb, which is that *above the Earth's atmosphere* about 10 000 photons arrive each second on each square centimetre in each nanometer of wavelength range from a zero magnitude B0 star. Now only about 20% of those photons which hit the photocathode give pulses which are detected. In addition some photons are absorbed in the Earth's atmosphere, or fall prey to inefficient reflection from two (or more) mirrors in the telescope, absorption by filters and any lenses within the photometer itself. These two factors together mean that one detects far fewer photons than are actually arriving from the star, above the Earth's atmosphere. So we need another rule of thumb. In a high-altitude good site an average figure is that approximately 5% of the photons are detected, that is, 500 counts per second per cm^2 per nm, rather than 10 000. (Incidentally, another equivalent rule of thumb is that at high altitude one counts half a million photons per second from a fifth magnitude star through a blue filter with a 20 inch (0.5 m) diameter telescope.) Of course, at sea level the numbers will be still less than this due to absorption in the lower part of the atmosphere. Suppose we take 100 counts per second per cm^2 per nm, and consider a mirror of 30 cm diameter—700 cm^2—and a bandwidth of 300 nm. That gives $100 \times 700 \times 300 = 2.1 \times 10^7$ counts per second for a zero magnitude star or 2.1×10^5 counts per second for fifth magnitude. This still means a healthy count rate at mag 5 but can we go as faint as 10 or 15 mag? In order to work out how faint a star can be measured it is important to know how the total number of photons counted limits the maximum possible accuracy. There is a statistical rule which applies to all counting processes, that the accuracy is equal to the inverse of the square root of the total number of pulses detected. This is the best possible accuracy, assuming perfect equipment, perfect timing, no misalignment or passing clouds, and so on. (Perhaps impossible accuracy would be a better name.) Therefore if one has detected 400 photons the accuracy attainable is $1/(400)^{1/2}$, that is, one part in 20. If 10 000 pulses are counted the accuracy is $1/(10\,000)^{1/2}$ or one part in one hundred (1%) and so on. Table 10.1 below summarizes these.

Table 10.1

Total counts	Maximum possible accuracy (fraction)	Maximum possible accuracy (magnitude)
$100 = 10^2$	± 0.1	± 0.11
$10\,000 = 10^4$	± 0.01	± 0.01
$1000\,000 = 10^6$	± 0.001	± 0.001

Note that the reason that the small values of errors in magnitude are equal to the small values of errors in fractions is that logarithmic scales are nearly linear for numbers close to 1 ($\log(1 + x) \approx x$ if $x \ll 1$).

Notice that in ideal conditions it makes no difference whether one has counted one million photons in one second or one minute, the accuracy is the same. Of course in practice the longer the time needed to reach the count, the more chance there will have been for inevitable variations due to atmospheric transparency changes, imperfect guiding and so on to have occurred. There are also spurious counts coming from the tube itself. Even a good, cooled photon counting tube is likely to produce between one and one hundred counts each second of 'dark current'. If we had ten dark counts each second and counted for one thousand seconds (~16 minutes) then ten thousand of the counts would be from the equipment and not from the star. From that point of view it would be nice to have a strong signal which makes the spurious counts insignificant. On the other hand, very high count rates ($\geqslant 10^6\,s^{-1}$) are not an unmixed blessing because the dead time correction becomes too big to handle. This is a lesser problem because it can be coped with using a neutral filter or by stopping down the mirror, which only needs a cardboard ring.

Since there will rarely be a problem with excessive signal with a small telescope, let us instead concentrate on how faint a star can be measured. If the sky were perfectly dark, one could detect arbitrarily low counts—say less than one per second. In practice the sky is far from perfectly dark. A realistic minimum value for a detectable count rate might be 200 per second. Since, for the example above, 2.1×10^5 counts corresponded to fifth magnitude, 200 counts being a factor of 1000 lower corresponds to a further 7.5 magnitudes ($\log_{2.512} 1000 = 7.5$), that is 12.5 magnitude. Since there are about a million stars of more than 12.5 apparent magnitude this gives quite a choice. Table 10.2 shows the *limiting magnitude* for telescopes of different sizes, right up to 200 inches, corresponding to the Palomar telescope. In fact Palomar gains a magnitude or two by being at high altitude, so its limiting magnitude is about 22. One advantage of a logarithmic scale is that a 6 inch telescope does not look so much worse than

Palomar! There is one sense in which this is genuinely true, which is that even with a small telescope shortage of stars is not a problem, for many classes of star. The exceptions are for intrinsically faint stars (i.e. large absolute magnitude) such as white dwarfs, for which choice is limited with a small telescope. A lot of work has been done on white dwarfs with large telescopes, leaving relatively little for amateurs.

Table 10.2

Diameter of telescope: inches	Diameter of telescope: cm	Approximate magnitude of star to give 2×10^6 counts s^{-1}	Approximate magnitude of star to give 200 counts s^{-1}
2	5	0	10
6	15	2.5	12.5
10	25	3.5	13.5
12	30	4	14
20	50	5	15
100	250	7.5	17.5
200	500	10	20

Table 10.2 also gives, for the various mirror sizes, the brightest stars which can be measured without taking steps to reduce the incoming light. Naturally this is more of a problem for Palomar than for an amateur.

The broad indications of table 10.2 are elaborated in figure 10.5 which

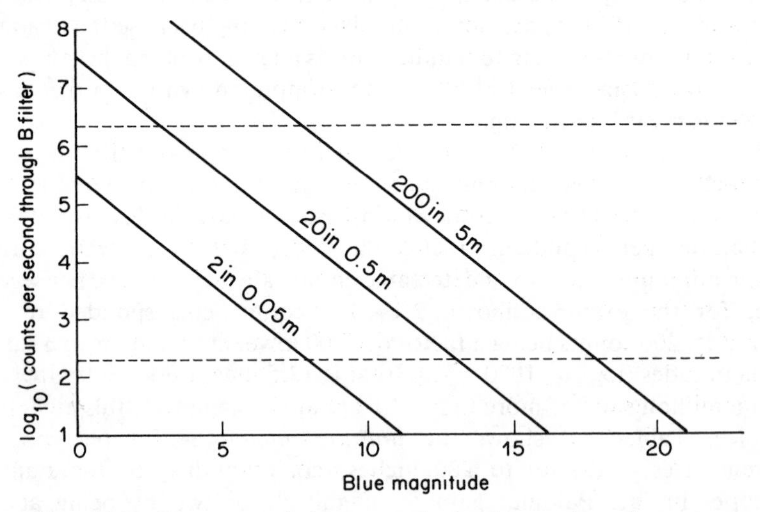

Figure 10.5 For a star of given magnitude, the number of counts per second to be expected for telescopes of three different sizes can be read off the vertical scale.

gives graphs which allow the intensity of light and the number of photons in various spectral ranges to be calculated. Notice that the intensity of light, that is the energy arriving per second, is the number of photons multiplied by the energy of each one—which depends on its wavelength. Hence the intensity graphs and number graphs have different shapes. This means that reference stars should, as far as possible, be of the same spectral class as the star of interest.

10.3.3 Scattering of light

Before describing a photometer in detail it is necessary to understand one more characteristic. This is the scattering of light from the various optical surfaces, mainly the telescope mirrors. It is often believed that the smaller the entrance aperture to the photometer which one uses, the greater is the potential accuracy of one's photometry. This at first sight seems sensible because the sky contributes light to the total signal, and so the less there is of this 'extra' light the better. The light from the sky can be due to a variety of causes such as scattering of nearby town lights, air glow, noctilucent clouds, etc, and the larger the aperture, the greater the area of sky contributing, and so the greater the extra, spurious, signal. Conversely, reducing the aperture size reduces this unwanted signal. However one must obviously not reduce the aperture to below the image size. The question is, therefore, how big is the image? Because of *diffraction*, it is surprisingly large—usually much larger than it appears to the unaided eye.

On a good night with a medium sized telescope, say a 25 cm (10 inch) diameter, one might look through the eyepiece and see the image of a seventh magnitude star with an apparent diameter of one arcsecond. Now move the telescope to a zero or first magnitude star with a similar elevation. You will now probably see an image of perhaps four or five arcseconds diameter. The sky has not become worse and your telescope has not changed. All that has happened is that the light surrounding the bright central core is now bright enough to be seen. The situation however is much worse than a five arcsecond visual image suggests. Look at any long exposure photograph on which both faint and bright stars have been recorded simultaneously. You will note that whereas the images of the faint stars are small, the images of the intermediate brightness stars are larger (due to scattering of light in the photographic emulsion), and the bright stars are still larger because the diffracted light is seen out to bigger radii. The images of bright stars are also often surrounded by crosses and haloes. The haloes are merely out of focus light reflected from the back of the glass photographic plate but the crosses are light scattered off the secondary mirror supports. These crosses have sizes of arcminutes not arcseconds.

There is scattering off the mirror, too. The following experiment is instructive. Shine a bright flash lamp (or, ideally a laser) at the surface of your primary mirror from the entrance to the telescope and then view it from the side. You will notice that you can see the area of the mirror which is being illuminated very clearly even when looking at grazing incidence across the mirror surface. The reason is that light is being scattered on particles of dust or polishing scratches from the surface of the mirror at up to ninety degrees to the direction of the incoming light. Of course, the amount of light reflected/scattered through these large angles is quite small per square arcsecond. If it were not, you would have a ground glass screen rather than a mirror! Even a perfectly clean mirror will probably have minor ripples on its surface which will contribute to this scattering of light. The problem arises because there is much more solid angle at large angles than there is within the one arcsecond or so of the obvious image. That is, there is a significant proportion of the total incoming light which is reflected/scattered well outside the few arcseconds of the apparent image. What has been found empirically is that for accurate work the photometer aperture should be about one arcminute. If apertures smaller than say 45 arcseconds diameter are used then errors due to faulty centring of the star in the aperture will be at about the 1%, i.e. 0.01 magnitude level, or worse, no matter how many photons are recorded. If your telescope mirrors are dusty or have poorly finished surfaces this will further degrade the results. Only if you are observing stars in a dense cluster, and need a small aperture to eliminate nearby stellar images, or are observing with much light pollution, either from the Moon or street lights, is it worth considering using small apertures. If this choice is forced upon you, then you will be restricted to observing programmes which do not require better than 0.01 magnitude accuracy.

10.3.4 Absorption of light in the atmosphere

Whilst it is obvious that mist and clouds absorb light, it is not so obvious that, even on a clear night, a significant amount of light is absorbed during its passage through the atmosphere. However, one only has to recall the dim red setting Sun, compared with its blinding brilliance at midday, to realize that clear air is by no means perfectly transparent. This absorption is called by astronomers atmospheric *extinction*, even though the stars are not extinguished. Because the effect is so marked, it would be foolhardy to try to compare the luminosity of a star at the horizon with one at the zenith. In fact important corrections must be made even if a reference star is only a few degrees from the star of interest. Clearly absorption is least at the zenith, and increases with the angle from the zenith, usually denoted by Z. If the Earth were flat and the atmosphere were a uniform layer above the Earth then, as can easily be seen from figure 10.6, the absorption would vary as $1/(\cos Z)$, written $\sec Z$. Since Z is the angle of the telescope with the vertical it can be measured directly. Figure 10.7 shows an arrangement of a pendulum and a protractor, pivoted parallel to its straight edge, which will do the job. One could, in fact, mark the scale with values of sec Z, rather than the angle Z. If you do so, it will be apparent that unless a very large protractor is used, this method is too crude, especially for stars near the horizon, to get accuracies better than a few per cent.

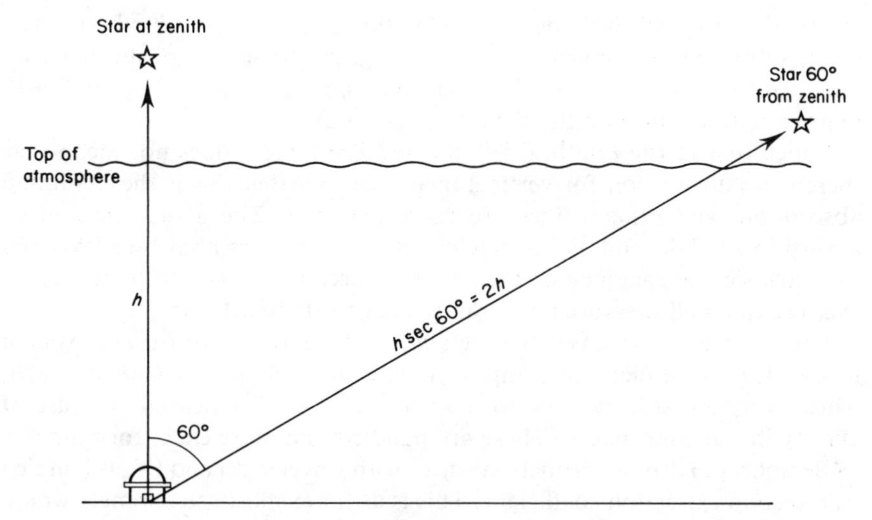

Figure 10.6 For a flat slab of atmosphere, the absorption would vary as the secant of the angle from the vertical, being a minimum (not zero) at the zenith.

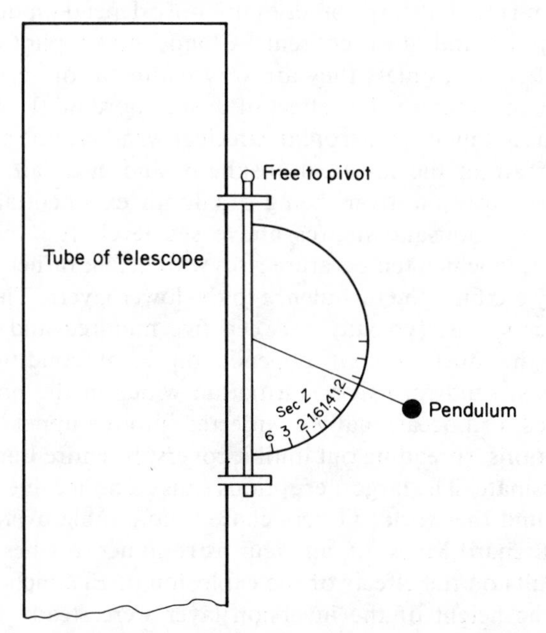

Figure 10.7 A protractor mounted on the telescope can be used to give the angle Z, or a sec Z scale could be constructed.

A more accurate alternative is to calculate the value of sec Z, using the values of the RA and dec for the stars involved, and the hour angle. The formula is

$$\sec Z = 1/(\sin \phi \sin \delta + \cos \phi \cos \delta \cos h)$$

where ϕ = observer's latitude, δ = declination of the star and h = hour angle of the star using the convention that it is negative to the east of the meridian and positive to the west. (The latitude enters, even in a flat Earth approximation, through the definition of δ.)

Notice that at the zenith $Z = 0$ and sec $Z = 1$. This does not mean that there is no absorption for vertical incidence, but that this is the minimum absorption, and other values are corrected to it. The actual amount of absorption will depend on the height of the observatory above sea level and the intrinsic transparency of the sky, but there is no need to worry about that provided all measurements are related to standard stars.

The sec Z formula arises from a simple flat Earth, uniform atmosphere model. It is not difficult to compute a better formula for a spherical Earth, which is of course very close to the truth, and an exponential decrease of density in the atmosphere. These are handled in a more exact formula due to Bemporad. Also the extinction varies with wavelength and this is handled by a second extinction coefficient. This is as far as most astronomers would go, but even with this improved calculation it is still essential to have nearby reference stars.

The problem is that absorption does not only depend on density, but also on water vapour and dust content. Clouds make photometry almost impossible, of course, unless they are very scattered, or very thin and you have a multichannel set up. The effect of dust is more subtle. Looking down from a high mountain or an aeroplane in clear weather, one sees a brownish dirty layer of air in the lower atmosphere, and clear air above it. The distribution of dust, far from being a smooth exponential, has a sharp cut-off at a few thousand metres above sea level. It is confined by an *inversion layer*, in which temperature rises with height rather than falls, and this effectively confines the turbulence to the lower layers. The height of the inversion layer varies, typically between five hundred and two thousand metres, and the dust content depends on local conditions—smog in industrial areas, sandstorms or Harmattan winds in the northern tropics and so on. Dust can occasionally be injected into the upper atmosphere by volcanic eruptions, spreading out until it covers the entire Earth, and taking months to dissipate. The largest eruptions cause a noticeable increase in red sunsets all round the world. Others cause a noticeable overall decrease in count rates. Richard Miles, an amateur astronomer in Cheshire, England, published results on the effects of the explosion of El Chichon in 1982, for instance. If the height of the inversion layer were steady, and its effects constant, it would have little effect on the *ratios* of count rates for stars.

Unfortunately, the layer drops during the night and this can affect not only rates, but also ratios of rates. As an extreme case, one can imagine an observer on a high hill who starts the night below the inversion layer and finishes up above it, in crystal clear air. One must simply be aware of the possible problems, and use reference stars.

10.4 PHOTOMETERS

10.4.1 Basic photometer design

We now outline the design of a typical photometer so as to pick out the various components and the reasons for their inclusion. Even if you buy a ready made photometer, an appreciation of the reason for design is helpful. There are several recent excellent books available which have a variety of designs and which are listed in Appendix 5. Figure 10.8 shows a typical layout rather than a specific design. We will list all the items and explain their purposes.

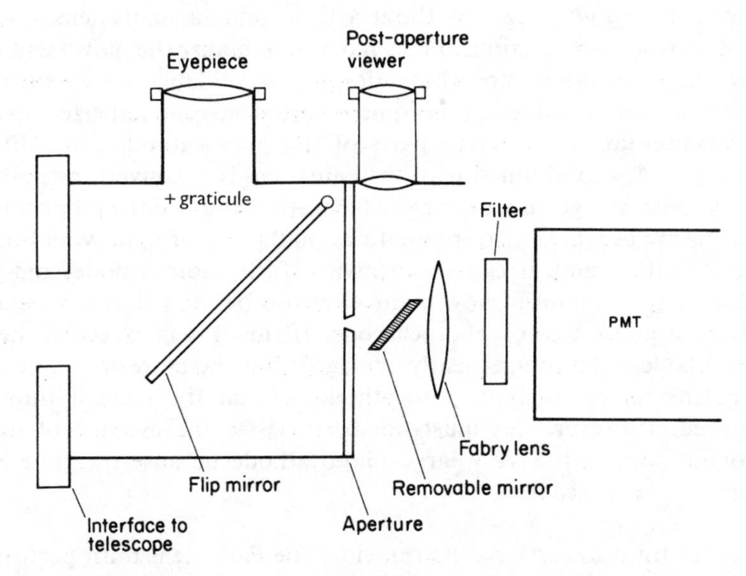

Figure 10.8 The layout, in schematic form, of the various components in a photometer using a photomultiplier. If a photodiode is used, the same components are needed, but the lens must concentrate the light onto the smaller sensitive area.

1. Previewer flip mirror, graticule and eyepiece

This assembly allows the observer to see a relatively large field at low magnification and to centre the star on the graticule. When the flip mirror is moved out of the optical path the light of the chosen object should pass through the aperture.

2. The aperture

The size of the aperture was discussed in §10.3.3. Its purpose is to limit the amount of sky which is seen by the detector whilst passing most of the light of the chosen object. It should have a diameter of approximately one arc minute and it should be accurately positioned in the focal plane of the telescope.

3. The post viewer flip mirror and eyepiece

The purpose of this assembly is to allow the accurate centring of the star in the aperture. Because the field of interest is restricted to about one arc minute it is normal to use a high magnification here. It is useful to have a microswitch arranged to ensure that measuring the stellar brightness cannot start until the flip mirror is removed.

4. The Fabry lens

In the JEAP design described in §10.4.3 the use of a fluid light guide avoids the need for a *Fabry lens*. Without a light guide a Fabry lens system is needed and has two functions. First, it has to minimize the movement of the stellar image on the photocathode despite unavoidable movement of the primary stellar image due to atmospheric turbulence and imperfect guiding. This is important as different parts of the photocathode have different sensitivities. A second function of the Fabry lens is to convert the point-like primary stellar image into an image which covers most of the photocathode. This is partly because the response to a small point of light would be very sensitive to the grainy, irregular sensitivity of the photocathode, and partly because there is a limit to how many electrons per unit time can be driven out from a given area of photocathode. If this flux is exceeded then the photocathode will be permanently damaged. It is therefore good practice to enlarge the image on the photocathode so that the intensity/unit area is reduced. However, this must not be taken to the extreme of using a photomultiplier with a very large photocathode because the tube's dark current will be larger too.

In order for it to carry out its functions the Fabry lens must perform the following functions:

(a) image the entrance pupil of the telescope on to the photocathode;
(b) turn the image in the focal plane into parallel light;
(c) produce an image of the required size on the photocathode.

Function (a) is satisfied when the distance from the Fabry lens to the photocathode is approximately equal to the focal length of the lens. This is because on all but the smallest telescopes the entrance pupil of the telescope is a large distance from the lens compared with its focal length and is therefore effectively almost at infinity. Function (b) is satisfied by placing the Fabry lens its own focal length away from the focal plane. Both functions (a) and (b) therefore require that the Fabry lens be approximately midway from the focal plane to the photocathode and that the total distance is about twice the focal length of the lens. Function (c) depends upon what area of the photocathode one wishes to illuminate. Now the size of the image produced by the Fabry lens is a function of the f ratio of the telescope and the focal length of the Fabry lens only. For example if the telescope is f/10 then a 2.5 cm focal length Fabry lens will give a Fabry image of 2.5 mm diameter and a 25 cm focal length will give a Fabry image of 2.5 cm diameter. An f/5 telescope with a 2.5 cm focal length Fabry lens would give a 5 mm diameter image and a 25 cm focal length would give a 5 cm diameter image. The final choice of Fabry lens therefore goes as follows. What size Fabry image is wanted and what is the f ratio of the telescope? These in turn decide the major characteristics of the lens and specify the focal plane to photocathode distance. They also govern the diameter of whatever filters are to be used in the system.

10.4.2 The filters

Before describing the filters in detail it will be useful to consider what filters do, and some of the characteristics of the various types. The purpose of a filter is to restrict the wavelength range of the system to one of particular interest, and to do so in a reproducible way, so that observations at different epochs can be meaningfully compared. Without a filter the range would be limited by the characteristics of the photocathode and/or by the transmission of the Earth's atmosphere. The former varies significantly from one tube to another and during the life of a tube, and the latter is not constant, being a function of pressure and water vapour content. These produce an ultraviolet cut-off which varies between 320 and 350 nm depending on atmospheric conditions. The infrared cut-off is near to one micron (1000 nm).

The simplest filters are low pass (high pass), that is they cut off above (below) a given frequency, having a specific wavelength, to the red (blue) of which they are transparent. The use of such a filter would mean that one side of the wavelength region to be observed was controlled by the filter and that the other side was controlled by the atmosphere or photomultiplier tube sensitivity. More normally a filter, or combinations of filters, is used which allow the transmission of only a fixed range of colours. It is easy to forget that these filter characteristics will only be repeatable provided that nothing else in the system, for example the red limit of the tube response, cuts into the transmission window of the filter.

Wide pass band filters can be simply pieces of coloured glass, or combinations of coloured glass. Typically such types of filters are used to give a transmission window ($\Delta\lambda$) of a few tens to about one hundred nanometers. They are generally very stable, i.e. their characteristics do not change much with time. They are, however, sensitive to temperature with a wavelength shift of about 0.1 nm for every 1 °C. The wavelength range transmitted by coloured glass filters is not sensitive to tilt, that is, they behave in the same way for rays passing through them at small angles to the perpendicular and hence do not have to be used in parallel light.

To achieve very narrow passbands (0.1 nm to 10 nm) one must use *interference filters*. These allow the light of particular elements to be isolated, which may pinpoint a significant feature of a star, see §10.5.1. However, interference filters age rapidly unless kept in conditions of stable temperature and humidity—filters even one or two years old often show serious departures from their original specification. They are not seriously sensitive to temperature but their multi-layer construction makes them sensitive to tilt, their nominal characteristics only being valid for rays which are perpendicular to them—'normal' rays. Rays passing through at any other angle will cause the filter characteristics to show both a shift of central wavelength and of transmission window width. They can also be very expensive. Whereas a set of UBV filters can be bought for about ten pounds, each interference filter is likely to cost ten to twenty times more than that. In choosing a project remember that one of the advantages of small telescopes is in being able to set down data bases over years or tens of years. To achieve this the equipment must be changed as little as possible, to alleviate problems with recalibration. This includes both photomultiplier tubes and filters. Therefore from considerations not only of cost but also of long term stability there are advantages for small observatories in using the UBVRI system. To be quite specific we suggest the *UBV system* developed by Harold Johnson in the USA in the years 1940–1960, and R and I modified by Gerald Kron in the USA and Alan Cousins in South Africa in the 1960s and 1970s.

This system therefore needs to be understood in some detail. It is based upon the use of broad band filters, several of which are combinations of two different filter types. Not every observer will have photomultipliers allowing the observation of all five of the different colours. Bialkali photocathodes cease to be sensitive at wavelengths longer than ~650 nm, that is the approximate red limit of the V filter. S11 photocathodes have a similar red cut-off. S20 photocathodes have some sensitivity out to 850 nm while extended S20 can reach 900 nm (see figure 10.2). Some exotic photocathode materials can cover much wider wavelength ranges than any above but suffer from severe practical disadvantages such as always having to be kept cool after manufacture, since warmth destroys their sensitivity. It is not normal to try to observe the whole UBVRI system with one photomultiplier tube, except with such very expensive systems. A tube which is red sensitive will be unnecessarily noisy for blue observations while tubes with good characteristics for the blue band will not detect red light. For UBV only, bialkali photocathodes are probably a good compromise, and S11 photocathodes are also used. For V, R and I an S20 or extended S20 will be needed.

In table 10.3 we list the filter or filter combinations required to provide the UBVR and I pass bands. The transmission curves shown in figure 10.9 make it clear why (and which) combinations are needed.

Table 10.3

Filter	Glass (Schott type)	Thickness
U	UG1	1 mm
B	GG385 +	2 mm
	BG12	1 mm
V	GG495 +	2 mm
	BG18	1 mm
R	OG570 +	2 mm
	BG38	1 mm
I	RG9	2 mm
	or WRATTEN 884	Thin film

Note that the U filter has a severe red leak between 700 and 800 nm and if it is used with a red sensitive device it must, in addition, have a copper sulphate filter in the system. This extra filter typically consists of 5 mm thickness of a 100% copper sulphate solution. This is one more reason to try to have a system where UBV are done with a blue sensitive tube and VR and I are done with a red sensitive tube or other detector. It should also be noted that the R and I of this system are similar to those of Kron or Cousins rather than Johnson whose system requires the extreme red sensitivity of an S1 photocathode or similar. If one does have access to a detector with such a red response then Johnson's I is centred at 900 nm and a Schott RG780 filter could provide the correct blue edge to the pass band. The red edge might have to be provided by the photocathode cut-off in sensitivity. Even PIN photodiodes will not allow one to go as far to the red as the next standard colour, J (which is centred at 1.25 microns), as the PIN photocathode cut-off is near 1.1 microns.

The difference between the intensities in different wavelength ranges is often very revealing. For instance, the difference between the brightness through B and through V filters gives an estimate of temperature. This quantity is used so often that it is given a shortened name, simply B – V.

One cannot discuss the choice of interference filters in the same way. Their uses are more specialized and must be related to a specific project.

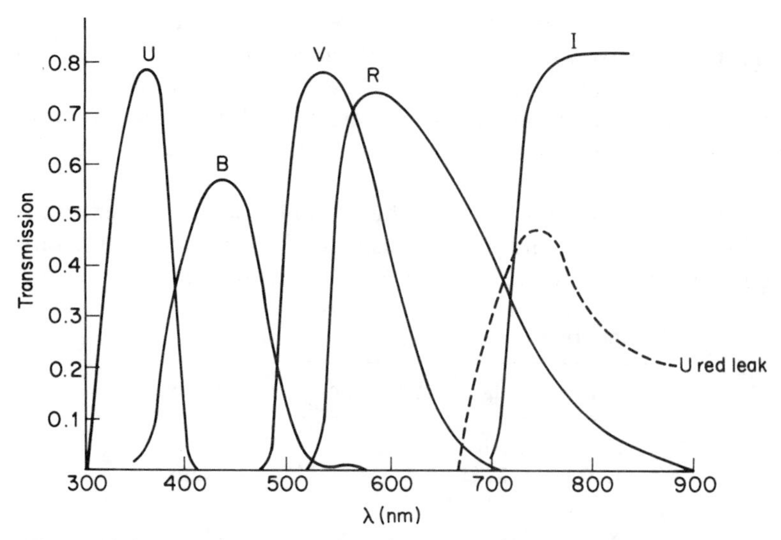

Figure 10.9 The variation with wavelength of the transparency of different filters. Broadly speaking, the filters pass the colour of light for which they are named—ultraviolet, blue, visible, red and infrared. However, the transparency strays into other areas. The response at a given wavelength of the (detector and filter) arrangement is the *product* of the transmission read off this curve, and the sensitivity read off either figure 10.2 or figure 10.4. All the other mirrors and lenses in the system will of course also cause some reduction in response. The effect of a combination of filters at a given wavelength is obtained by multiplying the transmission factors of the individual filters.

10.4.3 An overall design—the Joint European Amateur Photometer (JEAP)

Having explained several of the components which go to make up a photometer and the reason for their inclusion we will now give a brief description of a photometer designed for small telescopes and catering for widely differing f ratios and apertures. The need for such a photometer became obvious at a conference organized by one of us (ENW) in September 1984 for amateur and professional astronomers from all over Europe who were interested in collaborating on photometric projects. The requirement was for a lightweight system, capable of producing results equal, in good conditions, to the best available anywhere; capable of being used at from ~f3 to ~f30 on telescopes from 6 inches (15 cm) diameter to perhaps 40 inches (1 metre) diameter and being able to give on-line reduced photometry via a cheap microprocessor. A major requirement was of course a low price allowing as many people as possible to buy one. It was decided that cooling of the photomultiplier tube should be made available as an option and that, if possible, photon counting should be used. Considerable experience by one of us (ENW) in trying to do photometry in the southern parts of Europe showed that even at the best sites possibly no more than about one third of the time was suitable for single-channel photometry while another third has stars visible but with too much cloud to allow accurate photometry. In northern Europe the situation is much worse with possibly only a few or a few tens of nights each year being photometric. This led to the development for professional use of a four-channel fibre-optic-fed comparative photometer, known colloquially as the four-star photometer. This allows simultaneous measurement of the variable star, one or two reference stars and sky background, so strictly it should be called a three-star photometer. Experience with this has shown that even on cloudy nights it is still possible to produce comparative photometry results accurate to a few millimags. Figure 10.10 shows results from a 16 inch (40 cm) telescope in Nice using three channels only. It can be seen from the upper part of the figure, which shows the apparent magnitude corrected for an extinction coefficient five times the normal value, that the night was of poor quality with a hazy sky, stratus, passing bands of cirrus and some local cumulus which finally covered the whole sky. The observations were obtained despite much light pollution from the city. Nevertheless, the errors (Δm) in the lower part of the figure allow one to easily see a variation with a total range of only 0.01 magnitude superimposed on top of the previously known longer period variation of 0.02 magnitudes. Such experience led us to believe that although the JEAP must be optimized for stand alone use it should also be capable of being added to, one channel at a time as money becomes available, to allow it to be built up into a *multichannel comparative photometer*. Using two channels for star and sky is fairly easy since the sky channel need not be accurately positioned. Two simultaneous star channels adds considerably to the sophistication since they must both be accurately positioned on the same telescope—or even on separate telescopes.

The final design of the JEAP is housed in a small aluminium box of about 75 mm × 75 mm × 65 mm. This has a standard 11/4″ eyepiece tube on the front to allow it to be used as an eyepiece replacement. Inside it has a flip mirror with three positions: a 45° position to allow viewing of the focal plane and centring of the star on an eyepiece graticule; a position where it lies parallel to the optial axis and underneath the eyepiece/graticule assembly to stop the ingress of light whilst observing, and a third position when it lies perpendicular to the optical axis to stop light entering the rest of the system if, for example, the dome lights are switched on. At the rear of this box is a set of six interchangeable apertures ranging from 0.25 to 1.5 mm in steps of 0.25 mm. Into the back of this plugs a fluid light guide to transmit the light from behind the aperture to the filter and detector. The filter is a four-position wheel in front of the photomultiplier tube which is housed in a

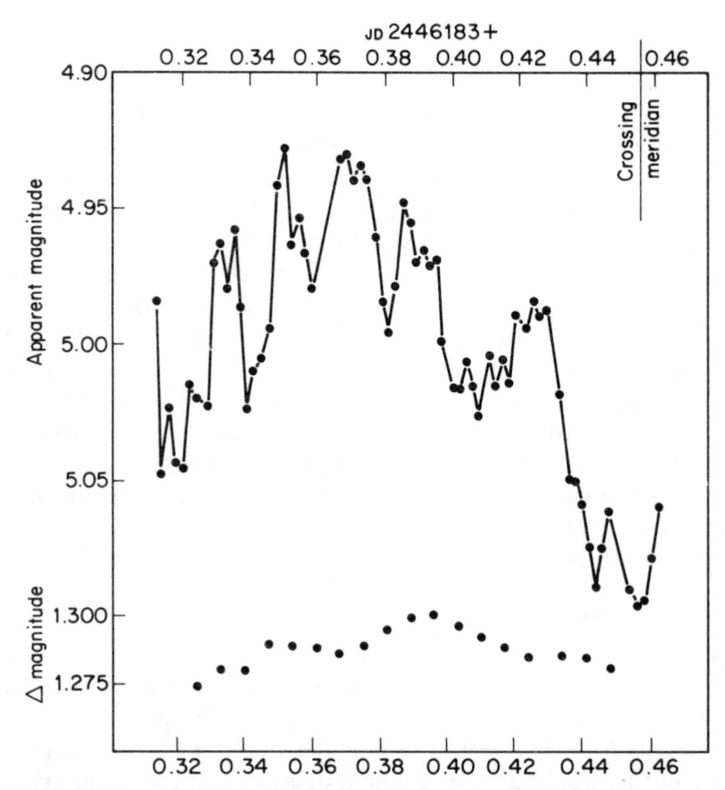

Figure 10.10 An indication of the way variations in sky brightness and haze are corrected out by taking measurements of a reference star and of the sky as well as the star of interest. In the case shown the measurements were simultaneous, thanks to a three-channel, comparative photometer system. With a single channel, the measurements have to be taken in sequence.

container machined from solid aluminium. (The whole assembly of filters and PMT can be cooled by a Peltier effect cooler which can maintain a low temperature, constant to $\sim 0.1\ ^{\circ}\mathrm{C}$. This, as mentioned above, is regarded as an option since it adds considerably to the price.) The result of this design is that the total weight put on the telescope is less than 0.85 kg and that when money or time permit it is possible to buy extra channels. Note that in this design there is no Fabry lens, as laboratory experiments have shown that the fraction of light which emerges from the fluid light guide is insensitive to guiding errors at the entrance. An EHT unit has been developed simultaneously with the rest of the photometer to ensure that a fully integrated and compatible set of parts is available. Figure 10.11 shows the general layout of the JEAP.

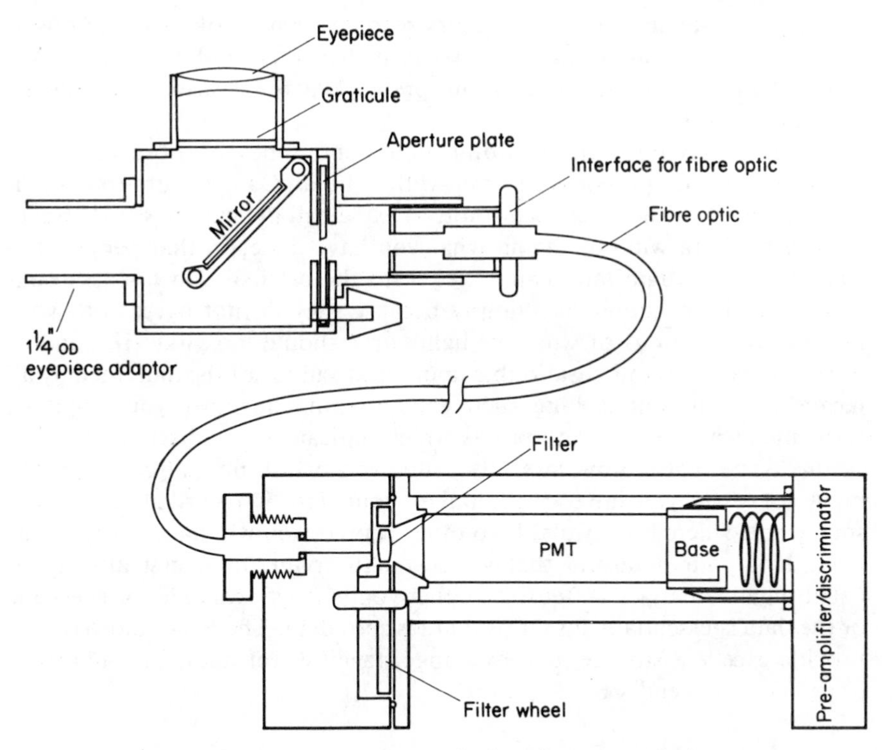

Figure 10.11 The layout of a particular photometer design, JEAP.

10.4.4 Recording data

In one way it is not too important which type of photometer you use. Whether you use your naked eye and a simple, small telescope, or binoculars, or a large, purpose-built, wide-field telescope and a comparative photometer will affect the type of project you can undertake, but you will always find suitable projects. With unaided eye observations you will have to confine yourself to variables with amplitudes greater than a few tenths of a magnitude. With sophisticated equipment you can look for variations of only a few millimagnitudes. What will finally decide whether your contribution is scientifically useful will be the *care* with which you observe. The most sophisticated equipment cannot stop a careless observer producing worthless data! Remember, you are laying down a database which might well be used years, or probably tens or even hundreds of years in the future. Asiatic records of 'guest stars' from one to two thousand years ago now shed light on nova and supernova events. Records exist of the passage of Halley's comet for over two thousand years and enable modern researchers to look for changes in its orbit since the long timespan makes up for limited accuracy. Contemporary astronomers reap the benefit of neolithic observatories such as Stonehenge from nearly five thousand years ago! We cannot know now what aspect of our present day observations will interest our descendants.

Therefore accuracy in recording dates and times of observations is important. If you present your data with a Julian Date be sure to make it clear whether it is heliocentric or not, as explained in detail in §11.1. Never modify the data without saying what you have done, so that people who want to use the data later can get back to the original. Do not leave out points which you don't like simply because they do not agree with your preconceived notions of what the light curve should look like. If cloud or instrumental problems cast doubts upon the accuracy of the data then by all means reject it; but lacking such good reason remember you might be throwing away the few data points which indicate a new discovery!

Finally, no matter how carefully you have worked, no matter how many nights or years you have spent, unless your data is recorded and stored somewhere where it is available to other people, you will have wasted your time. It is your data and that gives you the right for a first attempt at collation and analysis. After that, either you make it available for everyone or the data is essentially lost. Lost data is zero data. The BAA and AAVSO provide excellent storage, and exchange, facilities for data. The addresses are given in appendix 6.

10.5 OTHER TYPES OF OBSERVATION

10.5.1 Narrow band filters

If I wanted to make a map of a city showing the cinemas I might take an aerial photograph through a filter which only passed one of the lines of neon (say 640.2 nm). If I was lucky all the neon signs would show up, but the sodium street lights would be suppressed. The width of the neon line is < 1 nm so a filter which passed only 5 nm out of the total visible band of 300 nm (400–700 nm) could, ideally, suppress the general background light by 60 : 1, and cut out the sodium lines completely. Since hydrogen is the predominant element in the Universe, a filter which only passed the hydrogen lines would show images of emission regions, suppressing other light, including, in particular, local light pollution. A limited amount of spectral information can thus be obtained without the complication of changing the optics to include a spectrometer. The characteristics of interference filters, which are the natural choice for narrow pass bands, are outlined in §10.4.2.

Alternatively, instead of only transmitting a given line one may want to absorb a given line. Filters which absorb the main yellow line of sodium are available, to help combat the bad effects of street lights. This allows better, cleaner photographs.

10.5.2 Polarization

Polarization is best pictured in the wave model of radiation, in which there are electric and magnetic fields which vary jointly and periodically in value. Their direction is normally random in time, but for some sources of light there is a detectable pattern. For instance, the field may show a preference for a particular direction, called elliptical polarization. Another possibility is to find a field direction which rotates in a steady, rather than a random way, called circular polarization. The detection and measurement of polarization is fairly straightforward and can give new information about the source of radiation. For instance, strong polarization is often found in radio radiation from sources in magnetic fields, such as the Crab nebula and the Crab pulsar. To find the direction of strongest field in radio waves one has merely to rotate a (dipole) antenna about the direction of propagation to find the strongest signal. This is, of course, a familiar fact with television signals. Most TV transmitters in the UK transmit horizontally polarized beams. In the case of light, a dipole antenna would be only 250 nm long! This is not feasible, and light detectors are equally sensitive to all directions about the axis of propagation. However, polarizing filters are readily available, and these do indeed contain aligned dipoles, in the form of tiny rod-like crystals. One has only to place such a filter in front of a detector and rotate it about the axis of propagation to see if the signal strength varies. If so, the direction giving the strongest signal (I_{max}) is the direction of polarization. The strength of the polarization is defined as

$$\frac{I_{\text{max}} - I_{\text{min}}}{I_{\text{max}} + I_{\text{min}}}.$$

An alternative way of producing a variable polarization filter is to use two piezo-electric pistons at 90° to each other to compress a specially prepared polarizing filter. Compressing this first along one axis and then along the other causes the filter's plane of polarization to change through 90°. Such devices are commercially available.

It has been found that there is a whole class of irregular variables—archetype AM Her—which show polarization. They are binaries, and in AM Her itself the linear polarization varies by several per cent, in step with the rotation period of the binary (3.1 hours). The circular polarization varies with a bigger amplitude, but seems to arise from some more complicated source. Because of the varying linear polarization, rotating a sheet of polaroid in front of the photometer would produce changes from zero up to a few hundredths of a magnitude. This sounds small, but is not too difficult to measure, since the period of the variation (i.e. the rotational period of the filter) is in this case under your control and you will know exactly what curve to fit to the data. Finding the amplitude of a given well known frequency (and phase) is far easier and more sensitive than fitting an *unknown* curve to data. You can move the sheet of polaroid at a rate or in a sequence which best suits your recording equipment and the luminosity of the star.

10.5.3 Infrared detectors

A few years ago, infrared detectors counted as very specialized devices. Today you can get two-dimensional detectors in the near infrared which are used for detecting people coming to the door! Such burglar alarms cost only about £50! Perhaps infrared astronomy will soon be added to the activities of amateur astronomers.

Note 10.1 Comparison stars

Lists of reference stars are available from either the BAA, the AAVSO or your local or national astronomical society. These are stars which other observers have found to be of the correct colour and brightness to serve as useful comparison stars for the variable to be observed. The comparison stars need to cover a range of brightness similar to that which the variable possesses.

Chapter 11

Use of data

11.1 PRESENTATION OF DATA

Observations, and analyses of observations, need to be published or stored in such a form that they can be used, probably in conjunction with other data, at any time in the future. First, they need to include, or contain a reference to, a brief description of your equipment, type of photocathode used, filters, whether your data have been reduced to a standard system, which comparison stars were used, whether any of the data are of doubtful quality due to malfunctioning of the equipment or poor sky, and so on. Each observation must have an associated time. The accuracy with which you give this time will, to some extent, depend upon what type of data has been obtained and what sort of project you are involved with. The global availability of time signals which are accurate to better than one second means that start and ends of observing runs or individual integrations can have similar accuracy. Some of your integrations might take 100 seconds or more, and therefore it makes sense to quote the time of each of your data points by the centre time of your integration, i.e. start time $+\frac{1}{2}$ integration time. Somewhere you should say what your *integration times* are. The reason for this is that it is possible for shorter integrations to give more scatter than longer ones, even when the accuracy of the two should be the same. This may indicate short term variation in the star so that, say, 10 second integrations are seeing them whereas 3 minute integrations are smoothing them out.

It is normal to quote time in UT (universal time) which effectively is Greenwich Mean Time but with decimal fractions of a day. If for some reason you wish to quote in your own local time then make it clear how that relates to UT. It is much better if, instead of quoting UT (or local time), you give your date as a *Julian Date* (JD). This is a system of continuously and consecutively numbering the days so that, instead of having to work out how many years and months (which have 28, 30 or 31 days?) and days there are between two data points, all that is required is the simple subtraction of one JD from another. If you want to fold your data with some period or do some period analysis (see below) you will have to do this anyway. It is easier for you to do it than someone else later. The formula for converting from year, month, day, UT, etc, to JD is

$$\text{JD} = 367\,Y - 7[\,Y + (M+9)/12\,]/4 - 3\{\,[\,Y + (M-9)/7\,]/100 + 1\,\}/4 + 275M/9 + D + 1721\,029$$

where Y = full year number, e.g. 1988, M and D are month and day respectively, e.g. January = 1, June = 6, etc. However, in this equation (unlike a normal equation) only the integer parts of each term are used after each and every division. To make this clear we show an example. Let us take the date 4 January 1700. Thus $Y = 1700$, $M = 1$, $D = 4$. The terms in the equation are

$$\begin{aligned} &367\,Y && = 623\,900 \\ &-7[\,Y + (M+9)/12\,]/4 && = -2\,975 \\ &-3\{\,[\,Y + (M-9)/7\,]/100 + 1\,\}/4 && = -12 \\ &+275M/9 && = 30 \\ &+D && = 4 \\ &+1721\,029 && = 1721\,029 \\ &\text{JD total} && = 2341\,976. \end{aligned}$$

The table in appendix 7 gives the Julian Day for 1 January for the next few years. Intervening days can be worked out when you need them—remembering the extra day in Leap Years, of course. By convention, the change from one Julian Day to the next is at 1200 UT each mean time day. Thus the above worked example of Julian Date refers to 1200 UT on 4 January 1700. This convention arose when many of the world's observatories were in Europe and near the Greenwich meridian. It was thought that it was unlikely that many observations would be made during daylight hours and that it would be more convenient not to have to 'change the date' during any one night's observing run. Nowadays, with infrared observations possible in daylight and much of the observational effort moved well away from Europe, this is now an anachronism. The truth is, however, that wherever the zero longitude was moved to, someone would find the change of JD in the middle of their night.

Should you be sure that you are only ever going to use modern data, i.e. later than March 1900, then an abbreviated form of the above equation can be used. This is

$$\text{JD} = 367\,Y - 7[\,Y + (M-9)/12\,]/4 + 275M/9 + D + 1721\,014.$$

You will also need to calculate the fraction of a day at which your data were obtained. Remembering that the JD starts at 12 UT then the fraction of the day that would have occurred by, for instance, 3 h 12 m 12 s UT would be

$$\frac{15}{24} + \frac{12}{24 \times 60} + \frac{12}{24 \times 60^2}$$

$$= 0.625 + 0.008\,33 + 0.000\,12 = 0.633\,45.$$

You will see that if your times are really accurate to one second, then you will need to quote the time to 0.0001 day. However, quoting to 0.001 days is normally adequate. Should you wish to avoid remembering the 12 hour difference between the start of a mean time day and a JD for every calculation, it is possible to change the constants in the above two equations by 0.5 day so that they become 172 1028.5 and 172 1013.5 respectively.

For many stars such as, for example, long period variables, or wide binary systems with long periods, there will be very little need to quote times and accuracies of 0.001 day. Indeed, for spectroscopic observations, where integration times are often hours, it would be meaningless. However, in photometry, where integration times will be more typically 10 to 100 seconds, accuracy of 0.001 day is appropriate. This brings another problem. We make our observations from the surface of the Earth, circling about $8\frac{1}{2}$ light minutes away from the Sun. In principle, if we observed a star, exactly in the plane of the ecliptic, six months apart then on one occasion we could be receiving the light of the star 8 m 19 s before it gets to the Sun, and six months later 8 m 19 s after it reached the Sun. The Earth's motion around the Sun would therefore be adding a 1 year long, 16 m 38 s signature to the signal from the star. On the other hand, if the star was at the ecliptic pole then the Earth would not be moving alternately towards and away from the star, and no corrections have to be made. The equation to compute this correction is

$$\textit{heliocentric correction} = KR[\cos\theta\cos\alpha\cos\delta + \sin\theta(\sin\varepsilon\sin\delta + \cos\varepsilon\cos\delta\sin\alpha)]$$

where $K = 8$ m 19 s = 0.005 7755 day (light travel time for 1 AU), $R =$ the actual distance between the Earth and Sun in AU, $\theta =$ the longitude of the Sun, $\varepsilon =$ the obliquity of the ecliptic $= 23^\circ 27'$, $\alpha =$ the star's RA, and $\delta =$ the star's declination.

R and θ both depend on the date of the observation. R is variable because the Earth moves in a slightly eccentric orbit and is therefore sometimes nearer to the Sun than one astronomical unit (1 AU) and at other times further away. Both R and θ depend upon date, are available from the *Nautical Almanac* and *Astronomical Ephemeris* and could conveniently be programmed into a BASIC routine to calculate heliocentric corrections. It is normal to add the correction to the JD.

Therefore, when listing your data, you could use heliocentric JD epochs, geocentric JD, UT date and fraction of a day epochs, or even date and UT epochs. The nearer to the top of that list you go, the potentially more easily usable these epochs are for further analysis. On your own night to night records it is always good practice to write both the date of the day which is ending and the day that will begin during the night, to make any comments about the phase of the moon, cloud conditions and seeing. That way if anyone ever queries some of your data you are in the powerful position of being able to show them the actual record. A note such as '22/23 November 1987, no moon, light wind, some cirrus to NE 45° up sky, seeing 2–3 arc-seconds' is the kind of useful note for future use.

Now, having obtained a list of magnitudes, or Δ magnitudes, and epochs, what should be done with them? It is important to realize that the list should always be made available for future observers by being published or being lodged in the archives of the BAA, the AAVSO or some similar body. It is good practice to include a diagram also. If you have a microcomputer controlling your photometer, or doing your reductions, then it can also display your data against time so that you can look for trends or variations. The BASIC statements TAB(n) or PRINT AT make this easy.

The next stage depends very much on the type of star and the reason for observing it. Let us suppose that the star is in fact an eclipsing binary and that you wish to know what was the mid time of some eclipse which you have just observed and then to know where the time of this eclipse fitted on an $O - C$ diagram and whether it indicated that the period was constant, becoming longer or becoming shorter. We will start with a description of how to determine the time of a minimum or maximum and then discuss the $O - C$ diagram.

11.2 DETERMINING THE TIME OF A MINIMUM

Some eclipsing binaries have minima which are narrow, symmetrical and of the same shape every time. In that case a freehand line through the observations at one eclipse will give a reliable value, and it will be easy to find a period which fits the timing of successive eclipses. This can be done by plotting a graph of the estimated time of minimum against the number of the eclipse in sequence, or by superimposing data from several eclipses in a process known as folding the data. This will be discussed in more detail below, but the essential point is that, in order to superimpose the data from different eclipses, one must assume a value of the period so as to assign a phase within the eclipse cycle, to each reading. Only when the true period is chosen do the points fall closely round a mean light curve.

When any or all of the three desirable features listed above are absent, other techniques must be called into play. Let us take the case of a light curve which is symmetrical and constant, but not narrow. Unless the mathematical form of the curve is known or simple (which is rare for eclipse curves), hand fitting is needed, at least as an initial stage. One good way is to copy the curve onto tracing paper, rotating this 180° on the time axis and sliding it to get the best fit of the rising branch on the falling branch and vice versa. You can do the same trick mathematically but the human eye is very good at finding symmetries and superpositions. Even if the eclipse curve is severely asymmetric, but repeatable, it is still possible to obtain a time of minimum by obtaining a mean eclipse curve on tracing paper and sliding this, still with its *correct* orientation, on top of each minimum and picking out some epoch at which the individual curves best fit the mean. Mathematically an autocorrelation will do the same thing. However, the type of variably asymmetric curve to be found in, for example, CQ Cep (see figure 13.2) will not respond to this treatment and some other method must be found. Pogson suggested a method which is still used. A set of horizontal lines is drawn from ingress to egress of the eclipse. A mark is made at the centre of each line and these marks are connected up. Where this line intersects the eclipse minimum is probably as good an estimate of mid eclipse time as can be made. Figure 11.1 shows how this is done. Exactly the same treatment can be used if it is necessary to determine the time of a maximum. Should the curve to which you wish to assign an epoch have no symmetries and be variable then probably the best you can do is to obtain a mean curve and to slide each individual curve over this to get a best mean fit. This can be done mathematically using an autocorrelation fit. Once you have your newly determined epoch, what can you do with it? Below we explain the use of the *$O - C$ diagram*.

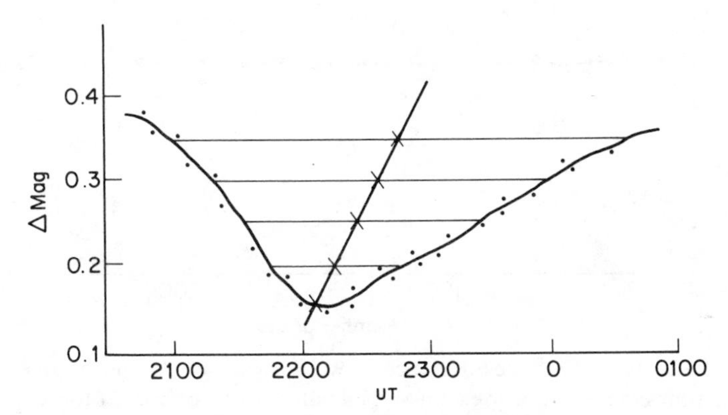

Figure 11.1 Pogson's method.

11.3 THE OBSERVED – CALCULATED ($O - C$) DIAGRAM

Suppose that there was some astronomical phenomenon which had a period of precisely 24 hours. This could, for example, be the period of an eclipsing binary star. In order to determine the stability and correctness of the period one might choose to determine the time of minimum by observing for 2–3 hours each night. Technically, it is more accurate to measure the time when the light intensity is changing fastest. This is not easy if the light curve shows variations from cycle to cycle. It will not be at a time of maximum or a minimum, but typically will be halfway up or down the light curve.

From these measurements the time of the minimum, or maximum, is deduced. Either way, one would have an observed time and this could be compared with the time predicted by calculating the product of the accepted mean period and the number of cycles since a given starting time. The difference between this observed and calculated value $(O-C)$ plotted against time or cycle number is the $O-C$ diagram. It is generally presented in the form shown in figure 11.2. The measurement will give approximately the same time each night. Then the $O-C$ diagram is a horizontal line, $O-C=0$. Note that, due to errors in determining the exact centre of the eclipse, there will be some scatter about the mean horizontal line. Unless the eclipse is unusually steep-sided these errors are likely to be several minutes (of time). It is not unknown for a few of these random errors to conspire to give the appearance of a quasiperiodicity in the $O-C$ values which disappears with the accumulation of more data.

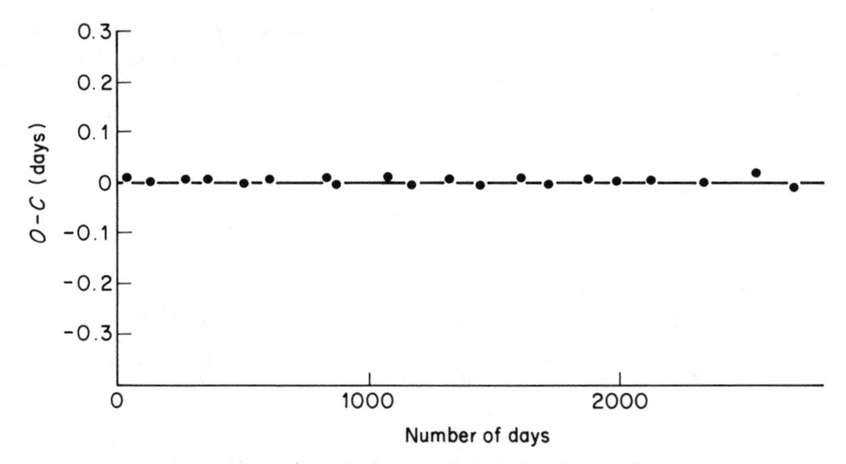

Figure 11.2 $(O-C)$ is the difference between the observed time of an eclipse and the time calculated, using an assumed value of the period. If the assumed value is correct, and the eclipses are perfectly regular, the $(O-C)$ will group around a horizontal line, with a scatter due to experimental error.

Suppose now that, unbeknown to us, the period of our imaginary eclipsing star is not exactly 24 hours but is 24 hours + 1 second. We still keep using 24 hours as the assumed period. How long will it be before the single extra second can be detected? Because of our inability to determine the time of mid eclipse with a precision greater than 1–2 minutes it will certainly not be clear after one night that the period is not precisely 24 hours.

However, after 120 days the $O-C$ value will be $+120$ seconds, or 2 minutes, which is becoming significant. After one year the difference is ~ 6 minutes or ~ 0.004 day. the curves on the $O-C$ diagram which result from either the underestimation or the overestimation of a 24 hour period by one second are shown in figure 11.3. It will be seen that although the precision with which the centre of any single eclipse can be determined might only be 1–2 minutes (i.e. ~ 1 part in 10^3) of the period, nevertheless, after one year the period itself is determined to better than one second, i.e. ~ 1 part in 10^5. There is no magic in this, it is the same as getting a more accurate value by repeated measurements. The $O - C$ diagram method is particularly suitable for astronomical measurements where, unlike the laboratory, gaps in data are endemic. It should be noted that at this stage we are still dealing with a case in which the star has a constant period, albeit different from our assumed value. The constancy is reflected in the fact that it produces a straight line, either horizontal or with a slope, on the $O - C$ diagram.

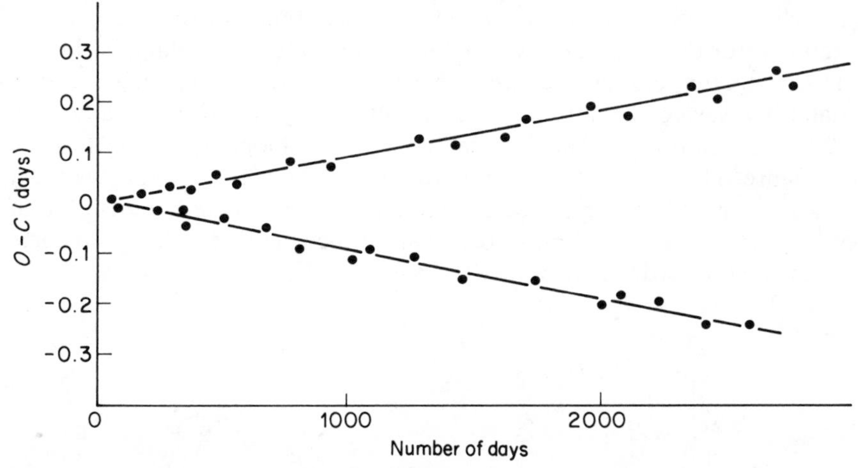

Figure 11.3 An $O - C$ graph for two different cases. In the upper plot, the assumed period is too short, and in the lower one it is too long.

What if the period is, in fact, changing? This shows up, as one might guess, as a curve on the $O - C$ diagram, but to say what curve we have to take a particular case. There are many possible physical causes for changing period but for illustration purposes we need only mention two. If we consider our hypothetical 24 hour eclipsing star again then we can imagine that as the two stars orbit around each other either mass is transferred from one star to the other, or that mass is slowly lost from the binary system. In the first case the mass ratio of the two stars is changing while in the second case the total mass of the system is decreasing. Depending upon how the mass is lost or transferred and how much angular momentum it possesses the period can either increase or decrease. Measurements of the changes in period will therefore be very useful in deriving an accurate model of the system. Observationally, we need to make a clear distinction between this case of a changing period and the former case of a constant but wrongly determined period. Before, each period really was of the same length as both preceding and following ones, but the estimated mean period was incorrect. With mass loss, each cycle is followed by one which is slightly longer (or shorter) which in turn is followed by one which is again longer (or shorter) and so on. Using our example of the 24 hour eclipsing binary again and supposing now that the period changes by + one second/cycle then we have successive cycle lengths of 24 hours, 24 hours + 1 second, 24 hours + 2 seconds, 24 hours + 3 seconds, etc.

Now, with this example, after 100 cycles the assumed period is wrong by 100 seconds. However, the $O-C$ value has changed by 100 + 99 + 98 + 97, etc, i.e. a massive 6050 seconds or +0.043 day. Figure 11.4 shows the resulting $O-C$ diagram. With this type of continuously changing period the exact form of the curve is very much a function of what period is adopted for the calculated part of the $O-C$ value. For example, in figure 11.4 a 24 hour period was assumed which was correct for the first part of the data. However, the lower curve would change its form totally had a different mean period and/or mean epoch been adopted. This is illustrated in figure 11.5. It is the degree of curvature which is the significant characteristic (from which the changing period can be determined) and not its arbitrary origin or mean slope. In algebraic language, it is the coefficient of the second-order term which is needed.

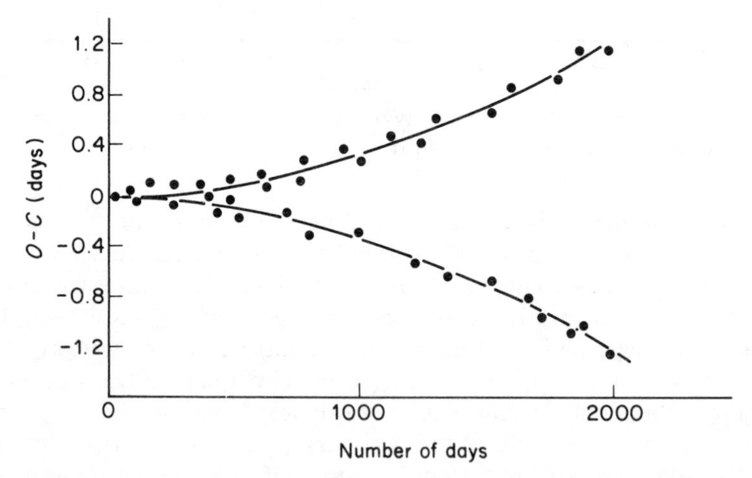

Figure 11.4 A varying period shows up as a curved line on the $O-C$ graph. A decreasing period would give the upper curve and an increasing period the lower curve. The amount of curvature depends on the rate of change of the period. In the case shown, the assumed period and phase were correct at time $T = 0$ (see figure 13.3).

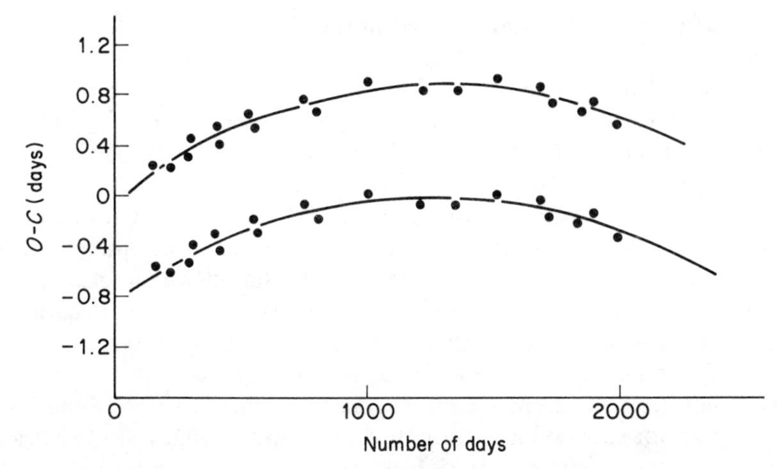

Figure 11.5 These curves correspond to the same (hypothetical) astronomical object as in the lower curve in figure 11.4, but with different assumptions. In the upper curve the assumed phase was correct at $T = 0$ but the assumed period was only true on day 50. In the lower curve the phase and period assumed for time $T = 0$ were in fact those for $T = 50$. The important point is that the curvature is the same in both these cases *and* in the lower curve in figure 11.4.

The coefficient of the second-order term can be easily determined numerically: it is just a matter of taking differences, as shown in note 11.1. Having done this, the difference between a constant and changing period is simple and clear cut. For a constant period the coefficient is zero. For a changing period it is non-zero. In a real case the error on the value of the coefficient must be determined to see if it is significantly different from zero.

This simple method has been used for some very fundamental research. For example, in 1987 the best evidence for gravitational radiation, as predicted by Einstein, was the slowly decreasing period of a close pair of pulsars. The changes are tiny ($\sim 10^{-18}$ per cycle) but pulsars are exceptionally good clocks, and have a short period so that very many cycles can be followed.

Wide binaries have long periods, which puts them in the court of amateur astronomers, with long sequences of data.

In the applications discussed above, the data are only needed at times which can be used to find the time of a given phase, e.g. data on each side of the minimum so as to determine its epoch. One way of doing this is by using Pogson's method (see figure 11.1). Alternatively, a mean curve for all the data can be derived (see §11.4) and this can be fitted to every individual observed minimum. Of course some feature other than a minimum might be chosen. Eventually, by whatever means, a fit is made to find the exact time of the chosen phase.

Thereafter, only that time is used in the analysis. Thus the analysis is a two-stage process, and this is a neat way of minimizing the amount of data being handled at any one time. As we have seen, the curve on an $O-C$ diagram is an accurate way of finding a period, whether constant or changing. In terms of sine waves, it distinguishes clearly between sin ωt, sin $[(\omega+\varepsilon)t]$ and sin $[(\omega+\varepsilon t)t]$ where ε is much less than ω.

Can the $O-C$ diagram tell us about the shape of the light curve? In an extreme case such as a spike which is always near the maximum, with measurements only near the crossing point, the answer is clearly no. In practice, secondary processes cause features that drift through the basic shape of the curve and so do affect the crossing points. So it is worth trying an $O-C$ diagram on a wide range of periodic processes.

The pulsation of the β Cephei stars, the light curves of binaries, the eclipsing of hot spots on accretion discs, the slow changes of the Miras and semiregular stars and the rotation of asteroids are all grist to the $O-C$ diagram's mill. Despite the attractive simplicity of the $O-C$ method, one can clearly get more information by measuring the shape of a complete light curve. The penalty is the need to handle large quantities of data, but this is now possible on home microcomputers.

11.4 FOLDING DATA

If there are no significant changes from cycle to cycle of the light curve, then it is appropriate to superimpose, or 'fold', these data with the period in order to produce a more accurate mean curve. There are two stages. To fold the data the period must be known. Then some objective way of 'averaging' the data is needed to produce a mean curve. In the simplest case, the period may be well known from previous observations. We will treat this case first and then discuss some ways to search for best fit periods in stars where this is not known.

We will assume that you have already organized your data so that it has a uniform set of epochs, either all Julian Dates (JD) or all Heliocentric Julian Dates (HJD). Note that if the periods are shorter than a few hours then, unless all the data were obtained in only a few nights or the star is near to the pole of the ecliptic, you must use HJD or the light travel time effects will ruin your attempts to superimpose the data. Ideally the data should all have the same reference stars. In some cases (e.g. asteroids) this is not possible, but the data must nevertheless all be normalized.

Suppose that some of your data is referred to a star with a catalogue brightness of $m_B = 6.27$ and some more to a star supposed to have $m_B = 7.32$. Let us now suppose that your mean Δm_B from the first star is $+0.73$ and from the second -0.30. They both suggest a value near to $m_B = 7.00$ but differ by $0.02 m_B$. Has one of your comparison stars varied, does it not have a well defined magnitude, have you incorrectly allowed for atmospheric extinction effects, is your equipment faulty? You will probably never know and you can only take an average. It may be that one of the reference stars has long term variations, which show up as the 0.02 mag discrepancy. In order to decide what you should do next, it must be made clear what the consequences are of any further modification of the data. You have two possibilities. One is to reduce every night's data to the same mean level and the other is to leave the data as they stand. If you are only going to search for the mean, short term light curve then the first possibility will do no harm unless the longer term variation is short enough, so that instead of removing a constant from each night's work you should have removed a variable value because the longer term variation shows significant differences within one night. If you were going to use this data set to do a period search then, as we will explain below, it would be disastrous to remove the night to night variations.

Let us now suppose that you have a *known period*, and data which do not show night to night mean light level changes. You need to 'fold' these data on to one cycle of the light curve. This is done by calculating the *phase* of each point from some epoch zero. The zero phase epoch could be totally arbitrary, e.g. JD = 0 or JD = 2440 000.0, or alternatively it could have some physical significance, such as being the time of mid eclipse. The *phase* of each point is simply the fractional part of

$$\text{(Epoch of observation} - \text{Epoch zero)/period.}$$

Plotting the data against phase has 'folded the data' and you now have only the problem of deciding how best to fit a mean curve to these data. What is needed is to divide the data into phase bins (0.05 or 0.1 are often used) and then to determine the centre of the distribution of that bin. If you have very many data points then it will not matter what you do next as there will probably be enough data points in each bin to make very little difference as to whether you calculate the mean or the median of each bin. However, if, as is more likely, there are only a few points in each bin (perhaps 5 to 10), or if the points show a large scatter, then one should use the *median* value. The median is that value at which there are as many points above the value as below it, and has the advantage that it is much less affected by odd points, way off the centre of a sparsely sampled distribution. For typically poorly sampled data, therefore, it is better than the mean as an estimator of the centre of the distribution. As an example, if you had seven points in one phase bin, the median would be the value of that point which was four from the top and also four from the bottom. If there were eight points then the median would be halfway between the point that was fourth from the top and that point fourth from the bottom (i.e. count the total number of points in each phase bin, divide by two and count that number down from the top, or up from the bottom). If, on the other hand, you have very many points in each phase bin, then the mean and standard error should be calculated. There will be little numerical difference between the mean and the median in this case and it is possible to put a meaningful error bar on each point, which clarifies the significance of the data.

You now have one average point in each phase bin, let us say in each 0.1. You could now just join up with a freehand curve (or even with straight lines) each of these ten points to make the mean curve. However, there are other possibilities. Overlapping phase bins are sometimes used—bins 0.1 wide but spaced every 0.05, for instance. Naturally the data in each average point are not independent but the proponents of the method can claim that it gives a smoother curve. If you do this, it must be made clear on the diagram. It might be worth trying 0.05 bins as well. Provided that you have a reasonably evenly covered phase diagram then you should choose the total number of phase bins so that you never have less than, say, five points in any one bin. If there are parts of the phase diagram which have not been covered, then you might have to settle for less than that number.

If your phase diagram, including a mean line, is to be published then it is good practice to put somewhere on the figure itself the name of the star and the period and zero epoch used. This information can also be put into the caption to the figure, but it is not unknown for captions to be put under the wrong diagram so annotating the figure itself is a safeguard.

Suppose now that there are cycle to cycle variations so that ‘folding’ is not appropriate. The data on the Wolf–Rayet star CQ Cephei shown in figure 13.2 is a case in point. For simplicity we will take a hypothetical star with a period of about 1 to 2 hours which is also thought to possess several other periods of similar, but unknown, length. If you have been lucky enough to have a spell of good weather then you might have a set of 6 or 7 hour observing runs on each night for a week. Suppose now that you find that in addition to the obvious variation with a period of one hour and an amplitude of 0.1 mag there is also a night to night mean brightness variation of a few hundredths of a magnitude. If there is good reason to suspect that there are other periods near to one hour present then the worst thing that you can do is to reduce each night's data to the same mean value. The reason is that whenever two similar periods are present there will be some *beating* between the adjacent periods; the more nearly equal are the periods the longer will be the period of the beats. The beat period is given by

$$\frac{1}{P_{\text{beat}}} = \frac{1}{P_1} - \frac{1}{P_2}$$

or, in frequency,

$$\nu_{\text{beat}} = \nu_1 - \nu_2.$$

Suppose that our hypothetical star with a period of one hour also has a period of 1.091 hours; then the beat period will be given by

$$\frac{1}{P_{\text{beat}}} = \frac{1}{1} - \frac{1}{1.091} = 1 - 0.9166 = 0.0825$$

so

$$P_{\text{beat}} = 12 \text{ hours.}$$

Now, as well as beats between the individual periods present there is also the complication of beats between beats and if other frequencies are present one can soon generate a system of rather slow variations superimposed upon the more obvious variations which not only are present because of the existence of several short periods but whose very existence aids in the discovery of the shorter periods. Suppressing the larger beats therefore directly detracts from one's abilities to discover what is happening with the shorter periods.

This hypothetical case emphasizes the difficulties. A 12 hour beat period means that there will be occasions when the star is slowly increasing in brightness, reaching a maximum and then decreasing over a six hour period. This coincides with the length of one night's observing run. If you have not used reference stars of rather similar colour and spectral type to the variable, and have not exactly compensated for the atmospheric extinction variations, then you might already have a spurious six hour trend in your data. If this effect is then complicated by the six hour variation due to the presence of a 12 hour beat and further compounded by the presence of a 12 to 18 hour period, as is found in some Be stars, you will soon realize just what care must be taken not to spoil the data.

In conclusion, if your only wish is to present folded data to obtain a mean light curve, then the removal of night to night drifts is acceptable. If you want to search for periods, then the greatest care must be taken and as a general rule night to night drifts must never be removed. Rather, every effort must be made to ensure that the night to night reference system is impeccable.

We will now describe ways in which a search for periods can be made.

11.5 CURVE FITTING

Any periodic curve can be synthesized from a set of sinusoidal curves of different amplitudes, periods and phases. If one has no idea of the processes producing a given light curve, such an analysis provides the only approach and curve fitting becomes a matter of finding the best set of such sinusoids. It may then be quite difficult to associate a particular component with a feature of the physical process. If, on the other hand, the physical process is well known and produces a well defined shape of light curve, then one would fit that shape. If an eclipse were known to give a square wave, one would fit square waves, not sine waves. In astrophysics, few problems produce a nice simple light curve, so most of this section is written in terms of sets of sinusoids.

The data to be used in the fitting will consist of one intensity value, and a time, for each (accepted) measurement of the object of interest. However, it will not normally be the raw data. There will be an initial stage in which corrections are made for atmospheric absorption. These corrections are made to the object of interest and to the reference star(s), before dividing one by the other to give the ratio of intensities, object/reference. This was described in §10.3.4. The resulting list of intensity ratios and times is best written on to a disc or tape file because it will need to be read many times as different fits are tried.

There are at least half a dozen different ways of computing the period(s) contributing to a measured curve, and curve fitting can become a sophisticated art. The basic methods are, in fact, based on very simple principles, but if used in their simplest form they are often very inefficient. The sophistication comes in making them fast and efficient, and therefore able to handle vast quantities of data. For an amateur astronomer questions of efficiency may be of little importance. The amount of data will probably not be enormous, and in any case if a home computer needs a couple of hours instead of a couple of seconds it does not matter much. We will therefore stick to basic methods.

First, if one can see a repetition period in plotted data, why bother with computed fits anyway? There are at least three reasons. First, it is more objective than hand fitting. You may well know the period(s) you would like to find, and this, probably unconsciously, will bias the curve you draw. Even the most highly trained observers cannot avoid such a bias. The computer will give an objective result. Secondly, most of the methods give an error estimate. Visual estimates of error are extremely unreliable, and often grossly underestimate the errors. The computation gives an error estimate using well defined, standard procedures. It is only in this way that results from different observers can be combined. Thirdly, although the eye is quite good at probing out a fundamental frequency, it is poor at identifying harmonics. Inspection of the plotted data will give a good starting point for a computation, but cannot be relied on to do more than that.

Unfortunately, a computer can be, but must not be, used blindly. This indeed is one of the limitations of fully automated photometry, with the telescope under computer control. Only *you* know which sections of data should be rejected because of spurious effects—clouds, an aeroplane, the EHT had drifted, and so on. It is best to work in photon counts, or intensities, throughout. Some programmes convert to magnitudes. However, a magnitude is a logarithmic scale and so distorts the curve and introduces spurious frequencies. In times past, it was easier to use magnitudes, because it is easier, by hand, to subtract than to divide. For a computer this is irrelevant, and using magnitudes will introduce an error, albeit a small error if the amplitudes are small (say, less than 0.1 mag). Let us suppose that the sky background subtraction works perfectly, and that the sky extinction is negligible. In this case the base line of the results will be constant. (In practice there will be some drift.) If the readings are evenly spaced in time the level of the base is computed simply by taking the average of all the readings. (If not, see note 11.3.) The baseline level is then subtracted from all the readings. The result should then be plotted, to see that it looks right. Even the simplest printer can plot out a curve. If it is plotted down the paper rather than across it needs only a single instruction, in a loop over the readings. At this point, one must decide which fitting procedure to adopt. Three types of procedure are given below, together with sample programmes, in BASIC.

11.5.1 The string length and phase dispersion methods

The *string-length method* is a simple and very effective method if only a single period is involved and the data extend over many periods. It may be very slow on a home computer but, as mentioned above, this is probably immaterial. The idea is to give the programme an initial guess for the period, from your inspection of the plot. Assuming this period, the phase of each reading is calculated and the readings sorted by increasing phase (this is what takes the time). Next the length of a jagged line joining all the points together in sequence is calculated. (This is the length of an imaginary piece of string joining all the points together, hence the name of the method.) If the initial guess of the period was wrong, there will be quite a few points lying badly off a smooth line, because they will have been assigned the wrong phase. Hence the line will be very jagged, and the total length unduly great. The method consists in trying a series of periods which bracket the original best estimate for the period, and seeing which gives the *shortest* total length for the piece of string. It could hardly be simpler.

Figure 11.6 shows the test results of a period determination of artificial data using this method. The program is shown in note 11.1. If you use this method you will immediately discover that you also need to consider scaling your data.

The phase dispersion method is simpler and quicker than the string method, but throws away some information. Again a trial frequency is chosen, and the phase of each observation time is calculated. However, instead of ordering the data, they are simply put in bins. Since phases run from 0 to 1, one might choose 10 bins, each 0.1 wide (0 to 0.1, 0.1 to 0.2, etc) or perhaps only 5 bins, each 0.2 wide. Next the spread of the Δm values within a bin is calculated. If the wrong frequency is used, the Δm values will be spread over a wide range, because there will be points that are wrongly binned. At the correct frequency, all the points are correctly binned, and only experimental error contributes to the spread. As an example of the statistician's craft, this method would not win prizes; it is too crude. It is, however, a fast way of making an objective estimate. To refine the estimate and calculate errors, the following methods are useful.

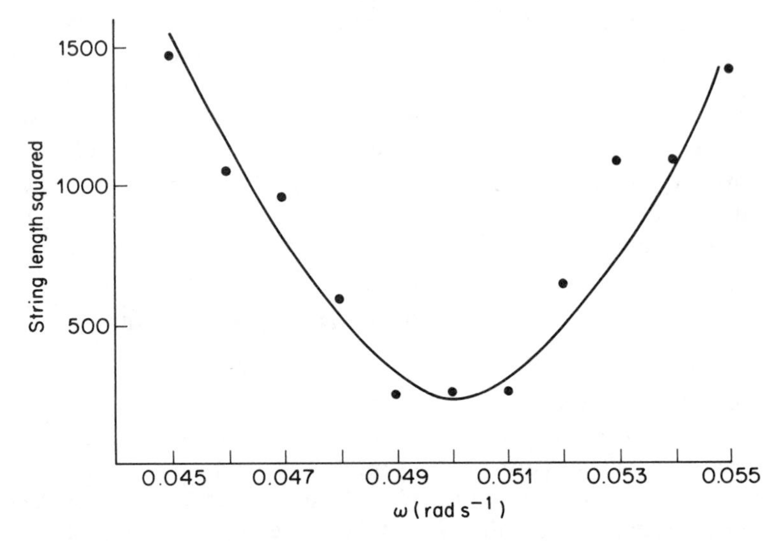

Figure 11.6 Simulated data, fitted by a sine wave whose amplitude, frequency and phase was found by the 'string' method listed in the text. Yes, the true frequency was 0.05 rad s^{-1}!

11.5.2 Method of least squares and periodograms

Again the starting point is a guessed value of the period(s). It is easy to try adding two or more waves, though the computation will become very slow. Each new parameter will lengthen the time by at least a factor of five! The 'squares' referred to in the name are the squares of the distances from each measured point to the curve drawn out by the guessed waves. There is no need to sort the data, so that time is saved. The problem is to guess a suitable function to match to the data—light curves are rarely a close approximation to a sine wave. The programme calculates the sum of all these squared distances, and the best fit is obviously the one which gives the smallest sum.

Figure 11.7 shows the test results of a *least squares* fit to artificial data using this method. The program is shown in note 11.1.

The least squares method operates by scanning over a range in each parameter. Suppose we take a scan over frequency, with all other parameters fixed. A plot of the value of (sum of squares of deviations) against frequency will show a pronounced narrow dip at the true frequency. If the reciprocal of the sum of squares is plotted, the true frequency will, of course, appear as a peak rather than a dip. This is, in effect, a periodogram but the standard (and faster) method, which gives a more useful vertical scale, is described below. There are various computational techniques for producing periodograms which are much more efficient than using a least squares programme, but it is useful to think of the plot in least square terms. If the square of distance is divided by the square of the error, the result is called 'chi squared'. If the errors vary significantly from point to point, it is better to make the program search for the minimum of chi squared rather than distance squared.

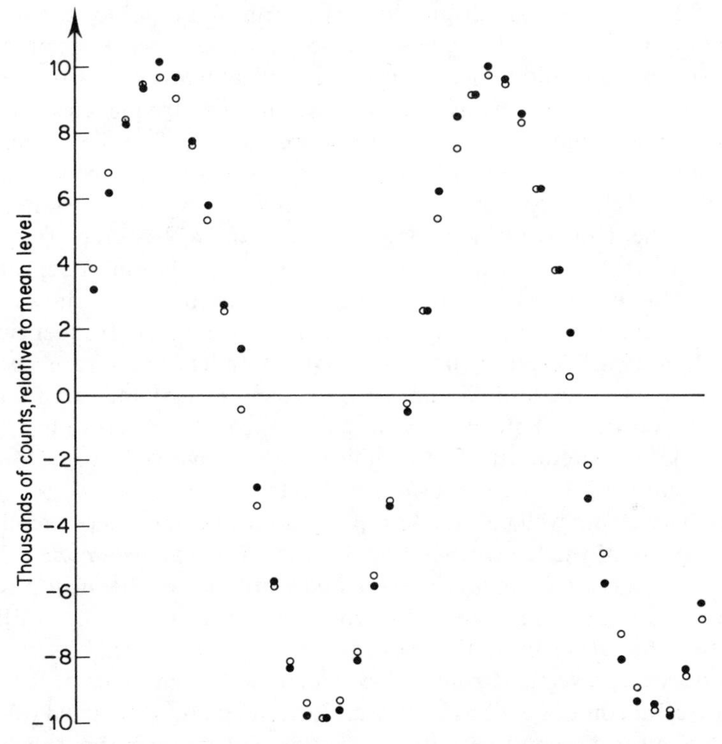

Figure 11.7 Simulated data (as in figure 11.6) fitted by a sine wave found by the least squares method. •, data; ○, fit.

11.5.3 Fourier decomposition

Any periodic signal can be regarded as a sum of harmonics, that is, a fundamental frequency and different amounts of any or all of integer multiples of the fundamental frequency. This avoids having to guess an analytical form for the light curve, and so is a very useful and convenient approach. Moreover, there are very efficient methods for finding how much of each harmonic is needed for the fit ('fast Fourier transform' routines). There are limitations, however. Such analyses ideally require observations to be equally spaced in time, over an integral number of periods. This is virtually impossible to achieve. One may fill in gaps by interpolation, but this always involves some guesswork, since there is no foolproof algorithm. To do this for more than a few points is unsatisfactory. (The same problem applies to maximum entropy methods.) Secondly, the harmonics do not always have direct physical significance because stellar oscillations do not necessarily have components with frequencies in integral ratios. It is therefore more revealing to find the frequencies of the components of the light curve. This is usually displayed in the form of a graph of the intensity of the component vertically, against the frequency along the horizontal axis. A pure sine wave would appear as a single vertical bar on a periodogram, a mixture of two sine waves as two bars at different frequencies and with different heights, and so on. The method used to calculate the strengths of the components from a given light curve is called power spectral analysis and is closely related to Fourier analysis. Both involve taking sums of the product of the data with sine and cosine functions with a given frequency, i.e. a given point on the horizontal axis. The calculation is repeated at points as closely spaced as one wishes. If the periodogram shows a few peaks standing out clearly, the components with those frequencies are clearly dominant. Usually, there are many smaller peaks, arising from errors in the data, the limited number of measurements, and from features of the method itself. To exploit a periodogram fully therefore calls for a more subtle interpretation. The technical points involved are outlined in note 11.3, and will be important if you use the method.

A very useful and well tested BASIC program for power spectral analysis is listed on p 289 of the September 1988 issue of *Sky and Telescope* (vol. 76, no 3). The program also gives the estimated amplitude of the most powerful frequency. The method is called 'discrete Fourier transform' but is different from Fourier analysis in that it does *not* need equally spaced observations, and produces a power spectrum rather than a set of components. When the most powerful component has been identified, it can be subtracted off and a search made for a second periodicity. (Try adding a subroutine to generate artificial data, replacing the data statements in the listed program. Do not forget to refer them to a mean level and mean time as explained in the article.)

11.5.4 Autocorrelation methods

If a copy of a perfect sinusoidal wave is moved along by exactly one period, it obviously matches itself perfectly. If a copy of measurements of a

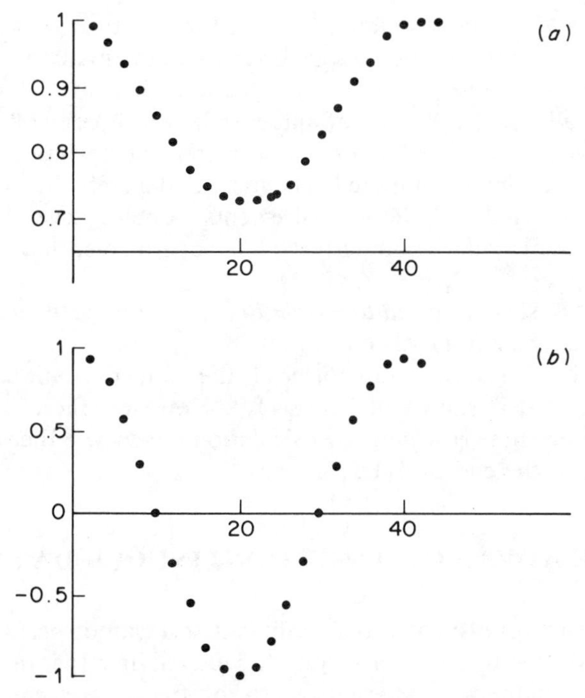

Figure 11.8 (*a*) Simulated data (as in figure 11.6) and (*b*) its autocorrelation function.

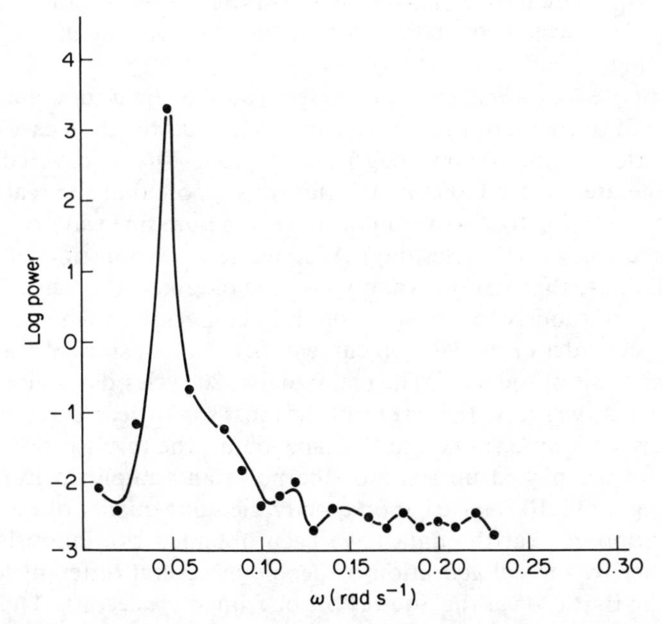

Figure 11.9 Fourier transform of the autocorrelation plot.

sinusoidal wave are moved along, the match at the true period will not be perfect because of errors, but will be better than the match for movements of less than (or more than) the true period. This is the basis of the computationally simple method of autocorrelation. A copy of the measurements is advanced by small steps. At each step the deviation between the original and the copy is computed. The first good match one comes to is the fundamental period. If the measurements contain several component frequencies there will be imperfect, but identifiable, matches at each component period.

Figure 11.8 shows an *autocorrelation* plot for artificial data. The programme is given in note 11.1.

Incidentally, a Fourier transform of the autocorrelation plot is the periodogram and is shown in figure 11.9. However, there is not usually much advantage in performing the calculation this way, rather than directly as described at the end of §11.5.3.

11.6 THE ADVANTAGES OF CONTINUOUS DATA RUNS

No matter how sophisticated the analysis it still cannot perform miracles. Inadequate data still give results which are uncertain in that they fit the data but are not unique, i.e. alternative sets of frequencies can fit the data equally well. The answer to this uncertainty is more data, or perhaps more accurate data. Gaps in the data due to daylight and poor weather are serious impediments to definitive analyses. We will illustrate this point below with synthetic data which are typical of the quality one might hope for in investigating δ Scuti stars (§14.4).

We suppose that a real signal is present, but having a total amplitude of only 10% of the noise of the measurements. In fact we take a case where the real variation is only 0.0006 mag peak to peak. This is covered by noise which generates $\Delta m \pm 0.003\,m$. We further suppose that the real variation which we are trying to discover in the noise is a pure sine wave of period 1.2 hours (frequency = 20 cycles/day). We give four demonstrations. For the first we suppose that an observation has been obtained at precise 7.2 minute intervals continuously for six hours on three consecutive nights, i.e. a rather good but short data run. The top curve in figure 11.10 shows the result of a Fourier analysis of the data. The real signal at 20 cycles/day is not detected. The second curve shows the results of the most sophisticated gap filling type of analysis we know, applied to the same data. The result is nonsense, the analysis having picked up some of the noise and amplified it. The third curve in figure 11.10 is based upon exactly the same quality of data but this time we suppose that the data have been obtained continuously for two days, either by the collaboration of people at several different longitudes around the Earth, or at the South Pole or from a spacecraft. This time the

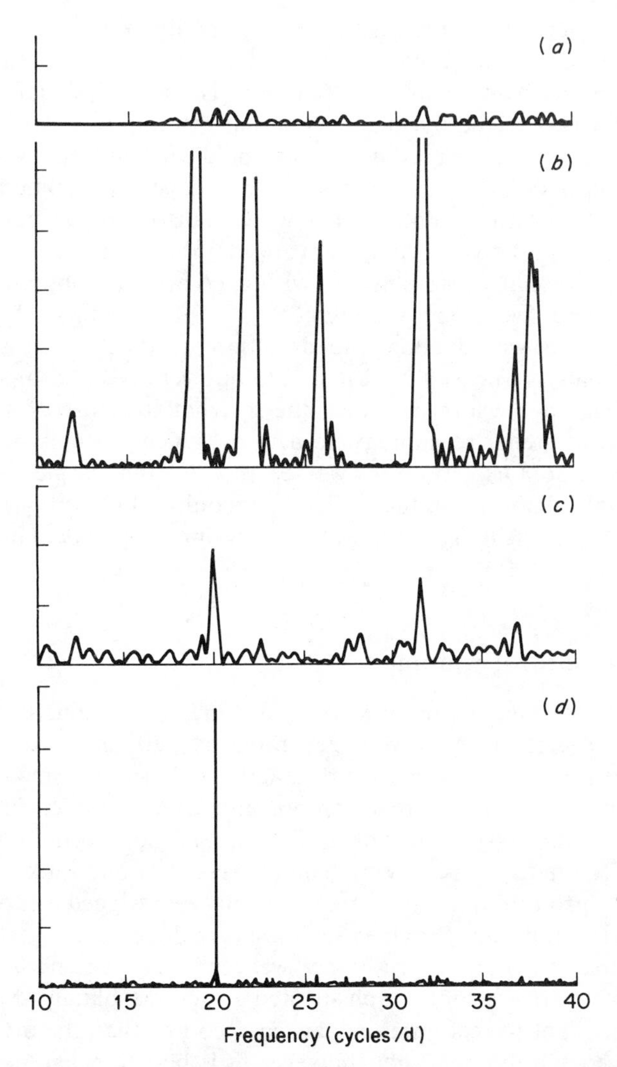

Figure 11.10 Telescopes on the ground can only take data at night and in good weather so there are inevitably substantial gaps in the data from a single telescope. These can introduce spurious periods into a fit, shown here with simulated data, in which the gaps are progressively filled. The graphs are of fitted amplitude against frequency. (*a*) 6 hours data on three consecutive nights using Fourier analyses. (*b*) Same data using 'gap-filling' techniques. (*c*) 2 days continuous data. (*d*) 20 days continuous data. It is only here that the time period in the simulated data, 20 cycles per day, shows up unambiguously. The simulated noise was ten times *stronger* than the signal, so to the eye the signal is lost in noise.

real signal is detected but it still cannot be unambiguously stated that it is the only period present. Finally we show what would be obtained from 20 days continuous data. Note that not only is the real signal clearly and unambiguously detected but that it is perhaps twenty times more powerful than the noise so the longer data run has paid off handsomely. Think now what this means to the user of a small telescope. We deliberately chose values for the synthetic data which were realistic for a small telescope (20–50 cm) on a relatively bright star (brighter than 7–8 mag) at a good site with a conventional photometer or with a comparative photometer on a poor site. The frequency was typical of δ Scuti stars. Twenty days continuous data would allow the detection of 0.0006 mag signals and possibly signals as low as 0.0001 mag. No one yet does photometry to that accuracy. This is the level at which the internal structure of stars can be investigated by stellar seismology, something which has only been possible with the Sun so far. However, we see that by the simple expedient of international collaboration small telescopes could transform our knowledge of this important field. Techniques for analysing gapped data are discussed further in note 11.3.

11.6.1 A word of warning

There are unfortunately many ways of picking up spurious frequencies in the fitting process. They may be generated by gaps in the data as in the above example, or may come from imagining that every small peak on a periodogram represents a real component. Even experienced observers make—and publish—such mistakes, so how can the newcomer to the craft survive? Fortunately it is easy to generate test data and look for spurious effects. For instance, one might try increasing the assigned errors or cutting out sections of the data. Provided such tests are done, it is arguable that it is better to use a program you know well, having written or adapted it yourself, than to use more sophisticated programs blindly. You will, of course, also want to read more advanced accounts than the merest outline given above. Before doing that, however, it is best to get some experience yourself on simple cases. Papers on statistics tend to hide their wisdom behind rather technical language, but if you have already worked out some simple cases their meaning is much clearer.

Note 11.1 Fitting programs

The following program generates artificial data, with errors, and then applies several fitting procedures. It is intended to indicate the steps involved in the procedures, but *not* to serve as a model program. Real data would require checks and tests to be added into the procedures before one could be confident of the results.

```
10 REM "FIT"
20 REM THIS PROGRAM SIMULATES DATA
30 REM AND THEN GOES THROUGH DIFFERENT
40 REM FITTING PROCEDURES IN RUDIMENTARY FORM
50 DIM Y(150)
60 DIM PS(150)
70 DIM T(150)
80 DIM XS(150)
90 DIM RP(150)
100 DIM PHI(150)
110 DIM L(150)
120 REM SF IS A SCALING FACTOR FOR THE "STRING" METHOD
130 SF=4E-05
140 PI=3.1415926535£
150 TP=2*PI
160 B=100000!
170 A=10000!
180 P=20
190 REM SO OMEGA=0.05
200 H=0
210 FOR J=1 TO 150
220 REM SYNTHESISE ONE COMPONENT
230 REM WITH ERRORS
240 YY=B+A*SIN(2*PI*J/P +H)
250 REM RECTANGULAR ERROR DISTRIBUTION FOR SIMPLICITY
260 Y(J)=YY+SQR(YY)*(RND(1)-.5)-B
270 GOTO 300
280 REM ADD BIAS SO ALL COUNTS +VE
290 Y(J)=Y(J)+20000
300 Y(J)=INT(Y(J))
310 T(J)=J
320 NEXT J
330 REM COULD ADD A SECOND COMPONENT
340 PRINT "AUTOCORRELATION"
350 I4=50
360 FOR K=1 TO I4
370 RR=0
380 RQ=0
390 FOR J=1 TO 90
400 RR=RR+Y(J)*Y(J+K)
410 RQ=RQ+Y(J)^2
420 NEXT J
430 RP(K)=RR/RQ
440 PRINT K,RP(K)
450 REM GRAPH HERE
460 LPRINT K,RP(K)
470 GOTO 500
480 GOTO 500
490 LPRINT K; TAB(30*RP(K));"*"
500 NEXT K
510 PRINT "POWER SPECTRUM"
520 R0=1
530 FOR P=1 TO I4
540 L(P)=0
550 NEXT P
560 FOR P=1 TO I4
570 FOR Q=1 TO (I4-1)
580 L(P)=L(P)+2*RP(Q)*COS(P*Q*PI/I4)
590 NEXT Q
600 L(P)=L(P)+RP(I4-1)*COS(P*PI)+R0
610 PRINT P,L(P)
```

Generate artificial data

Autocorrelation

Power spectrum

String length PDM

```
620 LPRINT
630 LPRINT P;TAB(.8*(11+L(P)));"*"
640 NEXT P
650 PRINT
660 PRINT
670 PRINT "STRING LENGTH"
680 REM STRING LENGTH
690 REM W IS A TRIAL FREQUENCY
700 FOR W=.045 TO .055 STEP 1E-03
710 FOR N=1 TO 50
720 TS=T(N)*W
730 PRINT "TS",TS
740 REM ORDER PHASES
750 PHI(N)=(TS-INT(TS))
760 NEXT N
770 FOR I1=1 TO 50
780 PM=-100
790 FOR I2=1 TO 50
800 IF PHI(I2)<PM THEN 850
810 PM=PHI(I2)
820 ID=I2
830 PS(I1)=PHI(I2)
840 XS(I1)=Y(I2)
850 NEXT I2
860 PHI(ID)=-100
870 NEXT I1
880 PRINT "X AND PHASE"
890 FOR I1=1 TO 50
900 PRINT XS(I1);PS(I1)
910 NEXT I1
920 REM CALC STRING LENGTH
930 SS=0
940 FOR N=2 TO 50
950 SS=SS+SQR(SF*(XS(N)-XS(N-1))^2+(PS(N)-PS(N-1))^2)
960 NEXT N
970 SS=SS+SQR(SF*(XS(1)-XS(50))^2+(PS(1)-PS(50))^2)
980 LPRINT "W,SS", W,SS
990 JJ=INT(.03*SS)
1000 PRINT "JJ=",JJ
1010 IF JJ>50 GOTO 1050
1020 IF JJ<1 GOTO 1050
1030 LPRINT TAB(JJ);"*"
1040 LPRINT W,SS
1050 NEXT W
1060 REM
```

Least squares

```
1070 REM TRIAL VALUES
1080 AQ=10000
1090 PQ=.3
1100 WQ=.05
1110 CZ=100000000!
1120 YQ=0
1130 PRINT WQ,XQ,PQ
1140 CZ=100000000!
1150 FOR WZ=.9*WQ TO 1.1*WQ STEP .04*WQ
1160 FOR AZ=.9*AQ TO 1.1*AQ STEP .04*AQ
1170 FOR YZ= -1000 TO 1000 STEP 500
1180 FOR PZ=-.1 TO .1 STEP .04
1190 REM TRY 50 POINTS TO START WITH
1200 REM THIS PART IS VERY SLOW
1210 FOR J=1 TO 50
1220 YA=YZ+AZ*SIN((TP*WZ*J)+PZ)
```

```
1230 CH=CH+(YA-Y(J))^2
1240 IF CH>CZ THEN 1340
1250 NEXT J
1260 REM SAVE IF MIN
1270 YS=YZ
1280 WS=WZ
1290 PS=PZ
1300 AS=AZ
1310 CZ=CH
1320 PRINT CH,WS,PS
1330 PRINT AS,YS
1340 CH=0
1350 NEXT PZ
1360 NEXT YZ
1370 NEXT AZ
1380 NEXT WZ
1390 PRINT
1400 PRINT"FITTED PARAMETERS"
1410 PRINT WS,PS
1420 PRINT YS,AS
1430 REM PRINT FITTED CURVE
1440 FOR J=1 TO 50
1450 F=YS+AS*SIN((TP*WS*J)+PS)
1460 PY=1.8E-03*(Y(J)+15000)
1470 PY=INT(PY)
1480 PF=1.8E-03*(F  +15000)
1490 PF=INT(PF)
1500 LPRINT F,Y(J)
1510 IF PF>50 GOTO 1630
1520 IF PF<1 GOTO 1630
1530 IF PY>50 GOTO 1630
1540 IF PY<1 GOTO 1630
1550 IF PY=PF GOTO 1610
1560 IF PY>PF GOTO 1590
1570 LPRINT TAB(PY);"+",TAB(PF);"*"
1580 GOTO 1640
1590 LPRINT TAB(PF);"*",TAB(PY);"+"
1600 GOTO 1640
1610 LPRINT TAB(PY);"="
1620 GOTO 1640
1630 LPRINT
1640 NEXT J
```

Note 11.2 Differences and derivatives

Four examples of the way a second difference is equivalent to a second derivative. The point to notice is that the second difference is equal to twice the coefficient of the x^2 term, whatever the coefficient of the x term. It is the x^2 term which gives curvature in the plot, so this helps in interpreting $O - C$ curves.

x	$1+2x+x^2$	Difference	Second difference	$1+2x+5x^2$	Difference	Second difference
1	4			8		
		5			17	
2	9		2	25		10
		7			27	
3	16		2	52		10
		9			37	
4	25		2	89		10
		11			47	
5	36		2	136		10
		13			57	
6	49		2	193		10
		15			67	
7	64			260		

	$1+3x+x^2$	Difference	Second difference	$1+7x+5x^2$	Difference	Second difference
1	5			13		
		6			21	
2	11		2	35		10
		8			32	
3	19		2	67		10
		10			42	
4	29		2	109		10
		12			52	
5	41		2	161		10
		14			62	
6	55		2	223		10
		16			72	
7	71			295		

Note 11.3 How to deal with gapped data

The periodogram method described in §§11.5.2 and 11.5.3 is the most widely used method of finding periods, but there are several pitfalls of which the potential user should be aware. We have shown in §11.6 how gaps in the data cause problems with the interpretation of an analysis, and conversely how long continuous data runs help to reveal periodic signals despite noise in the data. Even if we could reach that unattainable goal of obtaining error free data (i.e. data with no statistical noise), gaps in the data still cause spurious peaks. Periodograms do not avoid the problems, but they do make them easier to recognize. We will take as our starting point a hypothetical star with a pure sinusoidal variation and no data noise. Suppose first that its period is 24 hours (i.e. one cycle per day), and that it has been observed continuously for 6 hours in just one night. The data will reveal that it is variable, and give some idea of the period, but gives no information at all on what is happening in the other 75% of the cycle. Clearly there will be a lot of uncertainty.

Suppose now that the period of the star was six hours (4 cycles per day), exactly the same time as the duration of the observation. In one night we see one complete cycle, and measure its period, but can we be sure there are not other components? Figure 11.11(a) shows the periodogram for such a case. We have still not made any definitive determination of the components involved, or indeed of the exact frequency of the principal component, the reason being that the extra components shown in the periodogram would cancel out over the single cycle, and so it is impossible to tell if they are there or not. Could we rule them out if we have several cycles?

To find out look at the case of six hour observing runs on three consecutive nights. We have now not only observed three full cycles of the variations, but between the start of our first observing run and the end of the last run 13 cycles of the variation have occurred. Is this enough to give us an unambiguous value of the period of the variation? The periodogram shown in figure 11.11(*b*) still shows some spurious peaks, but more widely spaced. Finally, take a hypothetical experiment with perfect data obtained on five consecutive nights (figure 11.11(*c*)). The results are clearer, but it is beginning to be apparent that the spurious peaks will never disappear as long as there are any gaps in the data. Remember that there is no noise in these data so this is not the same problem as was dealt with in §11.6. The peaks are generated by the gaps in the data. If the gaps occurred at perfectly regular intervals one might identify the corresponding spurious peaks, but in practice one has irregular gaps which give a jumble of spurious peaks. However one could at least identify the *extra* peaks created by deliberately introducing one extra gap. This idea is the key to the analysis method described below.

The general shape of the periodogram distribution may look familiar, because it is similar to the diffraction pattern produced by a telescope mirror or indeed by any optical system. This is no coincidence. A mirror takes in a sample of the waves coming from a source, but there is a huge 'gap' in the 'data' from outside the mirror. A small mirror produces 'spurious peaks' (the diffraction peaks) spread over a wider range than does a larger mirror. Mathematically, the sampling of a set of waves, whether in measuring the light curve of a variable or in producing an image using a mirror, is similar and this shows up in the similar shapes. The shape of the pattern of spurious peaks in a periodogram can be calculated, just as the shape of a diffraction pattern can be. Its shape is called the 'window function', the name coming from the idea that the light curve is looked at through a few 'windows', rather than having a complete record.

The window function

Suppose that we had been looking at an absolutely constant source of light during our (perfect) observing runs, and so had got the same reading every time. Let us give this reading the value 1. What would we have obtained for the power spectrum of this constant source if we had observed it for six hours on each of five consecutive nights as in the last example above? The answer is shown in figure 11.11(*d*). The form of the curve is exactly the same as that in the power spectrum of our imaginary variable but the central spike is now displaced down to zero frequency, i.e. infinite period. Notice that the function extends to negative frequencies. This curve is called the window function of the data run. Each frequency component in the real data will in fact be seen not as a single spike, but as a window function, displaced along the horizontal axis to the particular frequency. However, since it is possible to calculate the shape of the window function this gives hope that its effects can be recognized and even removed.

Two more warnings are in order before we can confidently proceed to use the window function. The first concerns what happens to the window function when it is displaced from the zero frequency (infinite period) position. We have already seen what happens when it is centred on 4 cycles per day, but suppose it was centred at 1.5 cycles per day. The result is shown in figure 11.11(*e*). The part of the window function which would have remained on the negative side of the vertical axis is folded over as in a mirror and is superimposed on top of the rest of the 'window function'. The complexity of the resulting combined 'window function' can sometimes make it difficult to decide whether the highest peak in the power spectrum is real, or the chance superpositioning of two lower peaks.

The final note of caution concerns the additional complexities which are encountered when real data with noise and irregular gaps are used, as opposed to the 'perfect' data with regular gaps used above. Not only does the noise often have a similar amplitude to the true signal, but additionally it would be a fortunate observer who was able to observe continuously for a full six hours on five consecutive nights. When real data are considered it is often found that they are so irregularly sampled that, instead of the 'window function' being symmetric about a central point, rather the two sides of the pattern suffer different, irregular distortions. This will not occur when the pattern is centred on zero frequency but will become obvious once it moves from that position. The effect, when it occurs, is due to undersampling of the complete light curve, and the asymmetries and

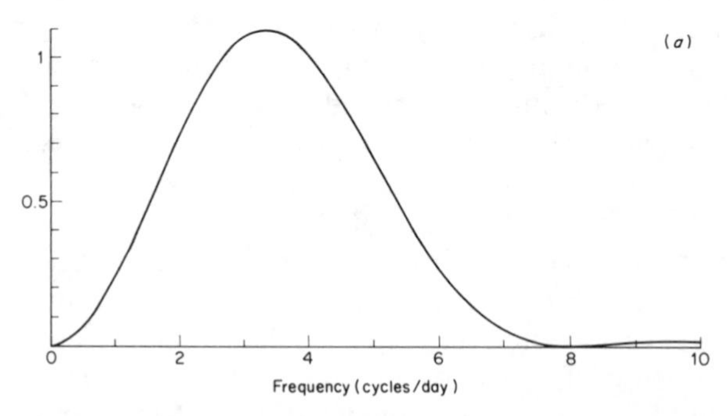

Figure 11.11 (*a*)–(*c*) show improving resolution as more data are obtained. (*d*) shows the window function for the data in (*c*). Finally, (*e*) shows the power spectrum for the same amount of data as in (*c*) but with a frequency of 1.5 cycles per day rather than 4 cycles per day (see text for more details).

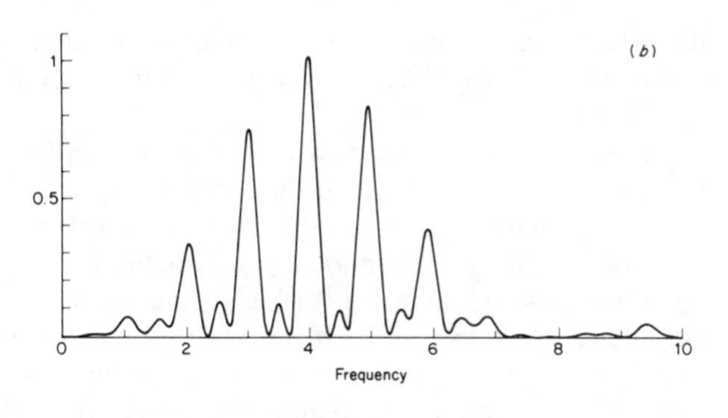
(b)
1
0.5
0
2
4
6
8
10
Frequency

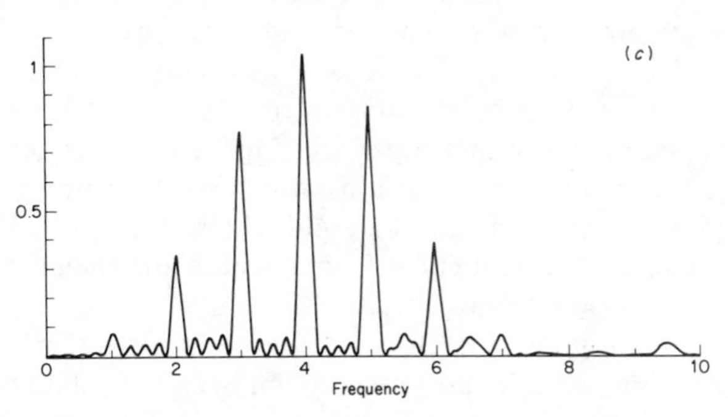
(c)
1
0.5
0
2
4
6
8
10
Frequency

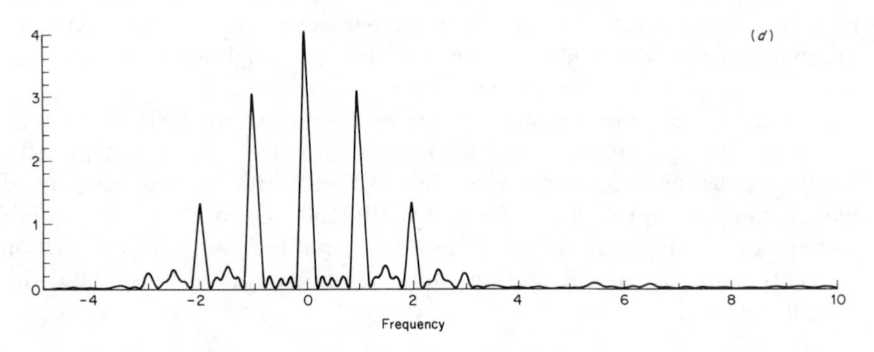
(d)
4
3
2
1
0
-4
-2
0
2
4
6
8
10
Frequency

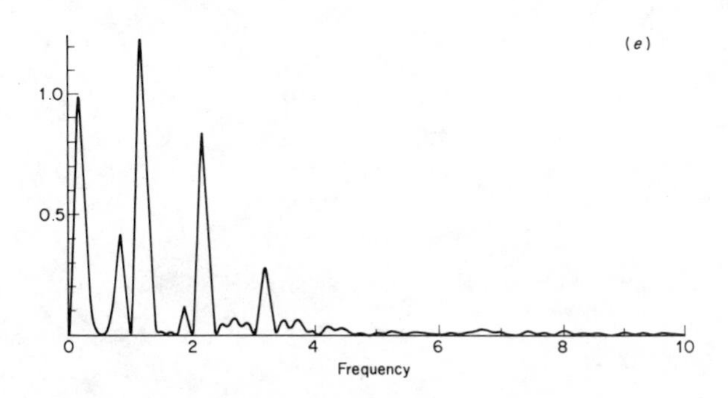
(e)
1.0
0.5
0
2
4
6
8
10
Frequency

distortion will change as the 'window function' is moved to different frequencies. These effects will be less troublesome if large amounts of data are available.

Given all these problems, what is the observer to do? First, it is clear that the more data you have, and the more nearly continuous it is, the less you will experience problems in finding the correct period. There is no substitute for large chunks of continuous, good quality data. But, whilst no analytical techniques can compensate for insufficient data, they can certainly increase confidence in one's results. It has been found that an effective procedure is as follows. First, analyse the real data and produce a periodogram. Next, synthetic data should be created, consisting of a single sine wave with the same frequency as that seeming to be present in the real data. To this is added some random noise and this noisy, synthetic data is then sampled at the real data epochs. Analysis of this is the same as calculating the window function except that, by displacing it to the frequency thought to be present in the data, the distortions which may be present are also displayed. If visual inspection shows a good match, then one can go ahead and subtract the predicted light curve from the data and then repeat the process to search for additional frequencies.

If no good match can be found between your displaced 'window function' and the power spectrum of the real data then almost certainly you have insufficient data to arrive at a definitive analysis. In this case the only remedies are either to go back to your telescope and get more data, or alternatively to try to find someone at a different longitude who either has data already or is prepared to work on the same star.

In summary, do not be easily satisfied with your first analysis of the data. The scientific literature is riddled with incorrectly determined periods, because someone simply chose the highest peak in a power spectrum of inadequately sampled data. It is not difficult to check your analysis techniques on simulated data and, as a further stage, the window function provides a way of subtracting out spurious peaks, even in noisy data with irregular gaps.

Part III

Projects

Chapter 12

Projects: introduction

We have made many general references to projects for which amateur astronomers, or others, who only have access to relatively small telescopes, might reasonably expect to be able to make useful contributions of data. We now come to specific proposals. Some of the projects which we will suggest could not reasonably have been expected to be undertaken by amateurs as little as fifteen years ago, due to the cost and complexity of the electronics involved, but are now possible. This advance will not cease. There are now amateurs using computer guided telescopes and, as we write, at least three different groups of amateur astronomers are experimenting with CCD (charge coupled devices). If such experiments and technical developments lead to low-cost two-dimensional detectors, then many new possibilities, such as the whole area of spectroscopy, might be brought within the range of the amateur. At the moment narrow band filters allow a toe-hold in that area, using photometric methods. We therefore concentrate on photometric projects in this book. In a further few years a Part IV on spectroscopic projects might be needed but at the moment the area in which amateurs are most easily able to contribute serious data is certainly by visual or photoelectric photometry using broad band filters which will be stable over decades. This allows both the laying down of long-term, well covered data bases over years, decades or longer, and intensive projects with observers at different longitudes giving continuous coverage for a few days or even tens of days. These two types of projects require very different commitments of time and effort. We hope that reading this book will suggest projects which not only appeal to individual personalities but which also fit in with the effort that you feel able to devote to astronomy. The types of projects to be chosen by someone who is only able to devote time once or twice a year to observing and, on the other hand, someone who can and would like to observe every time a clear night comes along will be very different. They can be equally important.

Our approach to most of the projects we suggest will be the same. We will identify the area of the HR diagram where the stars of interest lie, with a small diagram highlighting the appropriate area of figure 12.1. We will, wherever possible, use examples of research recently finished or currently underway to exemplify the type of data needed, the way in which it needs to be analysed, the constraints which will be placed upon the models by the data and some examples of objects that might be usefully worked on. For reference, figure 12.1 shows the HR diagram with the main types of variable star indicated.

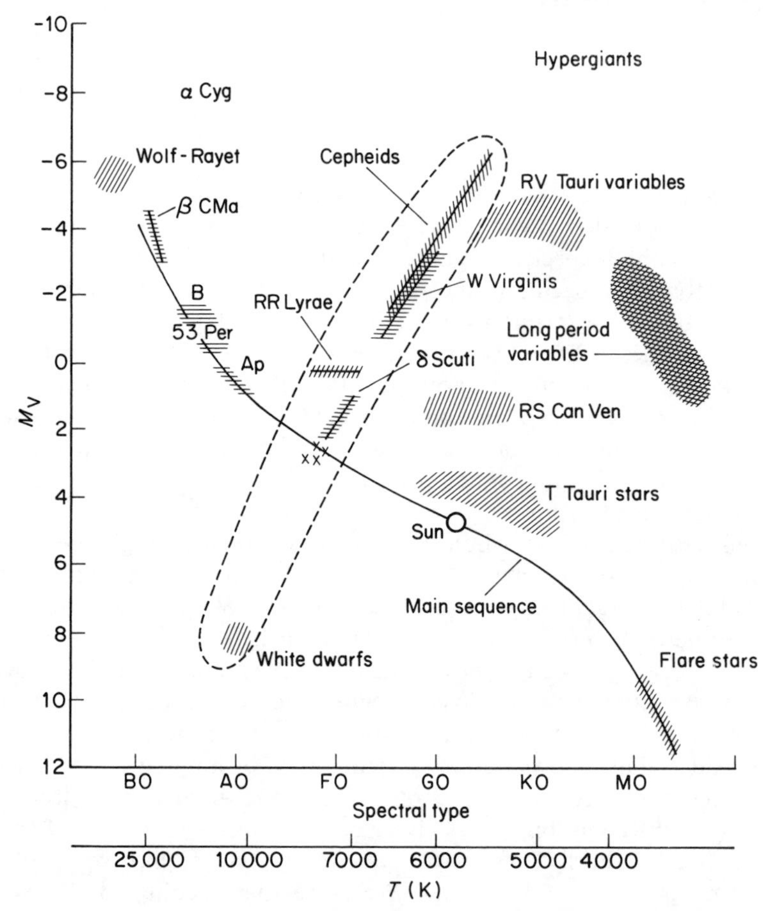

Figure 12.1 A Hertzsprung–Russell diagram showing the approximate location of various types of intrinsically variable star discussed in the text. Cataclysmic variables are binaries containing a compact star (usually a white dwarf) together with a red giant or main sequence star. However, these terms, which normally relate to stages in the development of an individual star, can only be used loosely in this context, since matter transfer can greatly affect the evolution of both component stars.

Although we cannot cover more than a small selection of projects, we have tried to give examples of the different observational challenges which can arise. Clearly it is easier to measure large amplitudes than small

Table 12.1

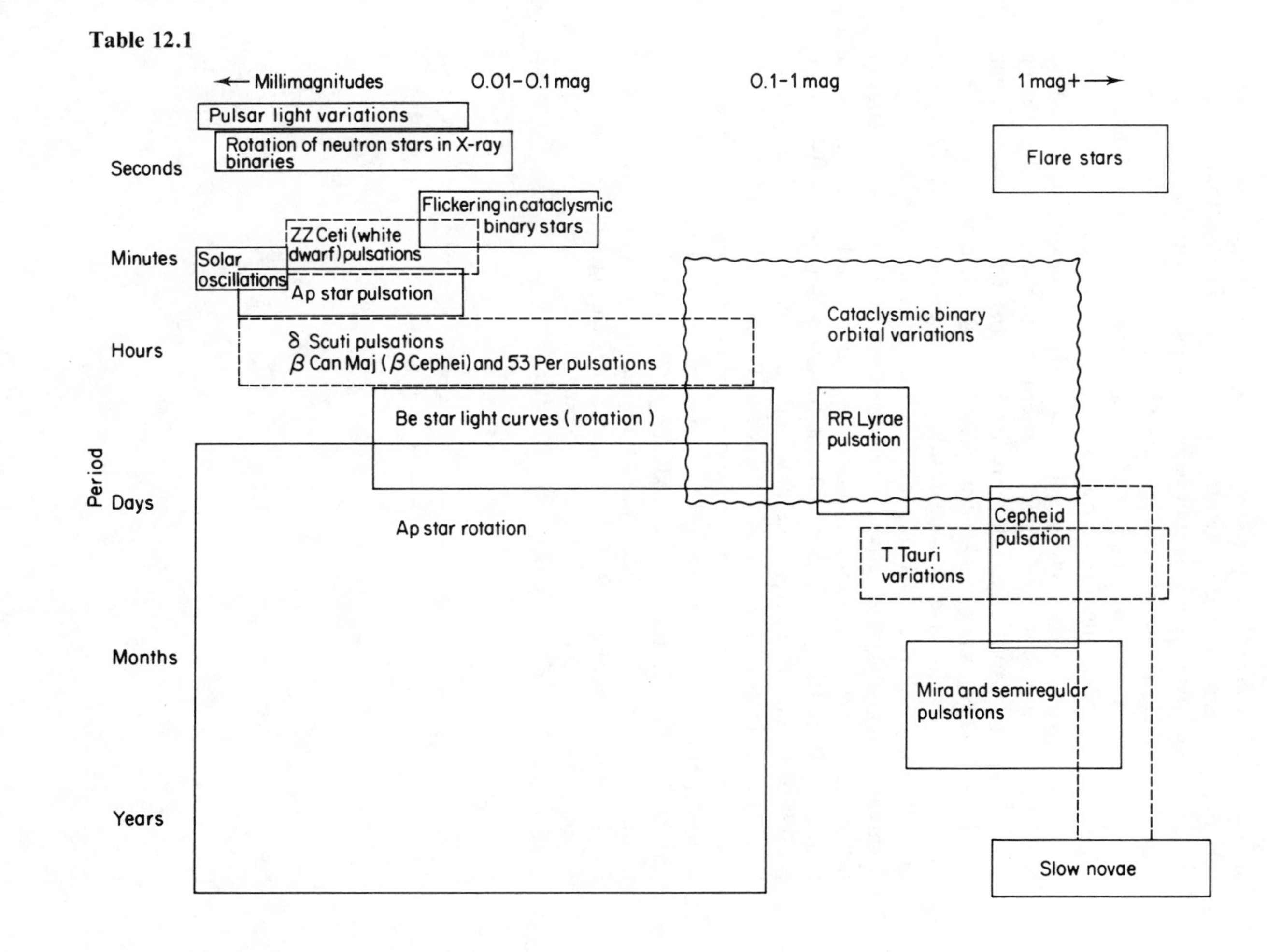
← Millimagnitudes
0.01–0.1 mag
0.1–1 mag
1 mag+ →
Period
Seconds
Minutes
Hours
Days
Months
Years
Pulsar light variations
Rotation of neutron stars in X-ray binaries
Flare stars
Flickering in cataclysmic binary stars
ZZ Ceti (white dwarf) pulsations
Solar oscillations
Ap star pulsation
δ Scuti pulsations
β Can Maj (β Cephei) and 53 Per pulsations
Cataclysmic binary orbital variations
Be star light curves (rotation)
RR Lyrae pulsation
Ap star rotation
Cepheid pulsation
T Tauri variations
Mira and semiregular pulsations
Slow novae

amplitudes, but also short periods are easier than long periods. Irregular variations demand a much greater investment of time than periodic variations, and may need to be split between members of a society. In purely observational terms, projects can therefore be divided up according to these three features, as shown in table 12.1. The nature of your equipment and available time may single out observations in one of the boxes as most appropriate. If it is the bottom left you will be breaking new ground! Naturally, we cannot give more than a very small selection from the existing data. Even the lists of types of objects in appendix 4 are very much abridged. You will therefore very soon find a need for more information. Books and journals give existing results, and the BAA or AAVSO can provide unpublished results and advice. But there is no substitute for discussion with colleagues with similar interests in astronomical societies or in the astronomical research institutions. It is useful to have a link with a professional and if you find it difficult to set this up yourself—not all professionals are amenable—you may find your local society already has a link. The authors could also help if you wish. When you have obtained data it is important that you send it to some central agency or individual for collation and/or analysis. The preservation of the data in central files is important for the future and the feedback you get from analysis from time to time is important to enable you to maintain your observing schedule in the most satisfying way. The addresses of such agencies are given in appendix 6.

Chapter 13

The brightest stars

13.1 BRIGHT RED STARS (HYPERGIANTS)

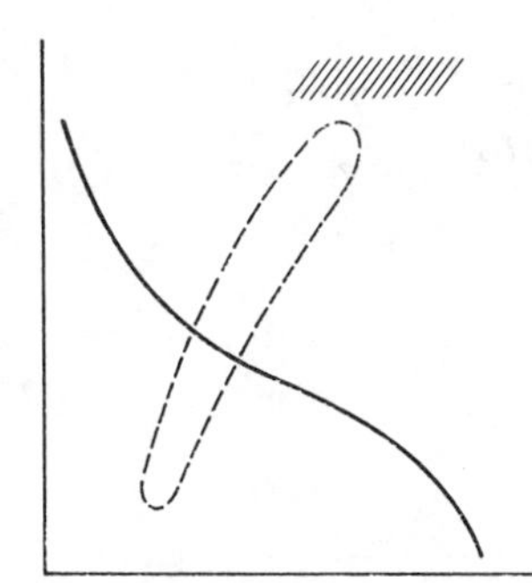

Across the top of the HR diagram lie the brightest stars. Some of these stars have a luminosity (intrinsic brightness) which is one hundred thousand (10^5) to one million (10^6) times greater than that of our Sun. The 'blue' stars of this luminosity are very massive, perhaps from 50 to 100 times the mass of our Sun, and they consume their fuel so quickly that their main sequence lifetimes are only of the order of one million years. These massive 'blue' stars will be dealt with in §15.2. Here we are concerned with *red* stars which are less massive, but so large that they emit as much energy as their massive blue counterparts. They do this with a surface temperature of only approximately 3000 K, while the blue massive stars have surface temperatures of approximately 30 000 K. Remembering that the total emission of thermal radiation varies as the fourth power of the effective temperature, T_e^4, and ignoring the complexities of stellar spectra for simplicity of argument, we can see that for these two types of stars to have similar absolute magnitudes it is necessary for the red stars to have surface areas 10^4 times greater than the hotter stars. As surface area varies as the square of the radius, these cooler stars must have radii 100 times larger than the bright blue stars. This makes them approximately 200 times larger than the Sun, hence their name '*hypergiants*'. Here is a story of one of them.

In 1950 HR 8752 was classified as a G0 star with a B – V of about 1.3. By 1958–60 it was 0.1 mag redder and by 1963–5 its B – V was near 1.6. In 1970 it was reclassified as G5 and in 1973 it was suggested that the spectral type was even 'later', i.e. cooler. Photometry by one of us (ENW) in the years 1976–8 showed that the star was changing on a timescale of years by over 0.3 mag in apparent visual magnitude and by 0.1 mag in B – V colour. Figure 13.1 shows a record of the star's movements in the HR diagram between 1950 and 1978. Percy and Welch observed the star during the 1979 and 1980 observing seasons while Zsoldos observed in 1978–81 and 1984–5.

Both groups of authors thought that a period of approx 400 days could be present. As explained in Chapter 6, a long period is a result of low density. So the value of the period is a constraint on the model for the structure and chemical composition of the star and hence for its mass and evolutionary status. The earlier data by one of us (ENW) suggests that a period longer than 400 days is present, and Zsoldos has suggested that the nature of the star's variability has changed since 1978.

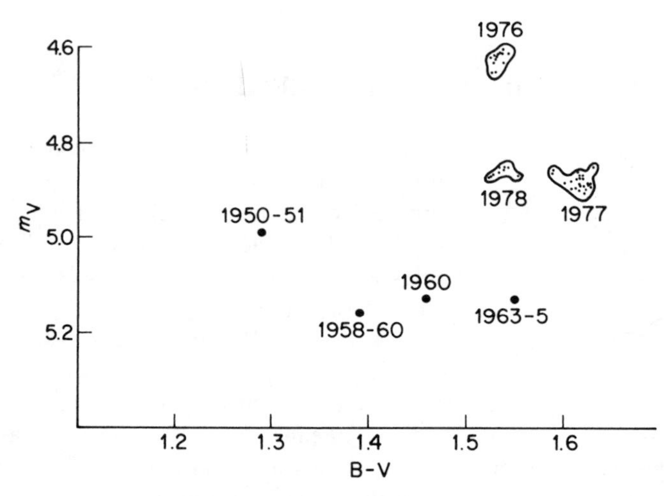

Figure 13.1 Variable stars change in colour, as well as in luminosity. Sometimes the colour changes are large enough to show clearly in the relative intensities measured through U, B and V filters. The star HR 8752 provides an example. The changes are shown here on a Hertzsprung–Russell type of diagram.

To investigate this suggestion the whole of the V based data from 1976 has been analysed by ENW. It is surprising that a good fit to the whole data set can be found by *three* approximately equally spaced frequencies. These are listed below.

Frequency (cycles/day)	Period (days)	Semi-amplitude (V mag)
0.000 470 97	2123.2	0.0789
0.004 955 6	201.79	0.1102
0.008 581 5	116.53	0.0416

Is the fundamental period 200 days, 400 days or is it as long as 2000 days, which would imply a very low density? A correct determination of such frequencies would act as a powerful constraint on models of the star. As yet, no-one knows what makes this type of star pulsate.

In addition to HR 8752, stars which need a similar long-term measurement and analysis are ρ Cas in the northern hemisphere and HR 6392, HR 4337 and HR 5171 in the south. These and others are listed in appendix 4. HR 8752 is merely a bright example of this class of star. For small telescopes in the northern hemisphere, it is an easy object with an apparent magnitude of about 5 and is only about 2 degrees in the sky away from a Johnson primary photometric standard, HR 8832, which is also about fifth magnitude. It is an example of how much work remains to be done with small telescopes that it is only since 1984 that this bright example has had (more or less) continuous coverage.

At the time of writing there is some indication that the B-band behaviour of HR 8752 differs from that in V. If this is confirmed then this might be due to the presence of a very hot companion star discovered by Stickland from spectra obtained by the International Ultraviolet Explorer (IUE). Therefore a (well organized) observing programme on this type of star should have as wide a wavelength coverage as possible. In appendix 4 we list ten stars which could be observed with relatively modest equipment in order to increase our knowledge of these largest of stars. We have arbitrarily split the table into stars hotter or cooler than K5.

So much for the story of one of them: what lessons are there to learn? Using our experience of HR 8752 as a guide, the following criteria should be built into the observing programme.

(1) Coverage to extend over several years with no instrumental changes.

(2) Coverage of B and V as a minimum, and as many of UBVR and I as is possible. Remember that if many of these stars have hot companions, then differences in behaviour between, say, the V, R and I bands (which will monitor the red star) and the U or B bands (which might be affected by a hot companion) will be important in suggesting which period, if any, is due to the binary nature of the stars. The use of filters greatly increases the value of photometry.

(3) The periods being sought are tens or, more likely, hundreds of days with amplitudes of tenths of a magnitude. If your observing equipment/site allows millimagnitude accuracy then you can detect changes from one week to the next and one good brightness determination each clear night will be adequate. If your potential accuracy is only 0.01 magnitude then it will take longer to see the changes, but frequent observations help with the statistics.

(4) If you cannot find comparison stars of a similar colour then it helps with the reduction if the stars are observed as near to the meridian each night as is possible. Always use two comparison stars, or even more if millimagnitude accuracy is sought, unless you are lucky enough to

have a primary photometric standard, or other demonstrably stable star nearby.

13.2 THE WOLF–RAYET STARS

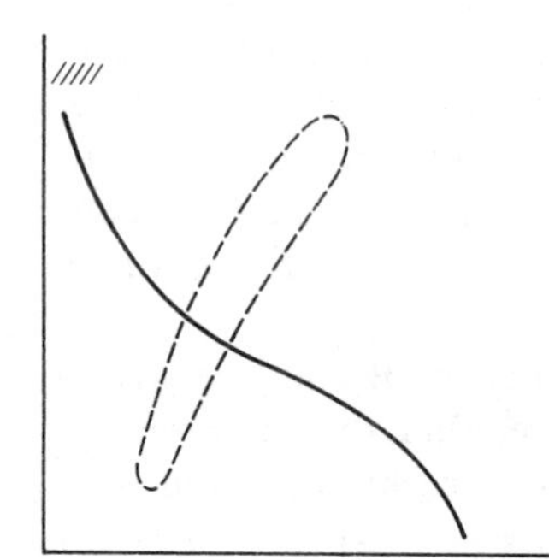

At the top of the HR diagram and only slightly less luminous than the 'hypergiants' are the Wolf–Rayet (WR) stars. The exact evolutionary status of these is still under some discussion but it seems likely that they are stars which, having exhausted the store of hydrogen in their cores, are now in the process of using helium or even carbon as nuclear fuel. It is the spectra of WR stars which identify them. They show very hot atmospheres (~20 000–40 000 K) and large amounts of mass being 'blown' off their surfaces, as evidenced by the shifted positions and distorted shapes of the numerous emission lines in their spectra. Two types of WR star are known. These are called the C (carbon) and N (nitrogen) types and it is the strength of spectral lines due to these elements which distinguishes between the two types of stars. Mass flow rates as high as 10^{-5} to 10^{-4} $M_\odot$/year have been derived from observations of widely different wavelengths (e.g. optical, infrared, radio). Most astronomers think that the speed of ejection of the matter is greater than the escape speed, so that it is lost completely from the star. If so, mass loss of 10^{-5} $M_\odot$/year is a dramatically rapid rate for a change on an astronomical timescale. Even in one human lifetime there may be detectable changes, especially if the star is an eclipsing binary, so that changes in period are a sensitive indicator of evolution.

Within our Galaxy about 160 WR stars are known, but the fraction of those which are double is still under discussion. In general, WR stars will be found to be photometrically variable. For those stars that are thought to be single, one could hope that, by monitoring the brightness over several months or years, some periodic signature of long-term secular variations might appear above the noise. This might be evidence for star spots on the surface—or perhaps it is a binary after all. Once it was thought that perhaps all WR stars were in binary systems, but now figures as low as 50% are mentioned. A small proportion of these will be eclipsing binaries. Four of the WR stars in our Galaxy are known to be eclipsing binaries—CQ Cephei, V444 Cygni, GP Cephei and CX Cephei. Possibly three more, V1676 Cygni, V1696 Cygni and CV Serpentis, should be added to this list. All seven could be well worth monitoring with small telescopes to look for period changes. Probably some of the other WR stars are close to critical stages in their evolution, but we cannot predict which ones.

The great strength and variability of the emission lines of WR stars reveals their nature but also poses an observational problem. Even with broad band UBVRI photometry these lines can significantly affect the fraction of light transmitted by any particular filter. Ideally, one really needs special, narrow band interference filters made, so that only the light of the star in between the emission bands (the continuum) is measured. Such filters are expensive and tend to be limited to professional use. If individual lines cannot be identified, some ambiguities may arise. For instance, a decrease in B – V, normally interpreted as an increase in temperature, may be due to the emergence of a strong line somewhere in the B pass band, rather than the general increase which would be associated with a temperature rise. Such a rise should be reported; it might encourage someone with a spectrograph, amateur or professional, to take a closer look.

Fortunately, varying line emission is less likely to affect measurements of orbital period, which in any case can be checked by independent determinations using different filters. If one finds a change in orbital period, one would want to know if it is correlated with any other change in the star.

Definitive results either way could be useful. For the binary WR stars, particularly the few known eclipsing ones, work of immediate importance is possible. To understand why, it helps to envisage the WR star in a binary system with the surface of the WR star slowly being blown away. Some of this material will transfer through the inner Lagrangian point (L1) onto the surface of the companion; some will be lost from the system. The material which leaves the WR star not only has mass but also angular momentum around the other star. Thus mass is transferred from the WR to the companion, and so is some angular momentum. Mass and angular momentum is also lost from the system. Clearly, if mass or angular momentum is being lost from the system, or transferred within the system, period changes will occur. Thus, the eclipsing WR stars are good candidates for investigation by the $O - C$ diagram. In order to demonstrate what can be looked for we present below the results obtained by one of us (ENW) on the eclipsing WR star CQ Cephei.

CQ Cephei is a spectroscopic binary with a period near to 1.64 days, of which one member is a WR star. It has been observed photometrically since the first years of the twentieth century, photographically until about 1945 and photoelectrically thereafter.

It was observed on four nights in September 1982 when it was predicted that the star would be in, or near, primary eclipse. Figure 13.2 shows the four nights observing runs (~6 hours each night) superimposed with the adopted period of $1^{d}.641\,436$, $t_0 =$ JD 2415 000.410. It will be seen that on the night 20/21 September we caught the star just starting to get fainter and on the other three nights we observed it near the centre of the eclipse. Two new results emerged from these data. First, the three light curves near minimum do not agree. One night, 12/13 September, emerges from mid eclipse much more slowly than the other nights, while on those other two nights the rise in the light is accompanied by 'bumps' on the light curve. The occasional presence of this 'bump' has now been confirmed by other observers. Possibly we are seeing structure in the 'wake' following the eclipsing star. This would be a fruitful study area.

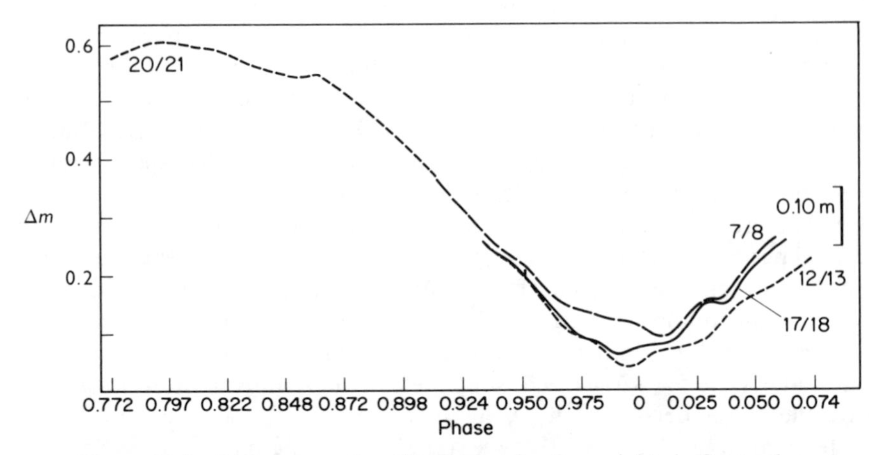

Figure 13.2 Light curves for CQ Cephei for four nights in September 1982. This is an eclipsing binary and the data show variations in the light curve between successive eclipses, which can be interpreted as rapid changes in a stream of matter being lost from the Wolf–Rayet member of the pair.

The second result came from adding the $O-C$ timing of the mean of the three mid eclipse timings to existing data. The $O-C$ diagram is shown in figure 13.3. Analysis of this $O-C$ curve showed that, following an apparent continuous period decrease in the early part of this century, the period has now stopped changing. Indeed, the data are compatible with no period change greater than ~ 1 part in 10^7 per year since 1927. If mass is really being lost or transferred as indicated by the results shown in figure 13.2, then only in a very special case will the period remain constant. It should be clear that very exciting results could be obtained in only a few years by more $O-C$ values and by closely investigating the detailed shape of individual light curves. In addition, pulsational variations might be present in some WR stars and departures from mean light curves should be looked at for evidence of such variations.

The later stages of stellar evolution are not well understood, and the relatively rapid changes in WR stars are particularly interesting. Correspondingly, one cannot be sure what observations would be most likely to produce interesting results. Careful work on even the apparently single stars, or on variation in the non-eclipsing binary systems might well shed new light on this interesting, but short lived, phase of stellar evolution.

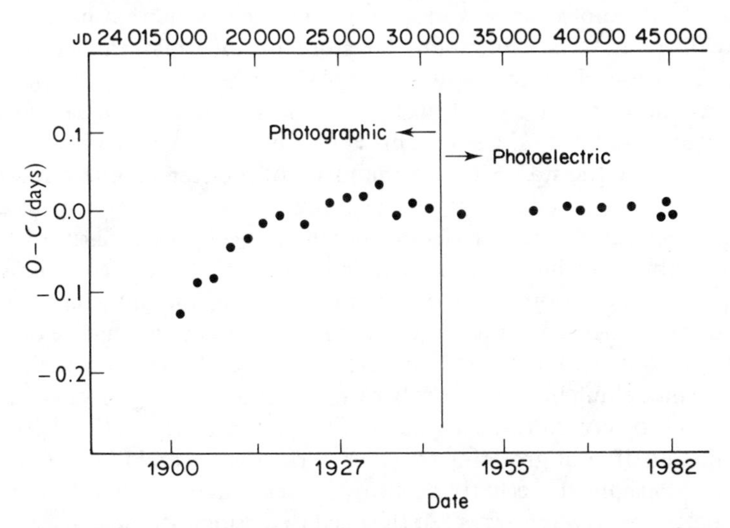

Figure 13.3 An $O - C$ diagram for CQ Cephei. The first part is curved, showing a changing period (compare with the schematic diagrams in figures 11.4 and 11.5). Since about 1927 the period looks constant.

13.3 B STARS WITH AND WITHOUT EMISSION LINES

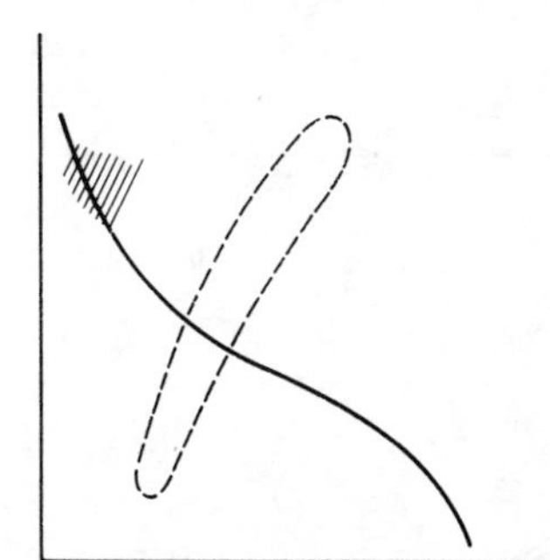

It is now time for us to work our way down the main sequence. It is on this part of the HR diagram that stars spend their time when they are in the full maturity of their hydrogen burning years, having condensed from the cold, diffuse interstellar medium, via their large, warm, red protostar stage and before they exhaust their stores of hydrogen fuel, start to burn helium and evolve off the main sequence towards the red giant branch. At the very top of the main sequence lie the O stars and then the B stars.

There are rather few O-type stars, although their great intrinsic luminosity renders them easily visible at great distances. Their observational requirements are similar to those of the B-type stars, so we may take them together. The coolest of the B stars have surface temperatures of over 10 000 K while the O stars and the hottest B stars have surfaces at 30 000 K or more. At these high temperatures the bulk of the radiation emitted by the star is not in the region to which our eyes are most sensitive, the 400–700 nanometer visual region, but in the far ultraviolet at about 100 nanometres. It is only the technological advances of the last few years that have allowed us to realize just what an untapped store of information lies in these regions. To the red of our visual region we have an excess of radiation (known as the infrared excess) over what one would have suspected from the black body curve of the star. This is probably caused by warm gas and dust surrounding the star. Since the majority of the star's emitted radiation is in the ultraviolet and infrared, one must beware of trying to draw conclusions about its structure from studying the very small part of its total energy distribution which is in the visible region. We have tried to illustrate this point in the cartoon where an explorer looks at the monster's tail and tries to decide exactly what it is that he is seeing. It is clear that he does not stand much chance of success. The advent of UV and infrared satellites has helped the modelling but for the bulk of work studying long-term variability we still have to rely on optical studies from the ground.

We must justify our grouping together of the B stars and the Be (emission line) stars. Perhaps as recently as 10 to 15 years ago this would not have been done, as there was always felt the need to distinguish those B-type stars which had only absorption lines in their spectra from those which had some emission lines. The range of emission lines runs the whole gamut from there being mild emission at the Hα Balmer line of hydrogen to the whole Balmer series and perhaps some other elements being in emission. Some of the stars have many very narrow emission or absorption lines while others possess very broad emission lines. This is not the place to enter into a detailed discussion as the causes of these different types of spectrum for the very

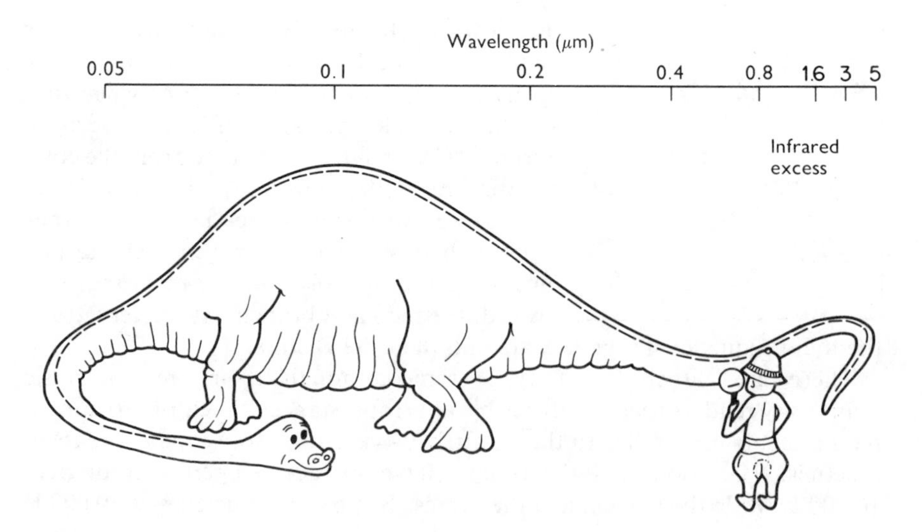

same reason that we are not choosing to distinguish between them in this chapter. The reason is that we now have observations of good quality extending back over several decades and these long data runs have shown that it is a normal part of these stars' behaviour to sometimes appear as normal and at other times to display emission lines. The stars γ Cas and Pleione are two of the better known examples, and it has been the continuous dedication of a rather small number of observers that has clarified this situation. It is now clear, therefore, that the distinction B or Be is somewhat arbitrary and does not tell us much about the underlying bulk of the star. Rather it is telling us something about the temporary nature of the outer layers of the star, perhaps whether the monster is shedding or growing a new skin.

There is a well known class of pulsating B stars known either as β Cephei or β Canis Majoris stars. Here we will call them β CMa stars. These are well established pulsators although the source of the pulsations is not well established. The other type of stars we will consider are the rest of the B stars, which contain some which might not vary, others which have short periods and yet others, known currently as 53 Persei stars, which certainly vary but with rather longer periods than the β CMa stars.

13.3.1 The β CMa stars

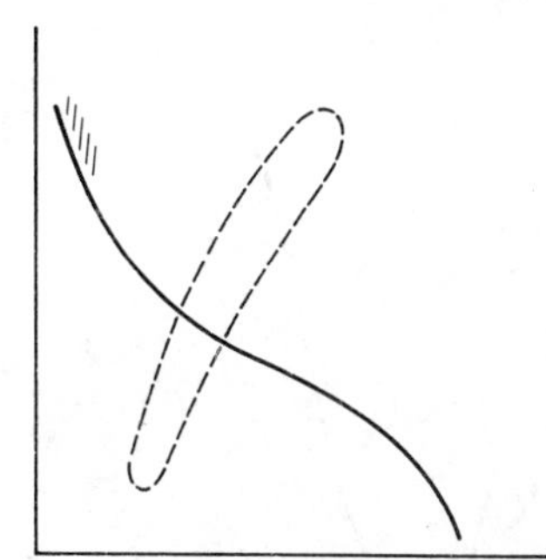

These are among the hotter of the B stars and typically have periods of only 0.15 and 0.25 of a day, suggesting non-radial oscillations. The amplitude of the photometric variations is generally low, from about 0.1 mag to below 0.01 mag and is larger towards the blue or ultraviolet than in the V, R or I bands. As photometric precision increases, it seems certain that many more of these stars will be discovered and for those capable of accuracies of a few millimags there almost certainly awaits the pleasure of adding yet more stars to this group.

Many of these stars have light curves which clearly contain more than one frequency, and the periods are such that the observer who is able to devote a whole night to observing can frequently obtain a full light curve. Given the nature of the stars, what data should the observer be trying to obtain and what questions will be addressed by that data?

Given good precision, the first requirement is ideally several light curves taken over a few closely spaced nights in order to generate a mean light curve, and so accurately determine the phase of the maxima or minima. For singly periodic stars this will allow another point to be put on an $O-C$ diagram and changing periods can be looked for. Exactly the same can be done with multiple-periodic stars with an $O - C$ value being calculated for each period present. This is the first occasion that we have had in this book to mention this possibility. Therefore, we will discuss in some detail what has been observed so far so that the newcomer to this field is aware of the possibilities and the traps which lie in wait.

The star σ Sco is a well known β CMa star with much work having been done on it. It was known to have an asymmetric light curve. However, a magnificent observing run in 1972 by Chris Sterken, a Belgian astronomer, produced 20 nights of data in a total duration of 50 days; 14 of these were in only a 19 day duration. Mike Jerzykiewicz, a Polish astronomer, and Sterken later analysed these data and they were kindly made available to one of us (ENW) in 1984. The star had been known to be a binary with a period near to 32 or 33 days since 1921 and in 1966 Van Hoof claimed that the period of 33.008 days was the correct one. The 1984 analysis was used to determine the amplitude of the main variation over Sterken's densely packed observing run. It was found that the semi-amplitude varied from about 13 millimags to 44 millimags with a clear signature of the orbital period. Figure 13.4 shows the results. In a star with several different periods present one has to be careful about what this means physically but it shows the nature of some of the changes which can be looked for.

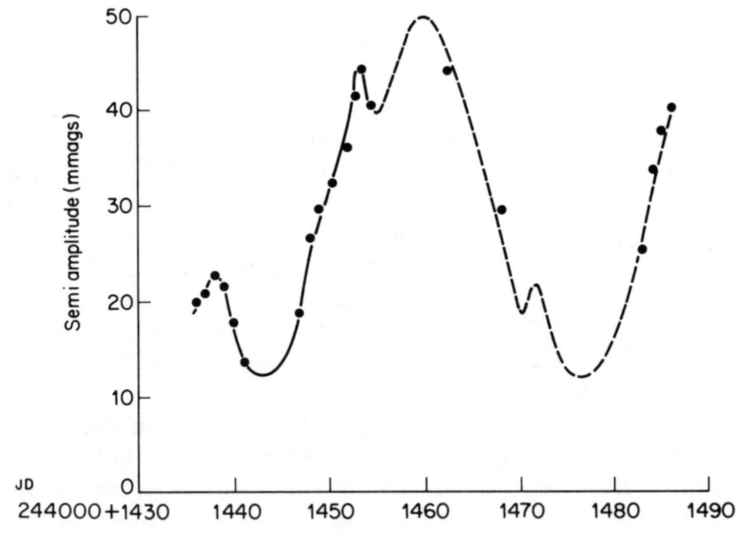

Figure 13.4 The semi-amplitude of the short-period variation in σ Sco throughout the 33 day orbital period.

The next type of research we will discuss is to look for period changes in these stars using the $O - C$ diagram. Some of these stars are nearly one million times brighter than our Sun and yet are only, perhaps, 10 to 30 times

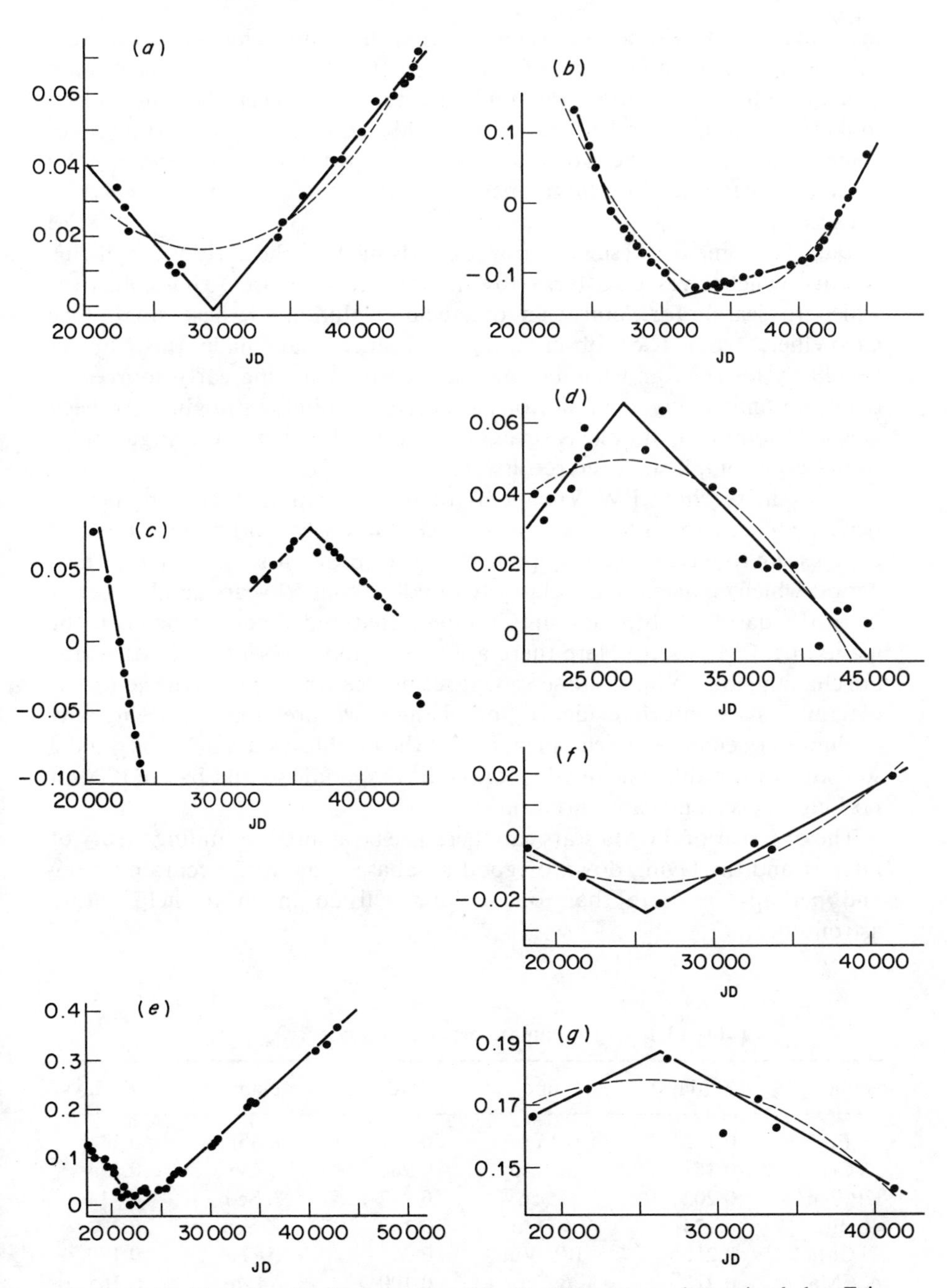

Figure 13.5 The $O-C$ diagrams for six β CMa stars determined by Eric Chapelton. (*a*) δ Cet, (*b*) BW Vul, (*c*) σ Sco, (*d*) 12 Lac, (*e*) β Cep, (*f*) β CMa, P_1 and (*g*) β CMa, P_2.

more massive. It will be clear therefore that they do not have the resources to keep up their flamboyant lifestyles and that they will only have main sequence lifetimes of about one million years. It has been thought that by looking at enough of these stars we should eventually be able to discover some which are in the process of evolving off the main sequence by detecting changing pulsational periods as the stars slowly change their densities.

Data for some of these stars now extends back to the early years of this century which gives us a large (by human life standards) timespan over which to search for some sign of stellar evolution in these stars. Eric Chapellier, from Nice observatory in France, has made this subject peculiarly his own and has become an expert in finding early sources of data. He finds many examples of period changes but, as might have been guessed, nothing as simple as straightforward stellar evolution. Figure 13.5 shows how complicated the results are.

You will see how BW Vul seems to have a continuously lengthening period while that of 12 Lac seems to be continuously shortening. δ Cet and β Cep seem to have increasing periods but for β Cep it is clear that a single period which changed once relatively rapidly about 50 years ago is a better fit to the data. σ Sco is not simply interpreted and, finally, what is to be made of β CMa itself? Here there are two periods present, both of which are changing but in opposite senses. It seems clear that, although the $O-C$ diagrams show much evidence for change, we are not yet seeing any evolutionary changes. Even Spica, one of the brightest stars in the sky and a well known variable, has slowly reduced its amplitude so that by the 1980s it no longer has significant variations.

The observer of β CMa stars can therefore be assured of finding plenty of interest and the laying down of good databases now with accurate timing and perhaps in more than one colour will do much to help future astronomers.

Table 13.1 Approximate periods of some β CMa stars.

Name	P (days)	Name	P (days)	Name	P (days)
γ Peg	0.152	τ_1 Lup	0.177	α Vir	0.174
δ Cet	0.161	α Lup	0.260	β Cer	0.157
KP Per	0.202	σ Sco	0.247	λ Sco	0.214
U Eri	0.174	θ Oph	0.140	K Scu	0.200
β CMa	0.250	BW Vul	0.201	ε Cen	0.170
ε_1 CMa	0.210	β Cep	0.190	δ Lup	0.165
15 CMa	0.185	12 Lac	0.193		
β Cru	0.160	16 Lac	0.169		

In table 13.1, we list some of the brighter β CMa stars and their periods. The observer wishing to work on any of these would do well to talk to a professional working in this field before starting to observe.

13.3.2 The rest of the B Stars

While it is clear that the β CMa stars pulsate, perhaps in a mixture of radial and non-radial modes, the source of this pulsation is not yet certain. However, on the same part of the HR diagram are other variables which are only now being identified as a group. They are called 53 Per variables and their main characteristics are still not well defined. Therefore, any thorough work on these stars will be useful. It is not really certain that they form a totally separate group from the β CMa stars. In general, they have longer periods than the well established variables, periods being thought to be typically an appreciable fraction of a day, say 0.7–0.8 of a day as opposed to 0.25 day or less.

The 53 Per stars are, almost by definition, 'normal' B stars and it is a point which has yet to be investigated as to whether most normal B stars would show this same effect if the observational accuracy was improved. As instrumental technology improves, many stars which were 'constant' with 1% accuracy are found to be variable at 0.1% accuracy. Users of small telescopes who can make observations approaching this accuracy and who have the time and dedication could transform this field in a few years.

What needs to be established is what fraction of 'normal' B stars vary at different amplitude levels and with what periods. For example, could it be true that at 0.1% accuracy all B stars are variables while at, say, 2% accuracy only 20% are noticeably variable? Is there any difference in period distribution between those stars with the larger and smaller amplitudes? Do the larger amplitude stars also have longer period, low-amplitude variations as well so that we have trouble seeing the longer period effects due to the larger amplitude of the shorter term variations?

Recently several groups have carried out careful observational runs on Be stars. It has been found that, as in the case of the 53 Per stars, variations with periods near to one day are common. The question which needs to be addressed is whether the 53 Per variations are the same as the Be star variations and, if so, is there any difference in amplitude, phase or the mix of harmonics in any one star between its normal B and Be phases. Could we be seeing stellar spots or magnetic fields causing the effects and if so how would these be changing during the transition from B to Be?

Periods close to one day are difficult to observe. From any one site it is impossible to cover more than, say, a six hour period, and conversely difficulties in correctly compensating for the sec z extinction coefficient changes at the few millimag level renders it very difficult for observers at different longitudes to combine their results if they use different photometers and different filters. What is needed is the use of identical photometers by observers at different longitudes, ideally with the longitudes less than 6 hours apart so that there is some overlap at the start and end of each observing run. This will allow accurate superpositioning of the various results.

A treasure trove of discoveries awaits the people who can organize that. Even if you feel that you could only organize one campaign each year on one star, provided that the results are accurate, you will be making a pioneering contribution to the subject.

Another group of variables among the B stars has also recently been identified. These have been called the ultra-short-period variables and they typically have periods of one hour or less, which makes them similar in this respect to the hotter of the δ Scuti stars discussed later. The amplitudes are only 1 or 2%, which is presumably why it has taken so long to discover them, and this type of variation should be kept in mind in any B star photometry. It is not yet sure in which types of stars it might be present and it may be superimposed upon longer period variations.

In summary, B star photometry can be a fertile field of research. It needs accurate equipment and people who can spend all night at the telescope. For the longer period stars it needs multi-longitude collaboration, not only to cover a single long period variation, but preferably to cover several contiguous ones, i.e. a continuous observing run of, say, a week. This is so that the variables with periods near to one day can be investigated for beats between several periods which might be present. The identification of these 'beats' is vital to an understanding of these stars. As such it is ideally suited to the small professional or college observatory which would value the international collaboration or to groups of dedicated amateurs, who might choose to make one or two observing campaigns each year.

13.4 THE A TYPE STARS

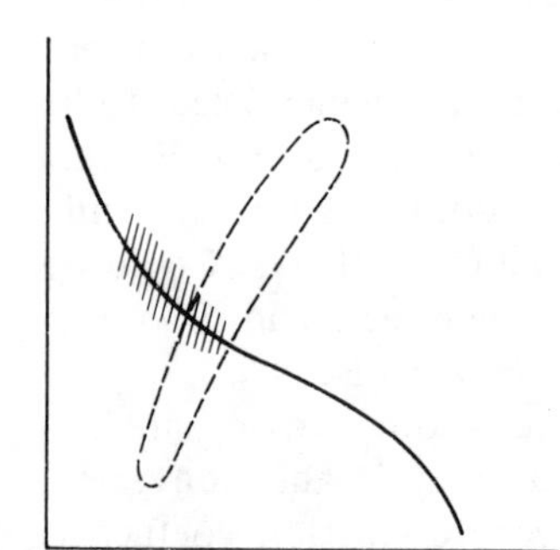

Moving down the main sequence from the B type stars we come to the A stars. The hottest of these have surface temperatures of about 10 000° and the coolest about 8000°. It is among the A stars that the instability strip crosses the main sequence and the hotter of the δ Scuti variables have spectral types near to A2. The cooler edge of the instability strip is among the F type stars and therefore for nearly the whole of the A type main sequence there is a mixture of well established variables and 'non-variable' stars. Also, about 20% of the A stars have chemical peculiarities on their surfaces; these are dealt with separately in the following section.

It has been clear for several years that advances in photometric accuracy have led to many previously 'non-variable' stars being reclassified as 'variable'. As described in the section on the δ Scuti stars, we are now able to look for periodic variations of about one millimagnitude (0.1%) or even less if long observing runs can be obtained. Similarly, the distinction between A and Ap stars has become increasingly blurred as technological advances have increased the accuracy of the spectroscopic data. As an example of this the star Sirius, the brightest star in the sky, was found a few years ago to show an enhanced metallic line spectrum. It seems at the time of writing that there can no longer be any clear distinction between A and Ap stars. Rather, there may be a continuum at one end of which there are clear examples of peculiarity while at the other end the peculiarities fade below the detectable limit. Possibly the same is true with the pulsational variations. While it is conjectured that there is some mutual exclusivity between chemical peculiarity and pulsation, the fact that both are sources of different types of photometric variation suggests that at some level it might well be possible to detect both.

We have already noted how some B variables have amplitudes which have died out over decades and among the A stars there is no good reason to believe that today's δ Scuti will not be tomorrow's constant star. The opposite of course is also true and if you are confident in your ability to produce very accurate results then even a good upper limit to the variations in some A stars might well be useful in the future if someone later discovers a star to be a variable. Note, however, that accuracy is of the essence here. A list of stars, all of which can be shown not to have varied by more than a few millimags over several years, might not seem very exciting. However, if it is later demonstrated that one of the stars then has a one hour period with an amplitude of 0.01 mag (i.e. 1%), it may be an important example of the detection of stellar evolution.

In the B stars the amplitudes of pulsationally induced photometric variations increase towards the blue and ultraviolet. For the A type variations this is generally, but not always, true. U band observation often suffers from variable water vapour extinction in the Earth's atmosphere, particularly at low altitudes, and therefore the B band is a good compromise for doing surveys of A stars. In an ideal world, of course, UBVRI data would be obtained, but if this is not possible then an interesting compromise is to try to get R or I data simultaneously with the B. This does require a special photometer which may be outside the reach of individual amateurs, but not of a society or small research group. The technique is to place a dichroic beam splitter in front of the photometer. This will deflect 'blue' light in one direction and 'red' light at 90° to it. It is then necessary to have another photometric channel to monitor this 'red' light through either the R or I filters. Many of the hotter stars show much reduced variation in the R and I bands when compared with the B band. One will not, of course, be measuring absolute brightnesses, but rather B – R or B – I colours. In many cases, however, these will be very near to the real intensity changes in the B band, because one is effectively comparing the star with itself. The results are very insensitive to transparency changes in the Earth's atmosphere and for those observers who work in photometrically poor sites, and who care to concentrate on the hotter stars, this is one alternative to comparative photometers.

13.5 CHEMICALLY PECULIAR B AND A STARS (Ap STARS)

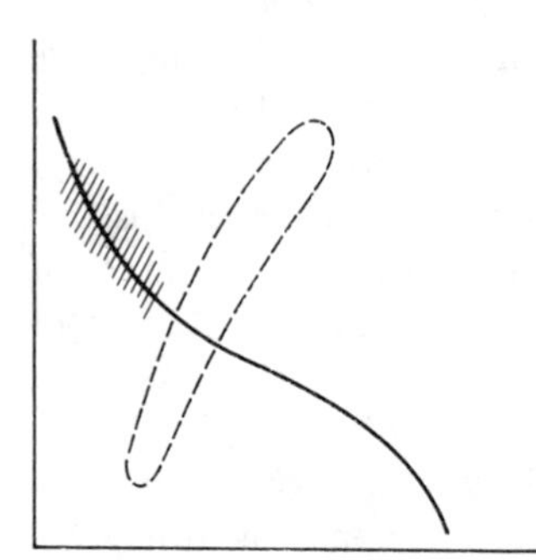

Imagine a star whose surface is severely depleted in helium (the second most common element in the Universe) except for a spot on one side which seems to contain all that missing helium. Imagine stars which have effective surface temperatures of 20 000 K to the red of 350 nm and 14 000 K to the blue of that colour. Imagine in that same star the whole surface being depleted in helium by at least a factor of 10 000, and yet iron, chromium and silicon being ten to one hundred times overabundant in spots or bands on the surface. Such effects are found among the stars known as Ap stars (although their spectral types range from B2 to F0 or even cooler). Others have the rare earths present in ten to one hundred times their value in the Sun, and some have lead, holmium and perhaps even uranium on their surface. Some Ap stars have magnetic fields of 3 Tesla (30 000 Gauss). The Sun's magnetic field for comparison is 10^{-4} T (1 Gauss) and it is only recently that fields as strong as 3 Tesla have been created in the laboratory and even that only for fractions of a second while a plasma is imploded. The Ap stars are bright (though less so than the hypergiants) and not on the instability strip, which is why they appear in this section. Not being on the strip is already an indication that different forces are involved—magnetic in this case.

13.5.1 History of Ap studies

Before describing current projects on these stars, it is worthwhile trying to put them into a historical perspective. Their existence has been known from the earliest days of spectral classification because in the more extreme cases their spectra appear bizarre even on low dispersion spectra. Typically, the spectral type derived from the hydrogen to helium line ratio is earlier (hotter) than that derived from the ionized calcium lines which are themselves hotter than that derived from the metal lines. Traditionally two sorts of peculiar stars were known. One sort was called Ap or Bp and ranged in spectral type from mid B to mid A. The others were called metallic line A stars (Am) and in general had spectral types from mid A to F. There was some overlap. Several criteria distinguished the two groups. The Ap stars often had strong and varying magnetic fields, while the Am stars either had weak or no magnetic fields. There appeared to be abundance differences ('*anomalies*') between the two groups and indeed several subsets of Ap were known, but the problem was compounded by the fact that lines of different elements are not visible over the whole temperature range. For example, in a normal mid B star lines of helium should be easily visible while lines of iron and chromium should not. Conversely, in a normal late A or early F star the temperature is low enough that lines of helium are not visible but lines of iron and chromium are. Both groups seemed to show abnormally low rates of rotation with some Ap stars going round so slowly that they take tens of years to rotate once. There are cases, however, of Ap stars rotating at near to the maximum possible velocity (i.e. less than a day, giving centripetal acceleration equal to the gravitational acceleration at the (bulging) equator). Am stars seemed to be generally members of binary systems while Ap stars seldom were. To further compound the mystery, among the B type stars some were helium strong and some were helium weak and it was not known how, if at all, these were related to the Ap stars.

The ideas to explain some of these peculiarities have been many and varied. The presence of strong, variable magnetic fields in the Ap stars together with their frequent spectral variability led to an analogy with sunspots. The idea is called the *oblique rotator model* and assumes that in Ap stars, as in the Earth, the magnetic axis is at an angle to the axis of rotation. It was assumed that spots occurred on the surface, say at the magnetic poles, and that the peculiarities occurred in the spots. Rotation of the star presented at times a more vertical and at other times a more horizontal view of these magnetic spots and hence caused the variations in observed magnetic field strength and spectrum. The abundance anomalies in the spots were thought, at various times, to be caused by element migration in a magnetic field or by variations of effective temperature. Some people thought that every Ap star perhaps had an evolved white dwarf companion, from which evolutionary products had spattered the surface of the Ap star. Another idea was that protoplanets containing fractionated material had fallen onto the surface. Currently the most popular idea is that under the stable conditions that pertain in the atmospheres of slowly rotating magnetic stars it is possible to get a gradual sifting of the material so that those elements which are light and have many strong spectral lines (and hence absorb radiation) are slowly moved upwards to the surface of the star. Conversely, heavy elements with few spectral lines are likely to sink and appear to be underabundant. Some people have put a lot of work into trying to demonstrate the feasibility of this idea and it cannot be dismissed lightly. However, we have already pointed out that not all Ap stars are slow rotators. If there is rapid rotation it normally induces internal streams of matter in the star which cause mixing and hence offset any element separation mechanism. The counter argument to this is that strong enough magnetic fields could perhaps suppress this mixing. Other worries are whether both Ap and Am stars can be explained by the same mechanism, whether it is realistic to get most of the helium in the surface of a star to migrate to one spot and, if the presence of uranium is confirmed, where does it come from?

What is clear is that the unusual spectra of these stars, particularly in the ultraviolet where many of the elements have many strong lines, causes a lot of absorption of the light which would otherwise be radiated from the surface of the star. This absorption (known as line blocking) can absorb up to about 30% of the light of the star in the visible and more in the ultraviolet. This absorption redistributes the energy of the star (this is called backwarming), and causes the distribution of energy against wavelength to depart from the black body curve. This is the reason that the star can appear to have different effective temperatures at different wavelengths, so any colour anomalies could be important. Attempts to look for correlations between the abundance anomaly and magnetic field strength, or other parameters, have generally led to a null result.

13.5.2 Measurements on Ap stars

Clearly there are many questions to work on, but if many of the anomalies appear in particular spectral lines, where does this leave the user of a small telescope without a spectroscope? The answer is 'in a very good position'. We can think of no other type of star so suitable to practice on. One of the reasons for this is that, so far as is known, the light curves are repeatable down to the millimagnitude level. Therefore if the newcomer to photoelectric photometry takes one hour or even all night to obtain one measurement in one colour it does not matter. One cycle later the star will behave in the same way in all colours and the light curve can slowly be 'filled in'. The light curve amplitude is often different in different wavelengths, but this significant information is not yet known for the majority of Ap stars.

Figure 13.6 shows amplitude in different colours for HR 3413, compared with the results for the Ap star HD 83368. Note how in this latter star there is a wavelength at which the amplitude goes to zero between B and V and for long wavelengths the amplitude has the opposite sign; in other words it is out of phase. What this means is that the maxima at short wavelengths are simultaneously the minima at long wavelengths. Many of the stars for which we have ultraviolet data (obtained from satellites) show that the amplitude trends continue with wavelength so that stars which have total amplitudes in the optical of a few hundredths of a magnitude might have several tenths of a magnitude variation in the far UV. Spaceborne telescopes for this extreme end of the ultraviolet spectrum are still in the planning stage, however.

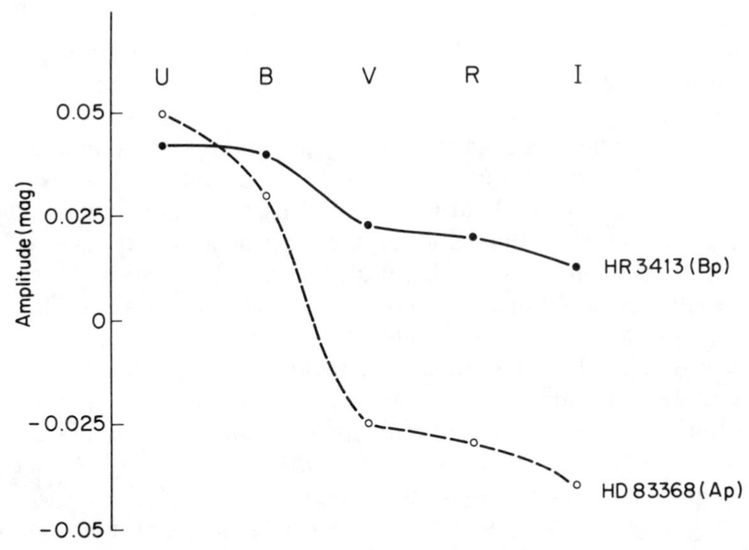

Figure 13.6 The amplitudes of light curves measured through different filters will not, in general, all be the same. The two stars HR 3413 and HD 83368 show very different trends of amplitude against colour. A negative amplitude means opposite in phase.

The fact that there is generally some wavelength at which the amplitude is zero and on either side of which the amplitude increases, but with opposite signs, has led to the concept of the '*null wavelength*'. Users of small telescopes able to do all five UBVRI colours are in a position to define the shape and gradient of the amplitude against colour curve and the position of the null wavelength. Many hundreds of Ap stars are known and if these data could be collected for many of them then correlations could be looked for between this behaviour in the visible and other parameters, derived from spectroscopy, some of which are: effective temperature (or even temperature differences for those stars which show different values at different colours); type or quantity of abundance anomaly; magnetic field strength; number, size or position of the spots or bands on the surface (many stars do not have just two diametrically opposite spots); stellar age or some other parameter. Common sense (which, having evolved for our physical survival on the surface of a planet, is no great arbitrator in matters astronomical!) suggests that eventually something should be found which will collate with something else. Even knowing accurate rotational periods for many of these stars would be a major advance. The shortest known period is under twelve hours. (Anything shorter would be remarkable because it is thought that an

A type star rotating with period of between eight and ten hours should be spinning itself to destruction.) The longest known period is over twenty years. Most of the stars have periods in the range of a few days to a few tens of days.

So far we have confined the suggested project to Ap stars to those essentially associated with the rotational period of the star. Should one look for binary orbit or pulsational variations? Very few Ap stars have been shown to be members of binary systems although not too long ago it was hypothesized that all of them had white dwarf companions. Any new discoveries of either orbital variations or eclipses in binary stars would of course be very valuable. Pulsation, on the other hand, is a topic of current and active interest in Ap stars. Until a few years ago almost all Ap stars were thought not to pulsate. The star 21 Comae was an exception but even here some people reported variations while others did not and the truth of the matter is still to be determined. In general, the theoreticians who would like to explain the abundance anomalies in Ap stars by radiative and gravitational sifting would have been happier if pulsations did not occur as they require these stars to have rather *stable* atmospheres. No-one was prepared for what was waiting in the wings.

The first sign was Przybylski's star. As a young man, Antonin Przybylski set off from his wartorn country of Poland and walked cross country, living rough and often having nothing to eat, until he reached Switzerland where he was interned for the remainder of World War II. He eventually ended up in Australia where, having obtained two PhDs, he became an observational astronomer, which he remained for the rest of his life. It was during a spectroscopic study that he drew attention to the remarkable star HD 101065 which now bears his name. Survey work in the early part of the century had classified the star as a late type object but higher resolution studies caused Przybylski to think otherwise. It took many years of study before the bizarre nature of the star was demonstrated and a less determined person might have been deflected to other projects. The spectrum of Przybylski's star is so full that it is difficult to decide if the continuum is seen, yet the lines of hydrogen are weak or absent. It finally transpired that many of the lines were due to abundant holmium which to us is one of the rarer of the rare earths. The effective temperature of the star is difficult to determine but it seems to be much hotter than its original G type classification and is now thought to be at least an F type star.

The chemical abnormalities of Przybylski's star were only the start of the story. The star was photometrically investigated by Don Kurtz (an American working in South Africa who does some of the most accurate photometry in the world) and he realized that this star was varying with a total amplitude of only 0.01 mag but with a surprisingly short period of about 12 minutes. Such a period can only be pulsational. In the few years that have passed since that original discovery Kurtz has made this field of low-amplitude short-period variations his own. About a dozen Ap stars, among the cooler ones, have now been shown to vary with periods in the six to twelve minute range, periods more normally associated with white dwarfs. Some of the stars are singly periodic, while others have more than one period present, the amplitudes varying with the rotational period of the star. It is assumed that these very short periods are due to the excitation of high overtone pulsations. The mode of vibration is governed by the forces involved, to which the magnetic field may contribute significantly. At the moment there is no consensus as to the exact nature of these pulsations. As more are studied some pattern will emerge. Since most of Kurtz's work has been confined to southern skies, there must be more of these variables still to be discovered by northern observers, so users of small telescopes could dramatically increase the amount of data.

In contrast to normal practice in variable star measurement we do not recommend the use of any comparison stars (unless you have a comparative photometer), when searching for this type of variation. There is simply not enough time to change the telescope's position and recentre the stars within a small fraction of a period of a few minutes. All one has to do is keep the telescope accurately centred on the star and let the computer log the data at, say, 20 second intervals. If you can arrange a real time display of the data you can watch the counts going up and down by about 1% every few minutes. Do not despair if after one night's run you have seen nothing. Remember that rotation of the star causes the amplitude of the variations to increase and decrease and it might take several nights before one has covered enough of the rotational period to bring the appropriate part of the stellar surface into view. Remember that, providing you are confident in the quality and stability of your observing site and equipment, even a well demonstrated null result could have future value. Note that, up to now, all the stars demonstrated to show this effect are the cooler Ap stars, so confine your initial searches to that spectral range.

If Ap stars are so promising, for both amateur and professional, then what of the Am stars? Although superficially similar to the Ap stars in their spectral peculiarities, and only slightly cooler than them in their position in the HR diagram, there are many significant differences. First, as a generality, most Am stars are in binary systems whereas at the moment most Ap stars seem not to be. Secondly, Am stars do not show the dramatic spectral changes which their Ap counterparts demonstrate as they rotate. Instead, it seems as though the whole Am star surface is uniformly covered in the same chemically peculiar atmosphere whereas Ap stars clearly have patchy surfaces. It has become almost dogma that Am stars do not vary and over recent years whenever an Am star has been found to show small fast variations it has generally been found that the star has evolved off the main sequence to luminosity class IV or III and the star has been reclassified. They are called δ Delphini stars after the prototype pulsating Am giant. A little reflection will show this is a rather curious state of affairs. On the one hand we have stars, most or all being in binary systems, many with periods in the 2–4 day range, and yet we do not see ellipsoidal variations, eclipses or any of the other host of variations that close binary stars are heir to. If you can find an Am star near in the sky to some other variable you are investigating, then generally it is thought that you have to look no further for a guaranteed, non-variable, comparison star. Presumably what is happening is that the variations caused by binary motion are of such small amplitude that they are able to fall below the typical, traditional 1% accuracy. If you can break this barrier, then perhaps you can demonstrate the binary effects.

Finally in this chapter, another word about the origin of the Am stars. Some of the proponents of the diffusion hypothesis for the origin of the abundance anomalies feel that this is perhaps even more applicable to Am stars with their very stable atmospheres than to the Ap stars. There are even those who claim that the chemical stratification will inhibit any incipient attempts at pulsation. Accurate observations at the millimagnitude level could help to differentiate amongst some of these ideas, but do naturally demand good observing conditions.

13.6 FLARE STARS

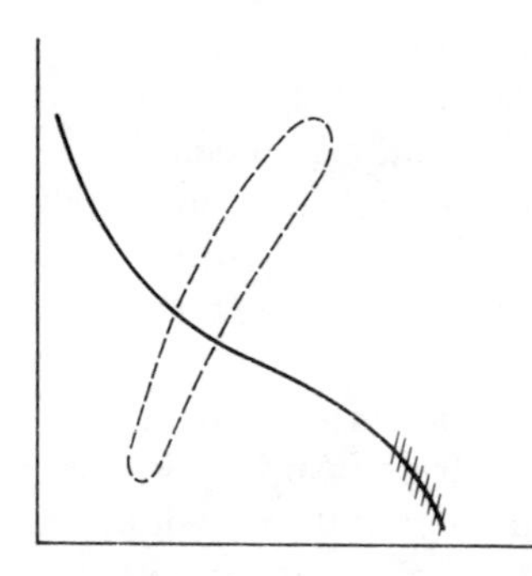

As mentioned in §4.1, flare stars were originally a nice neat group of dim red stars, the archetype being UV Ceti. However, some very fast flares and flares on RS CVn type binaries have also been discovered. Mathematical modelling of flares, even those on the Sun, is still at a rudimentary stage, so observations of flare stars would contribute to a state-of-the-art problem. Large amounts of time of the Jodrell Bank telescope have been devoted to the UV Ceti type, in recognition of their interest.

It needs great patience to keep records of a given star for a total of hundreds of hours, waiting for a flare that only lasts a few minutes. Some sort of innovative automation is called for. One way is a fully automated telescope, which will move periodically between the star of interest and a reference star, and keep taking measurements. A few amateurs have built such telescopes. However, with such a technique the telescope spends most of its time looking at the reference star and the sky background and could miss the flare. Thus a better method would be a telescope with a comparative photometer, or, one stage better still, with a CCD array which can monitor the star and the surrounding area continuously, except very briefly when the contents are read out into a computer. The computer can find the star of interest in the CCD image, which eases guiding, and a neighbouring reference star (or perhaps several reference) stars can be used to work out the apparent magnitude of the star of interest. The computer could also ring a bell and make a strong cup of coffee for the sleeping amateur astronomer when the star of interest shows activity! CCDs, previously the province of the professional, are coming into use by amateurs. (A CCD video camera could, in principle, be used as it is, but the 0.02 second integration time would be too short for all but the brightest stars.) The CCD method has been tried with limited success in supernova searches. The flare star case is, of course, much simpler because you know precisely where to look all the time.

Photography could also be used, with automatic advance at appropriate intervals, say 30 seconds. Film costs would soon mount up. The field of view would, of course, have to be wide enough to include a suitable reference star. Data analysis would be much more tedious than with a CCD. Flares would have to be searched for visually, probably with a blink comparator. A simple form of comparator using two standard 35 mm projectors can be made at home. There is no need, in this case, for the precise and expensive form of comparator used by professionals. Admittedly it is not easy to take precise absolute measurements from film, mainly because of its non-linear characteristics, but the main problem is to record the flare. It would be well worth the time afterwards to calibrate the film over the particular range covered by a particular flare.

It was difficult even for Bernard Lovell to get time on his radio telescope for studying flare stars, and it is scarcely imaginable that thousands of hours of time of a large optical telescope will be so used. The way is open for amateurs. Perhaps flare stars will give rise to a new observing technique among amateurs.

Chapter 14

Stars in and near the instability strip

As explained in Chapter 6, the cause of variability in many stars is a layer of ionized helium somewhere within them. The band which contains this type of variable star crosses the HR diagram from lower left to upper right. Where it crosses the main sequence the variables are known as δ Scuti stars. Moving up and across to brighter cool stars we find the RR Lyraes, the Cepheids and W Virginis stars and, finally, extending far to the right of the strip, the Mira and semiregular stars. We will start with the last type of stars first as their amplitudes are large (~ 2 magnitudes) making them relatively easy to observe. In addition, they are intrinsically so bright that they can be seen at great distances and hence very many are known. Their periods are long, of the order of hundreds of days, so good comparison stars are essential. These stars are at a late stage of evolution, and are at the cutting edge of advances in stellar astrophysics. This is the main reason for studying them. It is significant that a quarter of all requests from astrophysicists for data from the bank held by AAVSO (American Association of Variable Star Observers) concern this type of star. (Many of the other requests concern cataclysmic stars; see §15.1.)

14.1 MIRA AND SEMIREGULAR VARIABLE STARS

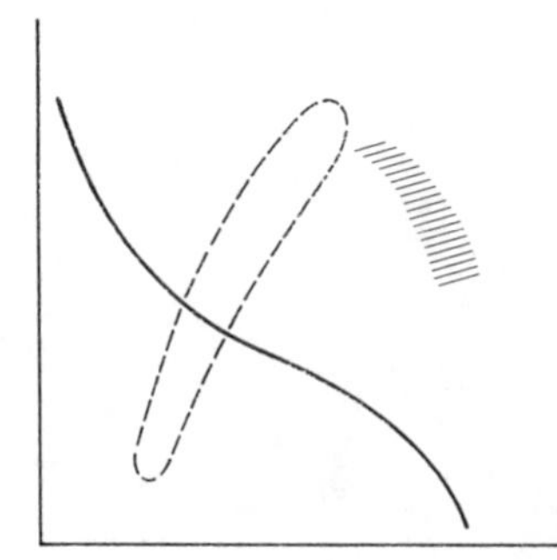

It is not common practice to group all these stars together, nor with those inside the strip, but from the observers' point of view there seems no great reason to divide them into various subgroups. Traditionally the name 'Mira' is reserved for stars whose light curves have the largest amplitudes and the most repeatable light curve. Miras show departures from strict periodicity and they merge into semiregulars, which themselves spread through a sequence showing progressive reduction in amplitude and increasing irregularities. We can discuss observations without getting into nomenclature.

The star Z Ursa Majoris is listed as an SRb with a period near 198 days and also a longer variation of period near 1560 days. Its spectral type is M6e. Members of the American Association of Variable Star Observers have been carrying out visual observations of this star for many years and in 1985 one of us (ENW) was given a computer listing of 7772 observations obtained between Julian Dates 2443 000.4 and 2445 706.7 (approximately between 1976 and 1983). An average of nearly three observations a day for seven years is a remarkable achievement by any standards and deserved to be analysed as thoroughly as possible. The data were analysed in several different ways and all analyses gave similar answers. The results were as follows.

Frequency (cycles per day)†	Period (days)	Amplitude (mags)
0.000 2010	4973.9	0.1973
0.005 1864	192.81	0.7004
0.010 2419	97.64	0.3269
0.015 5836	64.17	0.0882

† 1 cycle per day is 1.16×10^{-5} Hz, which is more correct but less convenient!

The lowest frequency cannot be determined accurately from these data. The fit suggests an approximately constant 0.005 cycle per day spacing between the various frequencies. A fit of these four frequencies to the data is shown in figure 14.1; it is good, though not yet perfect. This result was obtained from visual photometry with a probable accuracy of only about 0.1 magnitude, and because of this the logarithmic scale of magnitudes was not converted to the linear scale of intensities. The star was classified as semiregular from its behaviour over short periods, but shows a marked periodicity when viewed over seven years. If this were confirmed over still longer timescales then it would have an important effect upon our understanding of this type of variable. If we had the same quantity of photometry available but obtained with a photoelectric device, perhaps with accuracies of 0.01 or 0.001 magnitude, then what might we have found? More frequencies belonging to the same series, or one or more additional sets of frequencies? We cannot know but it is clear that much remains to be discovered by the acquisition of large amounts of high quality data.

In addition to the presence of many frequencies in these stars there is also an intriguing fact that many seem to show varying periods. Data runs obtained over many years on stars such as R Aql, T Cep, W Dra, R Hya and many other stars give $O - C$ diagrams showing large, continuous changes in period.

Whether such long data runs are best analysed by means of the $O-C$ diagram or by Fourier analysis is something that can only be decided empirically. First we need the data and then we can try to decide what it means. Remember that, because some of the periods which are present in these stars are many thousands of days long, the stability of the reference system is of vital importance. Notwithstanding the relatively large amplitudes of these variables it is still important to observe two, or even three, comparison stars in case one is not perfect. Only in this way can photometric results be reliable over many years.

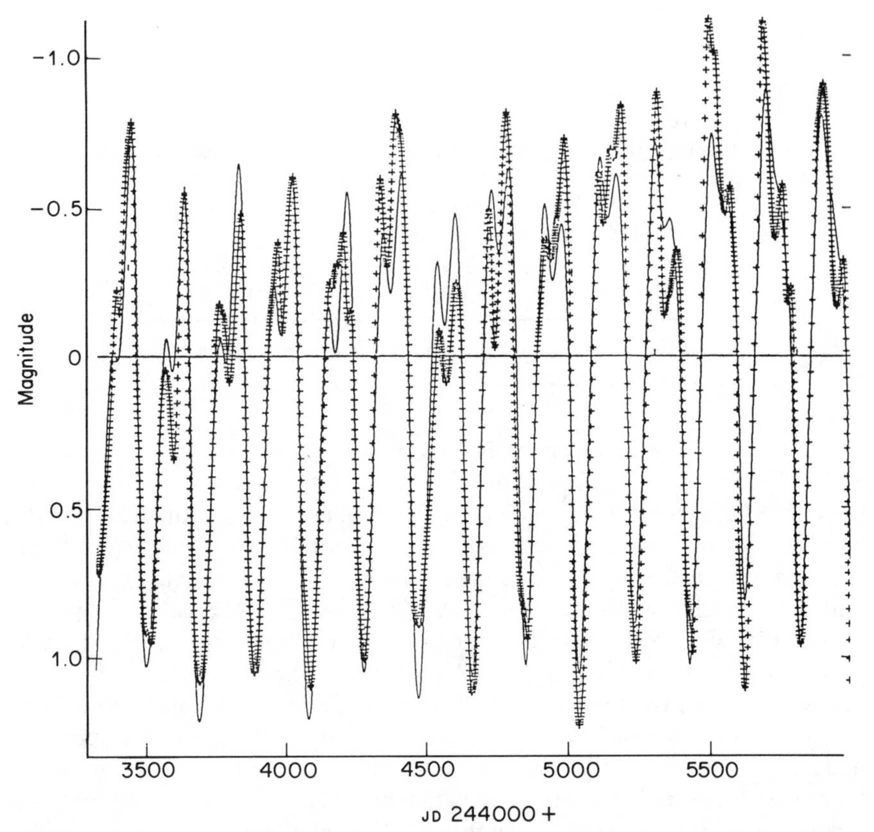

Figure 14.1 Seven years of data on Z UMa (crosses), and a fit using four superimposed frequencies. The fit is not perfect, and other frequencies may be involved. The 192 day component is obvious at a glance.

What kind of data will be most useful? First it is important to realize that the surfaces of these stars are so cool that they are bright in the red and infrared, but much of the data comes from visual observations by amateur astronomers, so that there is a tremendous amount to be gained by merely extending the observations photometrically towards the red. Note that colour blindness, a male linked genetic disorder, can cause serious loss of red sensitivity. The existence of subclinical disorders of this type enhances the case for photoelectric work on these stars, or for female observers. Amplitude and/or phase differences between the light curves at different wavelengths can tell us a lot about what type of pulsation we are seeing, radial, non-radial, which mode, etc. Even stars at the bright end of the instability strip will not have a significant amount of data available for each of the UBVR and I bands. Data runs of decades are needed and accuracies of 0.01 magnitude would transform what we understand about these types of star. The periods involved are sufficiently large that the fact that one cannot observe every night is not important. Working on circumpolar stars means that one does not have breaks in the data run for visibility reasons, although seasonal changes in the weather mean that from any one site some gaps are inevitable. This, of course, is why it is essential to add your results to those of others, in other parts of the world, by sending them to the AAVSO or BAA.

14.2 W VIRGINIS AND CEPHEID VARIABLES

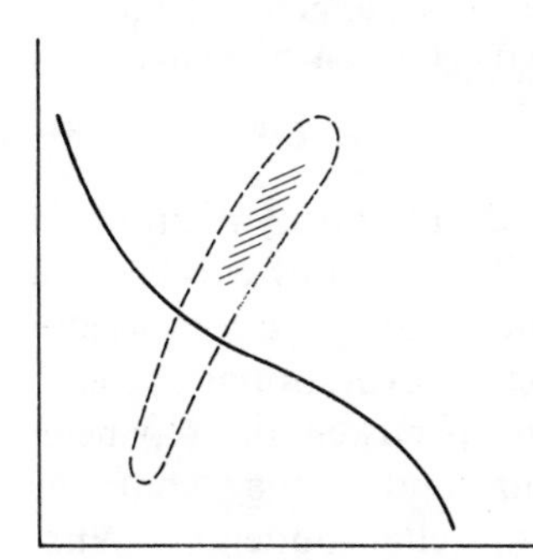

Passing down the instability strip from the large, cool, long period variables we come to the W Virginis stars and the Cepheids (named after the star δ Cephei). Although both groups of stars lie in roughly the same part of the HR diagram they are physically distinct and follow different period–luminosity relationships. The meaning of period–luminosity relationships was described on p 65. It is, briefly, analogous with the idea of a small bell vibrating more quickly than a large bell in that denser, more compact stars have shorter periods than less dense giant stars. The concept works so well that a rather tight relationship exists between the period and the absolute brightness of these variables. This relationship between period and luminosity is shown for both Cepheids and W Virginis stars in figure 14.2. It can be seen that the two types of stars are well separated on this diagram and that if the period is known accurately then, for either type of star, its absolute luminosity can be accurately predicted. This means that if the period and hence absolute luminosity of either a Cepheid or W Virginis star is known, its distance can be calculated from its apparent luminosity. This facet of W Virginis and Cepheid behaviour, in combination with their high intrinsic luminosity (they are bright enough to be seen in other galaxies), has led to them, together with the RR Lyrae stars (see §14.3) being some of the most important objects for determining intermediate distances in astronomy. Beyond our local cluster, however, it gets very difficult to resolve individual stars.

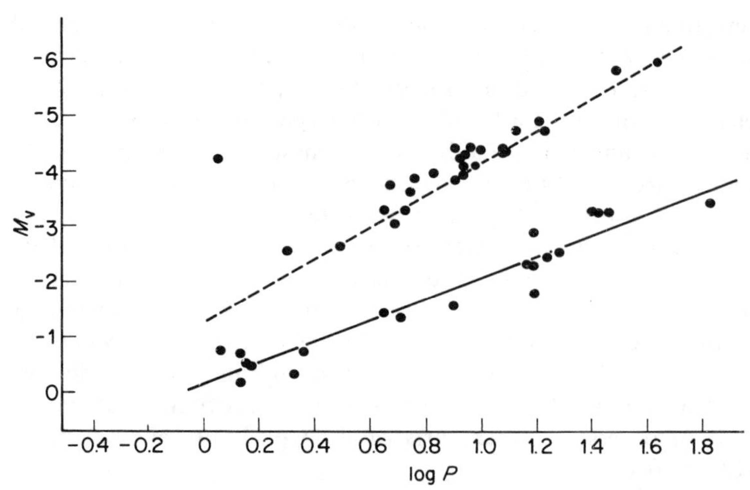

Figure 14.2 The period–luminosity relation for Cepheids (upper line) and W Virginis stars (lower line). Notice that magnitude is the logarithm of luminosity, so that *both* scales are logarithmic. The relation between absolute luminosity and period is a power law, $P \propto L^n$. n can be found from the slope of the line on a double logarithmic graph such as this.

Henrietta Leavitt discovered the empirical period–luminosity relation for which Cepheids are now famous. With the opacity data now available, it should be possible, as described in §6.1, to calculate the pulsation periods since the composition can be found by computing the evolutionary process all the way from the main sequence. There are some persistent discrepancies between the calculated and measured behaviour, and if they could be tracked down it would lead to improvements in all stellar modelling. What is needed is not so much more data on the relation itself, which is now firmly established from surveys of hundreds of Cepheids, but the detection of overtone frequencies and drifts in frequency as the star evolves. Overtones are a sensitive indicator of composition—the varnish on a Stradivarius violin is a nice analogy.

Currently very few double or triple mode Cepheids are known, but they are unlikely to be genuinely rare. (Remember the Sun has dozens of detectable overtones.) If a few dozen more were to be discovered and thoroughly investigated over the next few years it would be a major achievement. The observations are not difficult, as the amplitudes are large (approximately one magnitude), and the periods are in the convenient range of a few days to a few weeks. Observations on each star spread over a year or two could provide sufficient data for the detection of two or more overtones and, using the $O-C$ diagram, of any drift in frequency. (The prime requirement in looking for overtones is accuracy, and for drift it is a long span of observations. Naturally, both are desirable!)

Alain Figer and the GEOS (Groupe d'Etudes d'Observations Stellaires) observers recently discovered overtones from visual observations of the known Cepheid EW Scuti. These have been confirmed from photoelectric data. Remembering that these variables are going through a relatively rapid stage of their evolution it is quite reasonable to expect to see period changes within one person's observational lifetime. In δ Cephei itself the period decreases by about 10 seconds every century and yet, such is the power of the $O-C$ diagram that this small amount is rather easily detected. Delays as large as several tenths of a day now occur compared with predictions from the early part of this century. Some Cepheids are probably evolving right to left on the HR diagram, while others are evolving in the opposite direction because the evolutionary path is complicated in this region. Some should show increases in period, others decreases. Periodically varying $O-C$ values might indicate a binary Cepheid, as the light travel time varies. This is precisely the same effect as in Roemer's measurement of the speed of light from the satellites of Jupiter. In short, almost any new data on any Cepheid variable will be useful and the more intense the coverage and the longer the data run, the better the accuracy and the more homogeneous the data.

If the opportunity for new discoveries with Cepheids is great, then the W Virginis stars present an even more fertile field of endeavour. They seem to show variable light curves and perhaps changes in period. Their masses are not well established and they might perhaps contain several subgroups. Any thorough investigation of these variables can hardly fail to produce new discoveries.

Data in two colours, using filters, would be an added bonus. Differences in the light curves at different colours, in combination with radial velocity data obtained from Doppler shift measurements (by professionals) can lead to mode identification in the variables, allowing yet another intercomparison between observation and theory to be made.

14.3 THE RR LYRAE VARIABLES

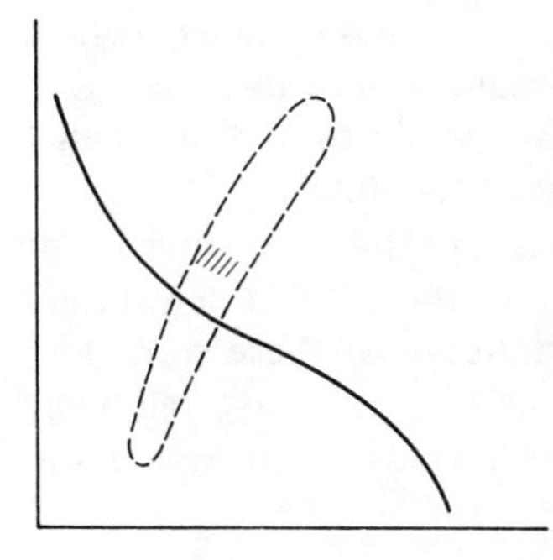

Proceeding down the instability strip we come to the RR Lyrae stars, sometimes known as cluster variables because of their abundance in globular clusters. They are intrinsically fainter and smaller than the Cepheids and W Virginis variables and hence have shorter periods; typically in the range 0.2 day to nearly one day. Earlier they were separated into three classes: a, b and c, following pioneering work by Bailey identifying various types of the light curve.

Later experience has shown that types a and b are sufficiently similar that they are now grouped together as type ab, but type c is still kept separate. Type c also has shorter periods, on average, than type ab. Figure 14.3 shows the different forms of typical light curves.

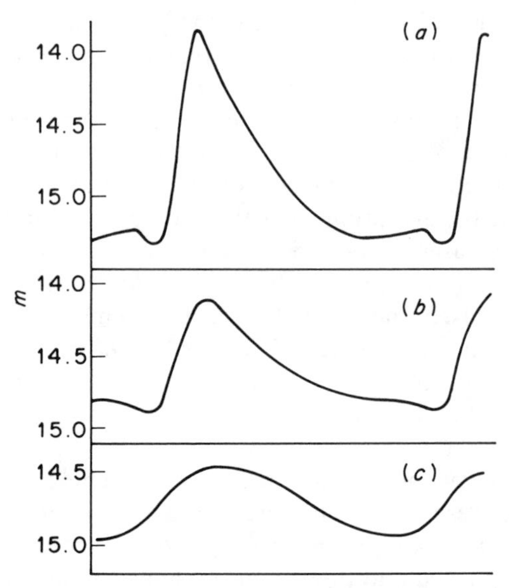

Figure 14.3 RR Lyrae stars show a range of shapes of light curve, and are categorized according to the similarity to the forms in the diagram.

Nearly 6000 RR Lyrae stars are known and, like the Cepheid and W Virginis stars, they have such a well established period–luminosity relationship that they have been very useful measures of astronomical distances. The RR Lyrae relation is much flatter, in fact their periods are mostly close to half a day. Their lower intrinsic brightness means that they are only useful in measuring smaller distances than the Cepheids but to counteract this they occur so frequently (about eight times more RR Lyraes are known than Cepheids and W Virginis stars) that they have been most valuable for determining distances within our Galaxy. Indeed it was their use in measuring distances to different clusters of stars in our Galaxy, where all the stars in any one cluster have approximately the same age, that allowed the changing pattern on HR diagrams to be discovered and hence contributed directly to our understanding of stellar evolution.

With so many RR Lyrae stars known there is certainly no shortage of stars to work on. The question is, therefore, are there still scientifically useful contributions to be made to our understanding of these stars with small telescopes? Even remembering that most of the stars have relatively accurately known magnitudes, UBV colours and periods the answer is still an unequivocal 'yes'. Although their fainter absolute magnitude means that there are fewer apparently bright RR Lyraes than Cepheids (there are several Cepheid variables visible to the unaided eye) there is still a sufficiently large number of relatively bright RR Lyraes to allow small telescopes to work on them for many years. All the observational projects listed for Cepheids apply equally to RR Lyraes. The search for changing periods (using $O - C$ diagrams), the search for multiperiodicity or even the discovery of previously incorrectly determined periods are all feasible projects and have similar significance to astrophysics.

However, some RR Lyrae light curves have an extra feature not found in Cepheid light curves. The effect is seen as a distortion of the light curve which can be visualized as a low-amplitude sine wave, superimposed on top of the mean curve, whose phase varies progressively with time. The distortion occurs on the rising part of the light curve and about the maximum. The effect is known as the *Blazhko effect*, after its discoverer, and perhaps as many as 20% of RR Lyraes show the phenomenon. The sine-like distortion of the light curve is periodic, with periods ranging from somewhat over ten days to over five hundred days but with most periods being in the range 20–40 days. In addition to this 'Blazhko period' some stars also show an additional period during which the amplitude of the Blazhko effect changes. For example, in RR Lyrae itself, the Blazhko period is 40.8 days while the amplitude of the effect varies with a period of about 4 years. The 40.8 day period is associated with the period of change in the observed magnetic field of the star.

It is thought that this star, like many other stars (and the Earth), has a magnetic field which is angled to the rotational axis of the star. Rotation of the star therefore sometimes presents us with a more normal and at other times a more oblique view of the magnetic field, and hence we observe a changing component of magnetic field strength. Since the magnetic forces affect the pulsation mode, this can give the superimposed variation. The exact way in which radial pulsations in a magnetic star might be affected and the way in which rotation of that star might cause distant observers to observe 41 day or 4 year variations is still under active discussion. What is needed is much more sound data which will constrain possible models of these effects and eventually lead to a correct understanding of their causes.

14.4 δ SCUTI STARS

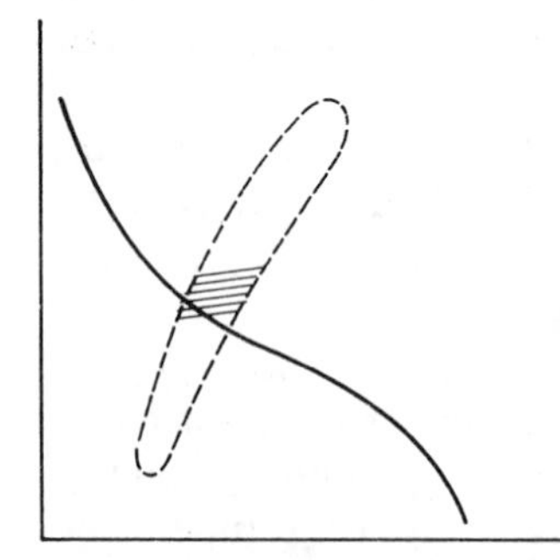

The area of the HR diagram where the instability strip crosses the main sequence contains the δ Scuti variables. Both on the main sequence and for one or two magnitudes above it, many of the stars are variable, but these are small, fast variations. The periods range from about 30 minutes up to nearly eight hours. Although the longer period δ Scutis overlap with the shorter period RR Lyrae stars, the stars seem to be physically different, as shown by the phase relationship between luminosity and radial velocity variations.

The amplitude of the variations ranges from a maximum of over half a magnitude down to the limits of detectability at the lower end, < 0.01 mag. Indeed it has been suggested that perhaps all stars in this part of the HR diagram are variable, but that some have such low amplitudes that they are classified as constant.

Most of the projects described so far require that observations over months or years be obtained for their maximum usefulness. δ Scuti stars, on the other hand, have periods short enough to include several cycles in a single night. The problem is to keep a cycle count from one night to the next. Multiperiodicity is common among δ Scuti stars. Unlike the Cepheids and RR Lyraes, which predominantly pulsate in the radial mode (although the Blazhko effect might be evidence for some non-radial variations), many δ Scutis seem to show non-radial modes or a mixture of radial and non-radial. To generalize, the longer period, larger amplitude, singly periodic δ Scuti stars are radial pulsators while for the shorter period stars our lack of detailed knowledge does not even allow us to decide whether we are dealing with complex mixtures of radial and non-radial modes or even non-periodic phenomena. A few of the approximately 200 known δ Scuti stars have been thoroughly investigated with measurements of period, amplitude differences at different colours and radial velocity variation. The vast majority, however, do not have such data available. In addition, even those stars which are thought to be well observed can still show surprising new results when new observations are obtained. As an example, the star 20 CVn has long been listed as a monoperiodic δ Scuti, with a period just under three hours and the low amplitude, for a singly periodic star, of 0.02 mag peak to peak. Observations by one of us (ENW) at Easter 1985, in Nice, France showed a newly discovered period of near 70 minutes which seems to be related to the previously known period by an exact ratio of 2 : 5 (see figure 10.10). It is entirely possible that many more discoveries wait to be made and certainly many δ Scuti stars require observations at wavelengths other than the U, B and V bands.

However, it is not just by extending wavelength coverage, or discovering the occasional extra frequency, that small telescopes can make important contributions. Of all the stars on the instability strip the δ Scuti stars are the least understood and the ones in which a properly organized observing campaign can add most to our knowledge. Note, however, that because the amplitudes are typically only a few hundredths of a magnitude photometry must be done either at good sites, or with comparative photometers.

The following is an example of real results on a δ Scuti, 29 Cygni, which has a period of 40 minutes. The data shown in figure 14.4 were obtained during a two week observing run in July 1983. Four consecutive nights produced continuous six hour runs; then there were four cloudy nights; then we had two consecutive nights on the first of which we had a continuous six hour run and on the second of which we had seven continuous hours, a half hour gap and finally nearly six more hours obtained in America. Tests on synthetic data using the same epochs as this real data set showed that our analytical techniques were well able to resolve up to six frequencies with values similar to those found in this data set. The upper curve of figure 14.5 shows the best fit power spectrum to the real data but in the lower curve of figure 14.5 we show why there are some worries about accepting this solution. We show an enlargement of the peak near 31.4 cycles per day from the upper part of the figure compared with what a pure sine wave would have produced. The real data do not fit a set of clearly defined single frequencies. To further demonstrate the uncertainty of the result various subsets of the whole data set were analysed independently. If the star was behaving in a predictable way and if the analysis was correct, then we would expect to get the same results from each subset. Figure 14.6 shows the results from three subsets of the data, each one of which would be capable of resolving up to six frequencies according to our tests on synthetic data.

The power spectra look significantly different and the conclusion must be that there are either complex power spectra containing many frequencies or alternatively the process is non-stationary. More data are needed!

Finally, lest it be thought that only continuous observations are useful on δ Scuti stars we would emphasize that there is still the possibility of comparing amplitude and phase with colour, and looking for period changes with an $O - C$ diagram for those stars which seem to be either monoperiodic or contain perhaps only two well defined periods.

As mentioned above, it seems likely that many of the variations to be discovered and identified in the δ Scuti stars will arise from non-radial modes. An early clue to their presence is when equally spaced frequencies are found. In multiperiodic radial pulsators, on the other hand, period–

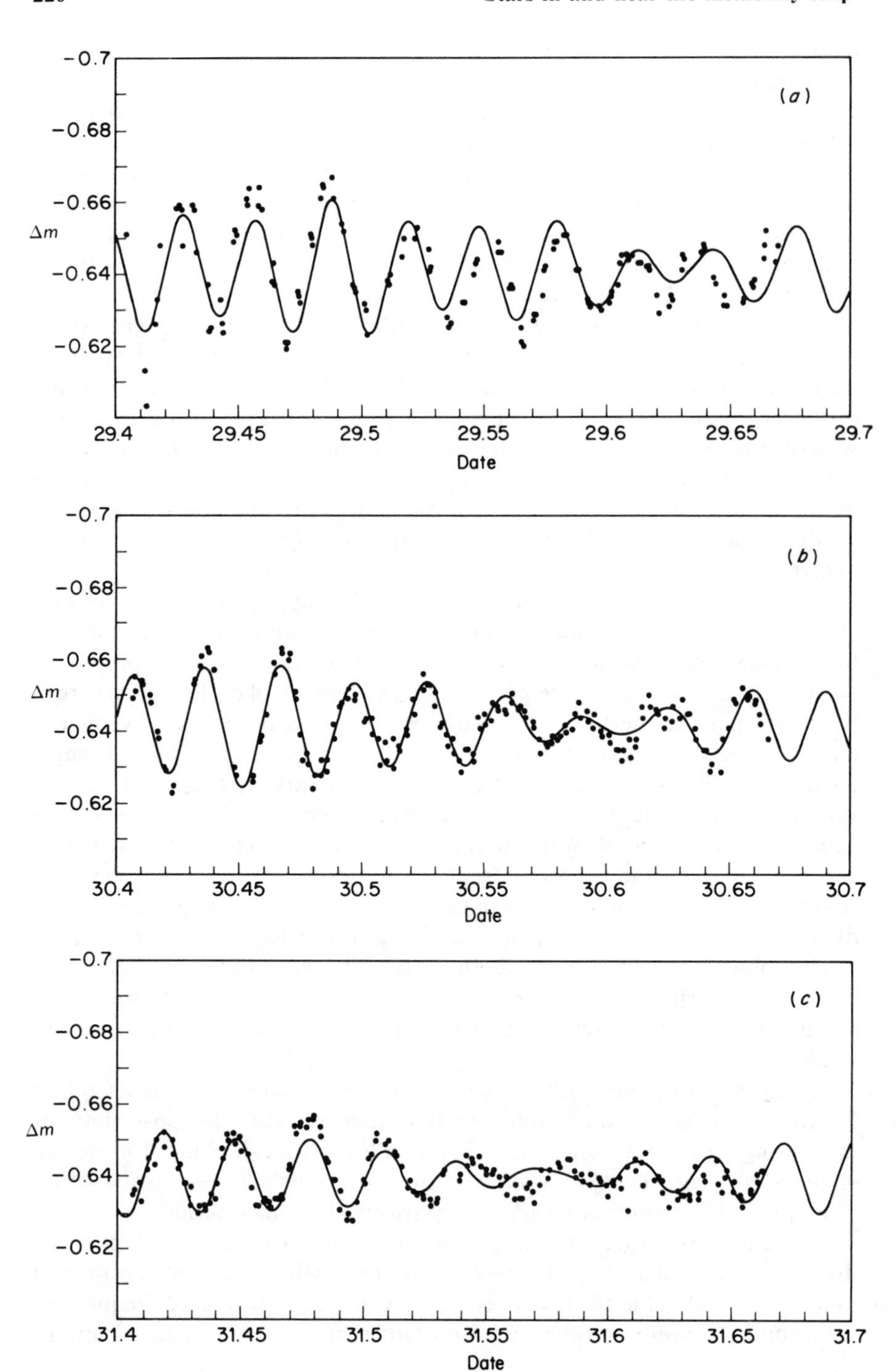
(a)
Δm
Date
(b)
Δm
Date
(c)
Δm
Date
−0.7
−0.68
−0.66
−0.64
−0.62
29.4
29.45
29.5
29.55
29.6
29.65
29.7
30.4
30.45
30.5
30.55
30.6
30.65
30.7
31.4
31.45
31.5
31.55
31.6
31.65
31.7

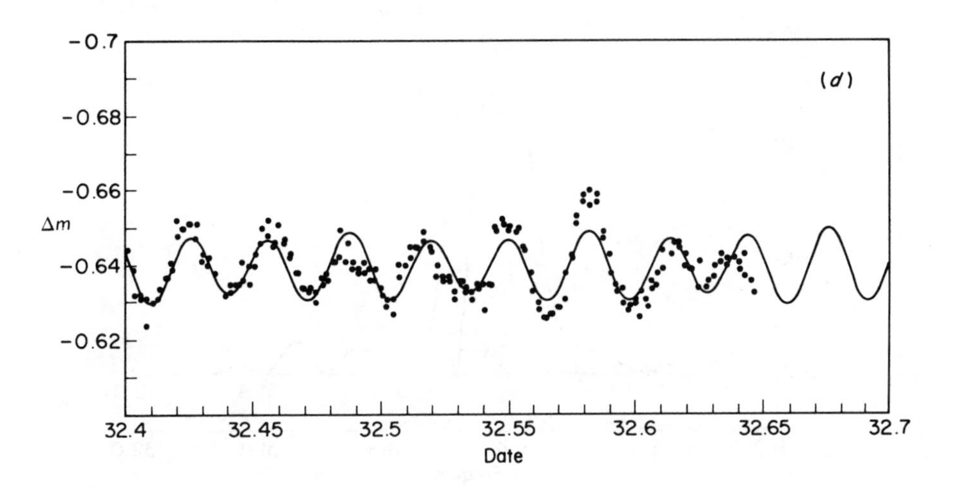

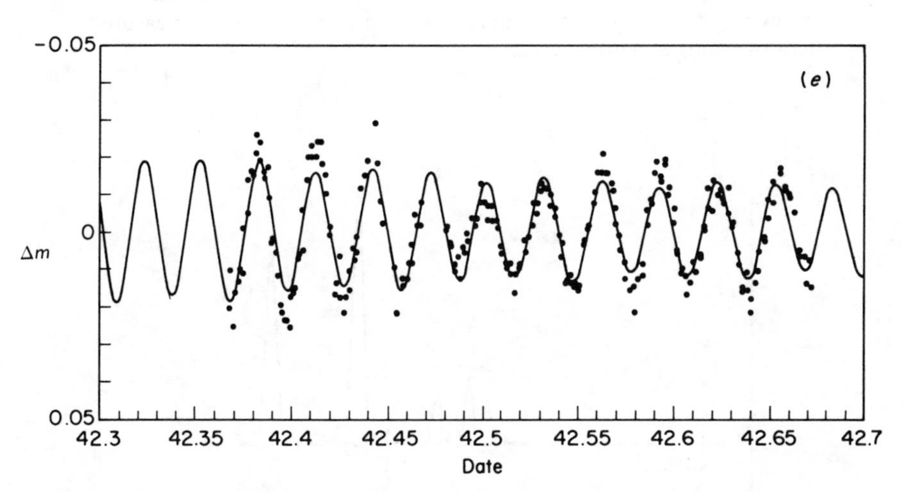

Figure 14.4 Data obtained in five nights of measurements on the δ Scuti type star 29 Cygni in July 1983. (*a*)–(*d*) 13–16 July; (*e*) 25 July.

frequency ratios near 0.7 are frequently found, generally identifying that fundamental and first overtone pulsations are present. These are only rules of thumb; to go further needs stellar modelling computations. Some of the more complex δ Scuti stars could well contain mixtures of radial and non-radial modes and probably many hundreds more of these stars, bright enough for small telescope observations, wait to be discovered and investigated.

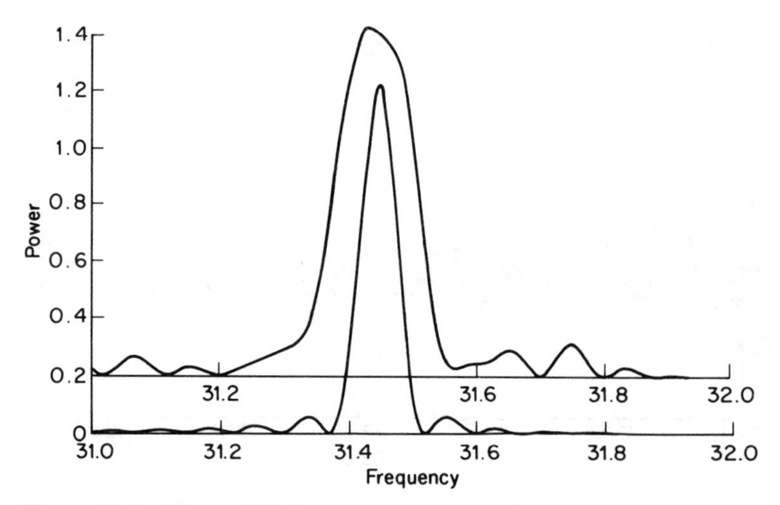

Figure 14.5 The peak in the power spectrum of the real data, near 31.4 cycles per day (=46 min) compared with a pure sine wave measured with the same experimental errors.

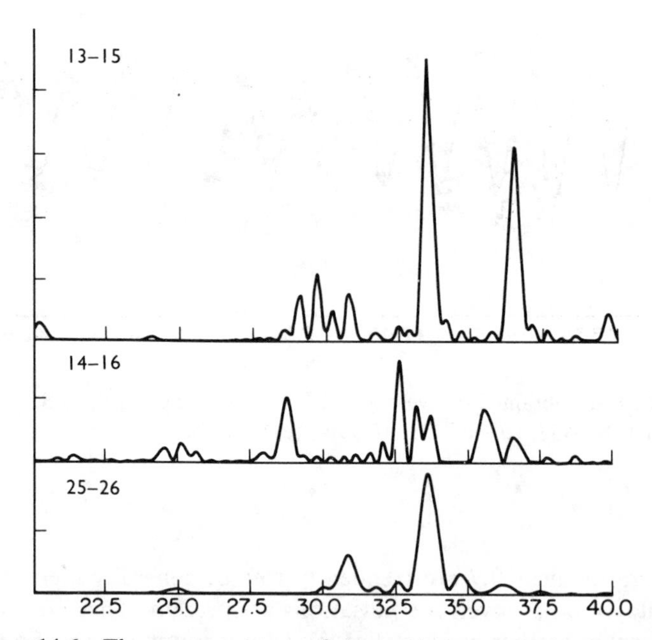

Figure 14.6 The power spectra of three subsets of the 29 Cygni data obtained on the data indicated in July 1983. Tests show that up to six separate frequencies could have been resolved from these data subsets and the power spectra should have looked identical if the variations were stable and composed of six or less frequencies.

14.5 THE LOWER END OF THE INSTABILITY STRIP

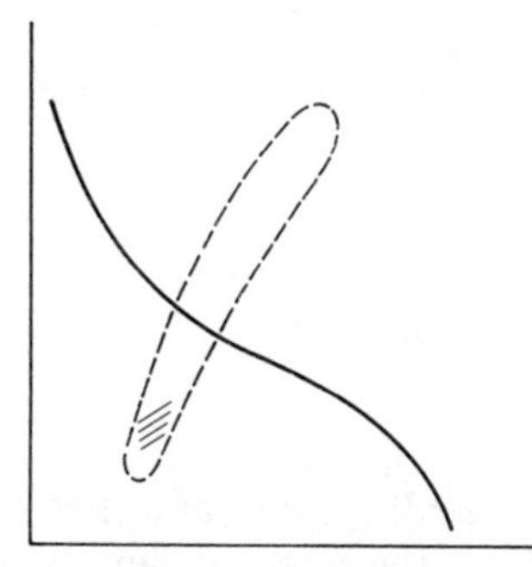

We have followed the instability strip across the HR diagram from upper right to lower left. Below the main sequence the HR diagram contains first the sub dwarfs and finally the white dwarfs. We have already used a variable dwarf as an example in §6.2.

The white dwarfs, one endproduct of stellar evolution, are five to ten magnitudes intrinsically fainter than our Sun and hence most are rather faint for small telescopes; only a handful have apparent luminosity < 10. The variable white dwarfs, which lie on an extension of the instability strip, are generally called ZZ Ceti stars. They have periods in the range of 2 minutes to perhaps 16 minutes and most (perhaps all) are multiperiodic. 20 to 30 different periods have ben found in one star! This should not be taken as an indication of extreme complexity, but as a reflection of the high quality data that can be obtained. The periods in the white dwarf pulsators are so short that very many cycles can be observed in a single 8 hour night, thus greatly helping the frequency analysis. Data runs of sufficient length in, say, the δ Scuti stars to give the same frequency resolution might well show similar complexities. (A 2 minute period observed continuously for 4 hours, which is quite possible, is equivalent to a 1 hour period observed continuously for 5 days which needs intense international cooperation.) The variations in the ZZ Ceti stars seem to be all of the non-radial type. We will not deal with these variables here in any great detail as their faintness means that, even with a 50 cm diameter telescope, photon rates of only a few hundred a second will be typical. Moreover, automated data recording is essential, given the short periods. White dwarfs would not be a first choice for most amateur astronomers.

Chapter 15

Variability in binaries

Many stars form part of a binary (or more complicated) system, and there are many ways in which the orbital motion affects the intensity and/or spectrum of the light. Some of these ways were indicated in Chapters 7 and 8. Observations of eclipsing binaries are popular, to the extent that *Sky and Telescope* publishes a monthly list of when to look for minima. From the observational point of view many such systems typically have comfortably large amplitudes, and convenient periods of days or weeks, so it would be merely repetitious to describe their measurement. The astrophysical interests for the amateur are in, for instance, mass determination, detection of third (or fourth) members of a system, and peculiar systems such as ε Aurigae, which has a long eclipse every 27 years. Close binaries in which matter is transferred from one star to another are of special interest, and also present new observational problems, so we shall concentrate on them.

15.1 CATACLYSMIC VARIABLES

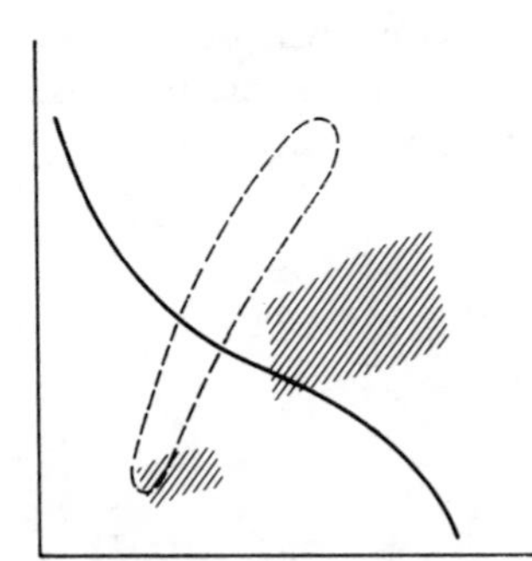

As mentioned in §14.1, many requests for data from AAVSO, the American Association of Variable Star Observers, concern cataclysmic stars. This is a rather general term but we will interpret it to mean large outbursts arising from mass transfer in binaries. (Flare stars, which are not binaries, are covered in §13.6.) As indicated in Chapter 8, the astrophysics of mass transfer in a binary is too complicated to handle from first principles. It has to be semi-empirical, so it must be guided by data which are good enough to sort out the effects of eclipses, mass transfer, intrinsic rotation and local hotspots. Any or all of these may be present simultaneously but if new physics of, say, thermonuclear reactions or black holes is to be extracted from these observations, the separation of the different processes must be on a sound base. Because of the irregularity of outbursts, cataclysmic stars present a more difficult task for the observer, for two reasons. Long periods of observation are desirable but, for different reasons, both professional and amateur find this difficult to achieve. Secondly, analysis is more difficult, without an underlying periodic variation to marshall the data. Unresolved binary stars generally show, and are recognized by some variability, of the types described in Chapter 8. It may be small periodically varying Doppler shifts in spectral lines, due to the orbital motion, or it may be striking dips in luminosity in eclipsing binaries. In close binaries, matter transfer adds a different type of variation which, to us as distant observers, can be very revealing of the processes involved. Close binaries are objects of exceptional interest. If astronomers were allowed to do experiments with stars a popular one would be to put a white dwarf or neutron star next to a star being studied, and watch as matter was pulled over to the accretion disc from which, due to its instabilities, some blobs of matter would become detached and would crash onto the surface of the compact star. This is indeed the case—or at least the most popular scenario for—the various types of cataclysmic star. The most spectacular type is the nova, which shows a sudden brightening of about 10 mag. Such a large increase is thought to be a violent energy release due to the ignition of nuclear fusion reactions in the transferred matter. The rate of this rapid rise is therefore of great interest. There are few measurements of it because the precursor star is often too dim to have been noticed. Of even more interest are the subsequent decay and, commonly, oscillations in the light curve of the nova as it returns to obscurity. The peak absolute luminosity of such events is thought to be $\sim M_v = -8$, and they have been seen down to $m_v \sim 14$, that is, out to ~ 1 kpc. Indeed, recurrent outbursts have been seen in some less violent systems. In general, however, it is not known which stars will become novae, so it is almost pointless in keeping up observations of a star, hoping it will be a nova one day.

There can also be spasmodic energy release in the *accretion disc*, not on the surface of the compact star. This is gravitational, not nuclear, and it is a more common process, there being perhaps a million such systems in our Galaxy, though only a few hundred are close enough for study. It is less spectacular, giving a Δm_v of perhaps 5, corresponding to a power range of 100 times less than that in novae. Nevertheless, it is easily measurable and such dwarf novae are, in fact, of greater scientific interest than novae since they can reveal the structure and physics of the accretion disc itself. There are enough known to make it relatively easy to choose a suitable one for measurement. Frequent measurements are needed during the whole duration of the pulse to find its shape. For instance, Z Cam stars have some feature in the process which causes a '*standstill*' (plateau) during the decay. SU UMa dwarf novae show two different types of outburst, a short (1–2 days) and a long (10–20 days) 'super outburst'. SS Cygni, of which a particularly fine series of measurements is shown in figure 8.4, exhibits a more subtle gradation of outbursts, which can be matched up with features in more sophisticated models of the disc. There are correlations between

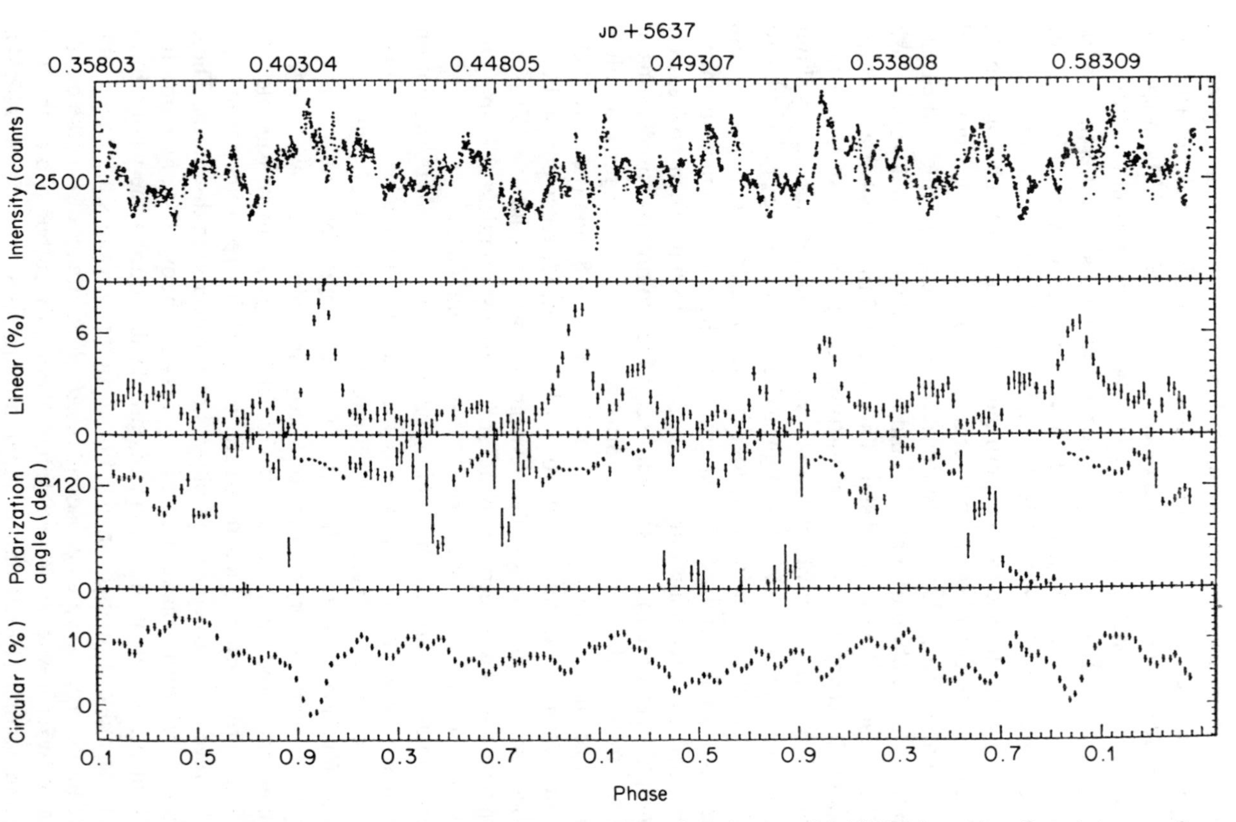

Figure 15.1 A variable star emitting polarized light, EF Eri. After Cropper M 1985 *Mon. Not. Royal Astron. Soc.* **212** 709. Reproduced by permission of Blackwell Scientific Publications Limited.

total burst energy, time between bursts, and decay times. The SS Cygni measurements were made at Harvard Observatory between 1896 and 1933. Notice that the magnitude scale is from 8 to 12. This is not a particularly difficult range—a 30 cm telescope, but with a modern photometer, could do as well as Harvard Observatory did 80 years ago. It is the continuity of the measurements over the long period which makes them so valuable. During dwarf nova *minima*, periodic oscillations, presumably reflecting the orbital motion, can sometimes be seen. They can be quite rapid. The fastest are the AM CVn stars, with periods of about 20 minutes which means of course that one can make a useful set of observations and still be in bed by midnight! During an *outburst* however, the light from the stars is masked, by a factor of several magnitudes, by the light from the accretion disc, so at that time one is, in effect, viewing pure fireball.

Some white dwarfs have very strong magnetic fields ($\sim 10^4$ T $= 10^8$ Gauss) and those in cataclysmic binaries are no exception. Such strong fields shape the plasma out to many stellar radii. In these cases there is not an accretion disc, rather one must imagine a tornado-like funnel of matter twisting down to the white dwarf (see figure 8.3). Whilst this introduces a further parameter into the description (the magnetic field) it also provides the observer with at least three more possible measurements—linear and circular polarization and Zeeman splitting of any emission lines (see figure 15.1). To measure the splitting needs an expensive high-resolution spectroscope but polarimetry is within range of the amateur. Polarization is imposed on radiation produced as magnetobremsstrahlung (synchrotron radiation). At orbital positions such that the spiralling particles are seen head on, linear polarization is observed, but at other positions the radiation is circularly polarized, to a degree dependent on the angle. The sense will change from left to right hand as the line of sight passes through the head-on position. The wavelength dependence of the polarization indicates the thickness (i.e. optical depth) of the accretion column. On balance, the extra complication of the magnetic model is more than compensated by the extra information! Figure 15.1 shows an actual case, EF Eri, measured at the South African Astronomical Observatory (SAAO). A polarimeter, as mentioned in §10.5.2, need not be a particularly complicated device. Their use could well open up many new fields of research for users of small telescopes. Polarizers and wave plates are commercially available, and can be rotated with a stepper motor. A cheap set could be obtained for less than £100. Whilst obviously not up to the quality and precision of the South African instrument, and perhaps needing more frequent calibration, they could yield extremely useful results. Readout, however, would have to be automatic. Both the angle and the count have to be recorded, and the integration time may be very short, since many of these magnetically active stars have short periods. Zeeman splitting can allow an absolute measurement of the magnetic field. However, in the violently turbulent conditions near a white dwarf, this is often masked by Doppler broadening.

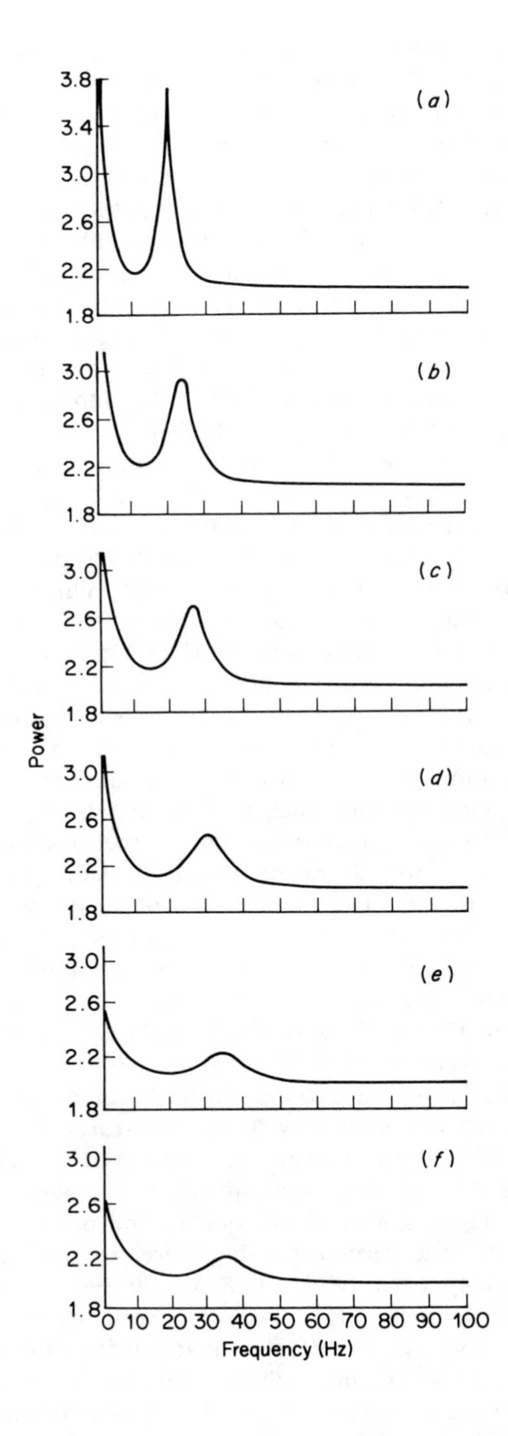
3.8
3.4
3.0
2.6
2.2
1.8
(a)
3.0
2.6
2.2
1.8
(b)
3.0
2.6
2.2
1.8
(c)
Power
3.0
2.6
2.2
1.8
(d)
3.0
2.6
2.2
1.8
(e)
3.0
2.6
2.2
1.8
(f)
0 10 20 30 40 50 60 70 80 90 100
Frequency (Hz)

Some cataclysmic variables vary so rapidly that it seems that they must be powered by accretion onto a neutron star rather than a white dwarf. The x-ray pulsars, as mentioned in Chapter 8, are examples of this. A class of particular interest shows 'quasiperiodic oscillations'. They are best revealed on a frequency power spectrum (see figure 15.2 and Chapter 11), where they appear as a definite peak on a continuum of other frequencies. In some cases the amplitude varies, and the peak frequency varies with it. Notice the high frequency—over 10 Hz. Such a fast variation obviously needs sophisticated automated recording. It may be necessary to record the arrival time of each detected photon.

15.2 MASSIVE BINARY STARS INCLUDING X-RAY BINARIES

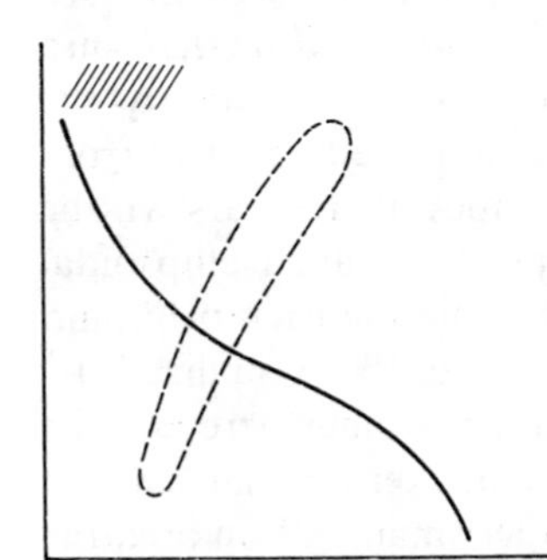

At the top left of the HR diagram, on or adjacent to the main sequence, lie other very luminous stars. They have similar absolute magnitudes to the brightest stars (see §§13.1 and 13.2) but the surface temperatures of these stars are ~ 30 000 K and they achieve these high luminosities with diameters of perhaps only 10–20 times that of the Sun, while the hypergiants are about 10 times larger again. We have taken the somewhat unorthodox route of choosing to group the two types of stars in the heading together and this needs to be explained. In order to describe the kind of work which can be done on these stars we are going to use as an example Cygnus X-1, a binary which contains a 'black hole' candidate. We will show that this star exhibits some unusual variations, but the question which is not answerable at present is which of these variations are linked to the x-ray emission (and therefore to the possible exotic nature of the secondary) and which of the

---

Figure 15.2 An x-ray source called GX5-1 was found to have rapid variations, but if all the data were taken together it was very difficult to pin down the frequency. However, there was a slow drift in the luminosity of the source and the analysis was tried again for different intensities. These curves show the power spectra at different intensities, increasing from the top to the bottom. Now there are clear peaks, but with the unusual feature that the frequency at the peak increases with intensity. It is this feature which gives rise to the name quasiperiodic oscillation. Such QPOs have been found in other sources. The underlying mechanism is thought to be a beat between the period of the disc and the compact star. Such striking features may lie hidden in optical data too. Data from van der Klis M *et al* 1985 *Nature* **316** 225.

effects would be seen in massive binaries which do not emit x rays. Professional telescope time is often allocated because some object is currently exciting or interesting but equivalent amounts of time are not easily granted to observe the 'controls', which is another example of how modestly sized telescopes can make significant contributions. It is not necessarily exciting to spend several decades observing 'normal' stars and large telescopes will not be used for it, but without such data we cannot know what aspects of the behaviour of the 'exotics' are normal and which are not. This is exactly the point made at the beginning of this chapter about needing data good enough to separate the various effects. For the rest of this section we will be concerned with subtle orbital effects rather than the dramatic outbursts.

The stars we are concerned with here will typically consist of one star with a mass of 10–30 $M_{\odot}$, with a secondary of similar or slightly lower mass. The orbital periods are days to tens of days and the proximity of the two stars to each other causes their surfaces to be *tidally distorted* into ellipsoidal (rugby football) or teardrop shapes. The effect is the same as the tides on the Earth, i.e. one tide towards the Moon and one tide away from it. The light variations caused by the elongated shapes of the stars will be present *whether or not eclipses occur as well*, and are called ellipsoidal variations. They will produce two maxima and two minima each orbit and the light variations will be smooth. The effect can be visualized by imagining a rugby football being rotated about an axis somewhere near its shortest diameter. One will see a large side-on area, then a small end-on area, another large area (side-on) and finally another small end-on area for each 360° rotation of the ball. (This and some of the other effects were described in Chapter 8.)

In all the binary systems that we are concerned with here, mass is being lost from the surface of the primary, and perhaps from both stars. The mass loss rates are much lower than in WR stars, perhaps 10^{-7} $M_{\odot}$/year, but can nevertheless cause observable effects. Some mass is lost from the system; some mass is transferred to the other star. Sometimes a third star will be involved, and it can be very difficult to distinguish between what seems to be a new physical process in a binary system and the relatively uninteresting complications of a third star! Sometimes, the transferred mass forms a stream which gathers in a ring about the receiving star. The incoming gas stream may cause a moving 'hot spot' where it hits the ring. Sometimes material spirals slowly out from the binary like material thrown out by a huge catherine wheel: there are endless combinations of all these effects.

The observer's task is to make observations which help us to visualize exactly what is happening in these systems and thus help understand how stars evolve. Once again, like the WR stars, many of the stars of this type might be intrinsically variable and thus patience and time are needed to distinguish between the intrinsic and rotational effects. If one wants to know the shape of the orbital light curve, but such variations are mixed with intrinsic pulsational changes, then clearly it is necessary to have very many observations. For instance, if the intrinsic pulsations are short one may average over them to eventually show the underlying orbital changes. Once one has an idea of the frequencies involved, a fit can be made to search for radial or non-radial modes of pulsation. The residuals may be genuine chaotic variations, indicating perhaps flaring activity in the system. This procedure of identifying individual underlying causes of variation and slowly peeling away each layer of knowledge to reveal yet a further source of variations is a 'reductionist approach' and before we describe the Cygnus X-1 results we will explain the procedure further. It is inherent in the reductionist approach to light curve analysis that one supposes that, at least to a first approximation, the various types of variability are *independent*. Thus, for example, if one has a pulsating, eclipsing star one is able to independently check the eclipse curve for period variations using an $O - C$ diagram and at the same time to remove the eclipse curve effects and look for changes in the phase or amplitude of the pulsations present. The alternative approach to the analysis is to put all the numbers into a period finding program and to search for strictly periodic phenomena. With some types of variability this can work well. If the processes are not independent because they interact with each other or are non-linear, it may not work. The problems of isolating basic mechanisms and the interactions between them, and the difficulty or impossibility of predicting non-linear processes, crop up repeatedly in all types of scientific analysis. It has been realized in the last few years that some basically simple non-linear processes, including oscillations, behave in an unpredictable way, now known by the name 'chaos'.

The optical counterpart of Cygnus X-1 is the star HD 226868. It is a ninth magnitude giant star with a surface temperature near 30 000 K (type B0 Ia). The mass of the primary component is 20 to 30 solar masses. The secondary is not visible but this is not unusual as the light from many normal stellar companions would be swamped by the brilliance of the primary. The mass of the secondary star is deduced to be in the range of from 5 to 15 solar masses from its effect upon the primary. The stars orbit each other every 5.6 days and there are orbital light variations of the type to be expected from tidal distortion of the primary by the close proximity of the secondary. The spectrum of the star shows no abnormalities and therefore the whole binary system is typical of many others which are known in our Galaxy. (Remember that stars such as this are so bright that they can be seen right across our own Galaxy and even in other galaxies.) Were it not for the fact that the star emits large amounts of x rays (in fact more energy in x rays than all the energy from our Sun), then it would have received the same scant attention that many similar stars have received. What particularly sets this star apart is that the mass of the secondary is so high. This in combination with the x-ray emission suggests that the secondary is a 'black hole'. If this could be proved it would of course be a most important discovery. 'Black holes' are still a theoretical prediction, able to give explanations of various astronomical phenomena, but which are still not proved to be the true explanations. The observations described below were made with a 30 cm (12 inch) diameter telescope and show that even ultimate forms of matter in the Universe such as 'black holes' can be investigated with modest equipment.

The observations are all B band using only one star, η Cygni, as a comparison (two other comparison stars were both found to vary, at the accuracy of these data). Observations began in 1972 and it was soon found that the star showed $\sim 4\%$ variations, two maxima and two minima, each with a 5.6 day orbit. Later it became clear that there were often significant departures from the mean light curve, but the data were not adequate to allow a decision as to whether these departures were periodic or were due to intrinsic irregular variations which are suspected to occur in many early type giant stars. It took several years before the mean orbital light curve was well enough defined to allow the departures from it to be accurately determined. Once this was possible the following questions could be addressed.

(1) What is the shape of the orbital light curve?
(2) Does an $O - C$ diagram based upon the mean light curve at different epochs show evidence for period change?
(3) Do the residuals from the mean light curve have any characteristics which would indicate their nature or their source of origin, e.g. primary star pulsations, gas streams, hot spots, etc.

Rather than work through the details we will only present the results, assuming a reductionist type analysis, based upon many years of data.

Figure 15.3 shows the mean light curve. Note how it departs from a sine wave and how the two minima are separated by $\sim 45\%$ and $\sim 55\%$ of the orbital period. Figure 15.4 shows the $O - C$ diagram from 12 years of data. There are two things to notice. One is that the period is decreasing, and the other is evidence on two separate occasions for a sudden discontinuity, or glitch, in the $O - C$ values. Both these glitches are associated with changes in the strength of the x rays given off by the star.

Next we show a series of light curves, all of which were obtained during a time when the x-ray strength was unusual and which were derived by looking at the residuals after the mean light curve had been removed. In order to emphasize the continuity and evolution of this residual light curve we have arbitrarily shifted these curves in orbital phase so that the deepest minima are aligned. In figure 15.5 you will see how this curve has a characteristic sharp minimum, broad maximum with a secondary minimum and how slowly the amplitude of this curve decreases. Finally, in figure 15.6 we show the real phases of both the deep and the shallow minima. You will see how they slowly drift in phase but over thirteen orbits return near to

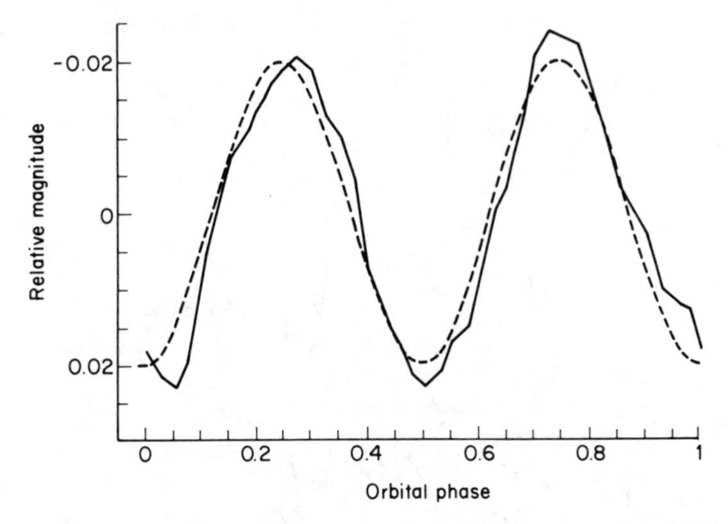

Figure 15.3 The mean light curve of the visible component of Cyg X-1. It is not quite sinusoidal in shape—the first peak is faster than the second.

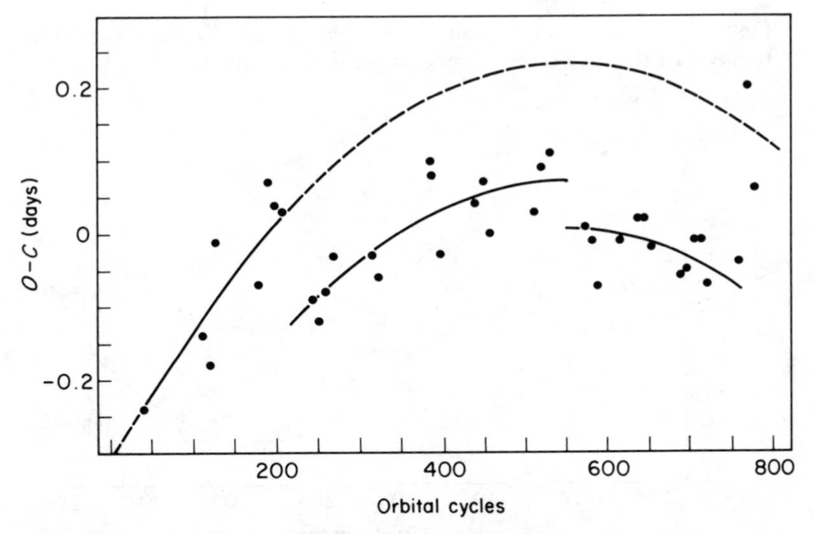

Figure 15.4 The $O-C$ diagram from data taken over 12 years. The period is decreasing, but there are also 'glitches' at which there is a slight but sudden increase. The timing of the glitches corresponds with changes in x-ray intensity of the other star.

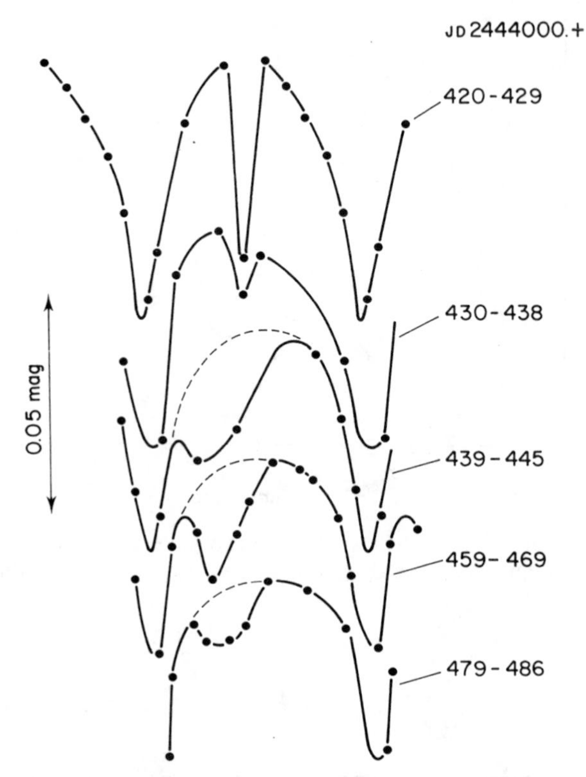

Figure 15.5 The changing shape of the residual light curve after removal of the long-term average shown in figure 15.3.

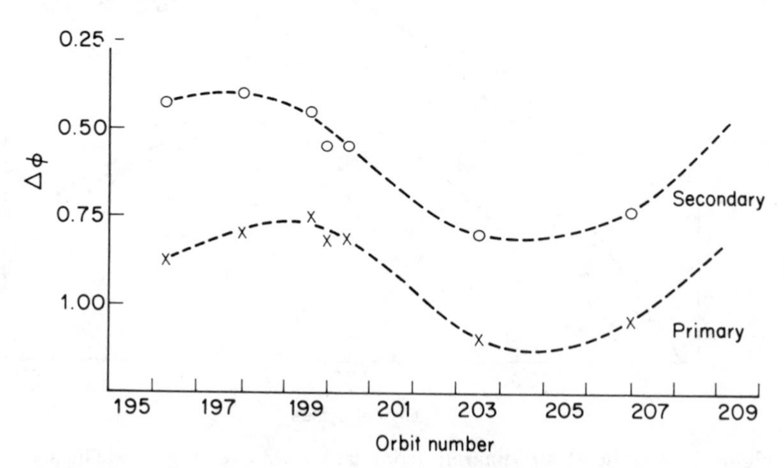

Figure 15.6 The phase of the minima which drift with respect to orbital phase. The diagram shows that the drift is steady and systematic.

their original positions. More features appear in the analysis but the above should suffice to show the kind of effects which can be looked for. It is important to understand that the only reason that this type of analysis has worked is that we are clearly dealing with several independent phenomena and that a change in one (e.g. the shape and phase of the residual light curve) has little effect on the underlying mean orbital light curve.

Several massive binary stars are known which emit x rays; many more are known which do not. Many of the non-x-ray binaries show 'ellipsoidal' orbital light variations, with or without the addition of eclipses, therefore readily lending themselves to an $O - C$ analysis. Whether or not glitches in the $O - C$ values are common even without x rays, whether periods increase or decrease, whether residuals show meaningful curves or random scatter, are all questions that can be answered by the patient acquisition of high-quality photometry over several years. This is necessary in order to identify which special effects are associated with the x-ray emission. If the data can be obtained in more than one colour then that would be a major bonus; for instance, it is frustrating to be able to see the residual light curve evolving in figure 15.5 but not to know whether this source of extra light is hot (blue) or cool (red)—or indeed whether its colour changes as its intensity decreases.

The spectral types of the massive x-ray binaries range from O down to the middle B types. In numbers, observers in the northern hemisphere are disadvantaged compared with their southern colleagues. The reason is that our planet's northern hemisphere faces away from the Galactic centre, while the nearest spiral arm and concentration of O and B stars lies between us and the Galactic centre. However, while the majority of the x-ray-emitting massive binaries are in the south there are still some in the north, including Cygnus X-1, and enough *non-x-ray* emitting binaries to allow work on the controls to take place in the north.

Work on this type of object is demanding and probably best arranged as a joint project between several members of a society. You and your colleagues should be prepared to maintain the project over several years (even decades). You will need to be able to work for about one hour per star on each photometrically clear night (only about a dozen a year in the UK if you do not have a comparative photometer) in order to get the high accuracy which is required because of the low amplitude of the light curves. If you can possibly set up a collaboration with observers at different longitudes it would be scientifically rewarding and personally satisfying. It is one stage better than putting data into a data bank, hoping that compatible data will appear.

15.3 RS CVn STARS

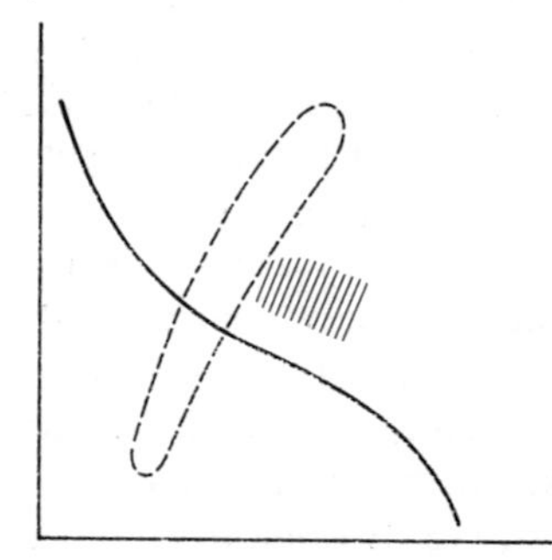

During total eclipses of the Sun it is clear that above the Sun's photosphere with its temperature of about 6000 K is a much less dense but hotter region of gas. This is called the chromosphere and has a temperature of about 20 000 K. Further out still exists the Sun's corona which is at a temperature of several million degrees. This could not happen if the only source of energy was in the centre of the Sun. It is thought that energy is being fed into these outer layers by shock waves, or streams of gas, which provide collisional energy.

We have already seen (in §13.3 for instance) that gas at 20 000 K emits most of its energy in the ultraviolet region of the spectrum. Gas at millions of degrees radiates x rays and it is therefore no surprise that the Sun emits both UV light and x rays, although the Earth's atmosphere shields us from most of these. The reason that the giant solar flares, already used as analogies in our description of the flare stars, are so easily visible in ultraviolet pictures of the Sun, taken from satellites above the Earth's atmosphere, is that they are so hot and emit so much of their energy in the UV.

However, the total amount of energy contained in the UV and x rays emitted by these outer layers of the Sun is very small compared to the visible light emitted. It was therefore a surprise to find, when UV and x-ray satellites were launched, that many K and M stars (i.e. with cool photospheres) are much more energetic at these short wavelengths than the Sun. It was soon found that many of these stars were double and in fact were known to be variable. Thanks largely to astronomer Doug Hall of Vanderbilt University, Nashville, these stars have become one of the star types most assiduously observed by American amateur astronomers, using photoelectric photometers. They have become known as RS CVn stars after one of the brightest of their members.

Typically, an RS CVn star consists of a late G or early K subgiant star with an F or G main sequence (luminosity class V) or slightly evolved (luminosity class IV) companion. Their orbital periods are typically in the range of from a few days to a few tens of days and spectroscopically they are noticeable for their strong emission lines from ionized calcium. Since the stars are binaries, one expects regular light variation due to orbital motion but it soon became clear that the variations were not repeatable. There was something else present which caused variation.

Visual observations from the ground combined with simultaneous UV and x-ray observations have now built up a fascinating picture explaining some of what is happening. Many of the stars show eclipses and this has allowed us to try to map the surfaces of these stars as the movement of one star across the surface of the other slowly obscures, and later reveals, details on their surfaces. The picture which emerges is bizarre; no wonder it has captured the imagination of many US amateur astronomers.

The lack of repeatability of the optical light curves can best be imagined as a sine-wave distortion which over a period of many years moves backwards through the underlying light curve. In RS CVn itself the orbital period is about 4.8 days, while it takes about 10 years for the distortions to move once through the orbital light curve. It seems that the surfaces of the stars are covered in huge, long-lived spots (or spotted regions). These should not be thought of as being like sunspots but rather as dark areas covering appreciable fractions of the star's surface.

The movement of the spot relative to the orbital period is thought to be due to the differential rotation of these stars. (This is well measured in the case of the Sun; lower latitudes rotate more slowly than higher latitudes.) The spots presumably occur in regions which rotate more slowly, but only about 0.1% (4.8 d/10 yr), than the orbital period of the close binary system.

The UV data show that above these regions are huge flares or prominences, much larger than those on the Sun. In some cases they reconnect back onto the surface of the same star while in other cases there is some evidence that the flare reaches from one star to the other in the binary. Imagine two white-hot balls of iron rotating about each other with a huge spark, like an electric welding arc, playing between one star and the other and you have some idea of what one would see from a planet going round them. The dark spot areas may be the footprints of these huge arcs.

In some of these stars the spots seem to be very stable and long-lived phenomena. In the star λ And it seems that one spot has remained identifiable for 50 years while in RT Lac there is some evidence that the same spot has persisted for 80 years. There seems to be clear evidence for change in a few months. Studies of some of these systems during eclipses has allowed the investigation of the size and temperature of the spots. In the Sun, spots are small (compared to its diameter) covering typically $< 1\%$ and have temperatures about 1800 K less than the brighter part of the star's surface. In the RS CVn stars the spots cover perhaps 10 to 20% of the star's surface but are only about 1000 K cooler than the rest of the stellar surface.

Given all this new data, what can the user of a small telescope contribute? Note first that these are rather cool stars and that therefore VR and I band observations would be expected to help, particularly with regards to identifying the even cooler spot regions (temperatures near to 3500 K). People with solid state detectors are therefore well placed to work on these stars. The amplitudes of the variations are large, typically several tenths of a magnitude, so that the intrinsically poorer accuracy of the solid state photometers will not be too much of a problem. In the USA several automatic telescopes, built by amateurs, are now working on these stars. A good light curve covering the orbital period of several days is required, and this needs to be repeated throughout several years to watch the slow migration. Note that only one, or a few, orbits should be taken to obtain the light curve as the slow migration of the spot will cause changes between light curves obtained months apart.

There have been some attempts to create computer models of the stars' surfaces and to use these to decide what type of data would be most useful in trying to make accurate models of the real stars. They confirm that VR and I photometry will be more useful than UB and V. It has also been discovered that at any one epoch it will be necessary to have complete phase coverage of the whole orbital light curve, which will mean superimposing the data taken over several nights. It is important that the light curve is defined everywhere to an accuracy of better than 10% of the total amplitude of the light curve. Finally, only those stars with inclinations of their orbital axis of about 70° to our line of sight will provide good models. Both above and below that value the models will become progressively less well constrained.

It would be wise to contact a professional expert on these stars before starting work so that your work can complement those already working in this area. A good first contact would be Doug Hall, Dyer Observatory, Vanderbilt University, Nashville, Tennessee 37235, USA.

Chapter 16

Minor planets and comets

Both asteroids and comets store information about the origin of the Solar System. Asteroids are a storage place because they have no atmosphere to alter their surfaces, and because they have few collisions. As was explained in Chapter 9, such collisions that they do have are a potential source of information. It can be unravelled by looking for the different spin rates about the three different axes which an irregular body such as an asteroid can have. Most of the comets we see on the other hand are short lived. Their tail, an atmosphere of sorts, only appears near the Sun, and is very variable. It is now known from satellite observations that comets crash to their death in the Sun. How then can they store information? Only if, as is currently generally believed, the vast majority of them reside in an '*Oort cloud*' very distant from the Sun (thousands of AU), with an occasional comet being perturbed into an elongated orbit with a small perihelion distance. The comets in the Oort cloud are, in this model, primeval. In this sense the asteroids and comets both carry primeval information. The methods of observation are, however, quite different, so we will take them in turn.

Since the orbits of asteroids lie (with a very few exceptions) wholly outside the Earth's orbit, their reflected luminosity varies with position, being a maximum when the Sun, Earth and asteroid are in line, in that order. Then the asteroid shows its full, albeit crinkled and irregular, face to us though unfortunately we only see a point image, since all asteroids are small. The (absolute) luminosity of an asteroid is defined as the luminosity it would have if viewed from the Sun (i.e. in opposition at a distance of 1 AU). It is therefore always necessary to apply a correction factor equal to the square of its distance in astronomical units. A second correction factor is needed for the angle of viewing, usually called the *phase angle* and denoted by α (see figure 16.1). Since an asteroid is rarely seen exactly head on, the next best thing is to measure the apparent luminosity at various angles α and then, numerically or graphically, extrapolate back to $\alpha = 0$. A complication arises, because at very small angles the curve is non-linear, rising sharply by about 0.32 mags. This is assumed to be because the surface is granular, and we see the net effect of shadows of myriads of grains, except in the exactly head-on position. If the head-on position happens to be the pole-on position then no variation of reflected light will be seen as the

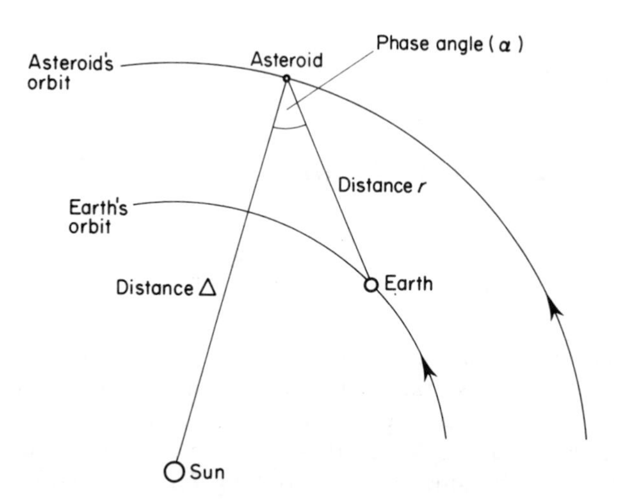

Figure 16.1 The definition of the phase angle for asteroids (and planets). The apparent luminosity depends on the phase angle—as is obvious in the corresponding case of our moon.

asteroid rotates. As an example this appeared to be the case with 8 Flora on its 1954 opposition. However, by 1955 this had changed and its rotation was evident by a periodic variation of amplitude 0.04 mags. Measurements in 1984 by Andrew Hollis, an English amateur astronomer, gave a value of 0.070 ± 0.005 mag (8 Flora). Clearly asteroid observations need to be repeated at as many oppositions as possible. The relation between the period of the light curve and the period of rotation depends on the orientation of the axis. Imagine an ellipsoidal (or more realistically, potato shaped!) asteroid rotating about the long axis which happens to be perpendicular to the line of sight. If there is a relatively dark (low reflectivity) area on part of the asteroid it will cause a dip in the light curve, once per revolution. The period of the light curve and the period of revolution will be the same. If, however, the rotation is about the short axis and that axis is perpendicular to the line of sight, then the potato will show a maximum surface area twice per revolution, so that the rotation period will correspond to twice the light curve period. Since the asteroid is not resolved, these two different conditions have to be distinguished by the shape of the light curve. Simulated light curves calculated for various assumed shapes and axis orientations are a help. Ambiguities can in principle be resolved by the form of the variation of the light curve as the phase angle changes, but it is not easy. Periodic variations of isolated asteroids can only be due to rotation, bringing different faces into view. They can have up to three different periods, corresponding to the three axes, as explained in Chapter 9. A purely observational problem arises because the asteroids move through different star fields during the course of a few days or weeks, and this means that different comparison stars have to be used. In stellar photometry one seeks to use the same reference stars, but in asteroid photometry this is unfortunately not, in general, possible. The reference stars should, if possible, be of spectral class G since that, after all, is the type of light being reflected by the asteroid.

The differences between the reflected spectrum and that of the Sun give information on the asteroid surface. It is assumed that there are three basic types of asteroid, mirroring meteorite types; stony, stony iron and carbonaceous. More advanced classification schemes have, however, been devised. Identification of different properties for different classes would give valuable clues to the formation mechanisms.

The fact that the most common principal period of rotation is near 8 hours means that spurious periods, or confusion between the period and a half period, is liable to arise. A good way to avoid it is to achieve near continuous observations, pooling the efforts of astronomers at different longitudes. (An ideal way would be to make observations from the appropriate pole of the Earth or from a satellite.)

No convincing cases of binary asteroids have yet been reported. (A genuine case would be very exciting, yielding, as with a binary star, radius and mass estimates.) The only direct way of measuring radius, therefore, is by *occultations* of stars by asteroids. Several potentially useful occultations are predicted to occur each year. The problem is that the shadow area is both narrow and difficult to predict exactly. For instance, to be sure of seeing an occultation by an asteroid 25 km in diameter, its orbit would need to be known to within 25 km and this is rarely the case. Nevertheless, several occultations have been effectively observed. It needs a lot of people spaced out in a line across the predicted path with many of them resigned to missing the event! Accurate timings from those who do see the occultation gives the lengths of a set of chords across the asteroid, after correction for the motion of the Earth and asteroid. Combined with the luminosity this gives the albedo.

Although asteroid astronomy is often regarded as the 'Cinderella' of the subject, its importance has been recognized by the fact that there are several proposals for satellites to fly by, or even orbit, asteroids. One proposal, accepted in principle, would land a probe on an asteroid. Even if this were to lead to asteroid mining operations, which is conceivable, only a very few of the 3000 known asteroids will ever be visited, so statistical information will always have to come from ground-based astronomy. Because of the Cinderella effect, this means amateur astronomers. A visit from a space probe can be relied upon to increase the level of interest in asteroids.

This has already happened for comets. The visit of Giotto, in close cooperation with Russian and Japanese spacecraft, to Halley's comet in 1986/7 certainly stimulated a great deal of public attention. While a comet is in the Oort cloud, it is rather like a small asteroid. It has no tail at that stage, this only being generated during the passage by the Sun. It does have different composition from a minor planet, containing a lot of water and other ices, but in appearance it could be mistaken for an asteroid. It is the evaporation of these components that generates the tail. An interesting mechanical similarity is that comets are thought to have a similar spin period to asteroids—about eight hours again. If comets could be detected in the Oort cloud, everything said above about measuring asteroid spin rates would apply to comets. However, we only see comets when they are near the Sun and the cometary nucleus is masked by the light reflected by the gases and dust forming the tail or tails. This makes it very difficult to detect the light curve due to the rotating nucleus. A good technique is to take measurements with several different apertures, looking for a more marked periodic variation with the smaller apertures. Ideally these measurements should be simultaneous (using a beam splitter and two channels), since there are also changes of brightness caused by sudden release of gas. The apertures should be small, say 10 arcsec and 100 arcsec so this demands very good guiding. Again this is not easy since the comet moves through the stellar background.

The gas and dust in the tail have noticeably different colours, so broad band spectrometry is adequate to distinguish them. The gas and dust tails separate under the influence of the pressure of the solar light and the solar wind. This can be a very satisfying area for amateur study. High-resolution spectroscopy can give a great deal of information about the changing composition of the tail, but we have considered the equipment required to be beyond the scope of an amateur observatory.

Finally one can look for new comets, rather than studying known comets. This has long been the province of the amateur. Large aperture binoculars with modest magnification give a bright image which makes searching easier. This is the only equipment needed, apart from the means to keep warm and comfortable for many hours, perhaps in a semireclining position. In searching for comets one may look for fuzzy objects. There are, of course, fuzzy images of other objects; nebulae, galaxies or star clusters for instance. Messier's catalogue was originally produced as a list of objects which comet hunters should ignore, because they were known not to be comets. Nowadays a good atlas performs this service. In the case of doubt, one has only to wait to see if the fuzzy image moves. The motion of a comet will be visible in a few days, or perhaps even an hour or two if it is close. Alternatively, one may look for unexpected point sources, which could be a comet before the tail is formed. Such an object could be a nova (which would be just as interesting!) if no movement is seen. If it does move, it could be a minor planet. Even a rough orbit determination could decide, since minor planets are nearly all in low eccentricity orbits between Mars and Jupiter, whereas a comet is never in such an orbit. An orbit can be calculated as soon as several positions have been reported, and it is important to report probable comets to the address given in appendix 6.

Chapter 17

Two more challenging projects

We wondered what to suggest if you want a more challenging project, perhaps demanding more equipment. How should one choose? Which is more rewarding, the technical challenge of finding effects others have not detected, or gambling on a project that might yield very important physics? This is a subjective choice, so we will give an example of each. The first is on finding whether the Sun is typical of its class, and the second is about tests of general relativity.

17.1 THE F AND SOLAR TYPE STARS

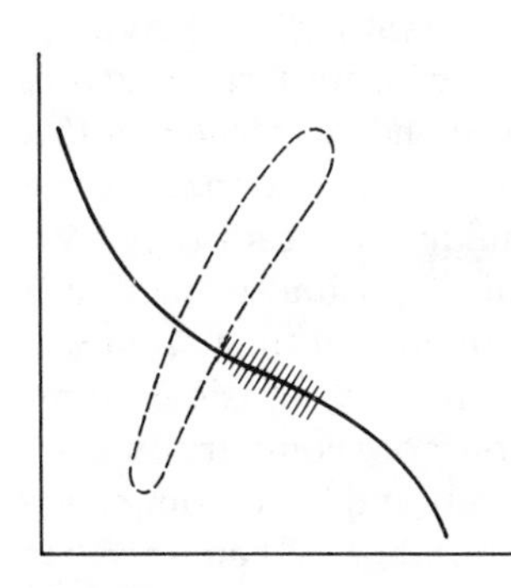

Upon continuing down the main sequence from the late A and early F type stars, which is where the instability strip crosses the main sequence, we enter the realm of the late F and G stars. Our Sun is a main sequence G star (G2V) and of course we know more about our Sun than we do about any other star. We might hope, therefore, that we could extrapolate from our knowledge of the Sun to be able to predict what kind of behaviour we should expect with these similar stars.

It might come as a surprise that the Sun is a 'variable' star. However, the variations are so small in amplitude that they are not obvious from changes in climate, in fact they are not easily detected. There are three types of variation which lead to photometric changes. The first is due to the '5 minute' solar oscillations. This is really a set of many hundreds of different pulsations, with periods ~ five minutes, which have been discovered over the last 20 years. The photometric amplitude of these oscillations is only about one part in one hundred thousand (10^{-5}). However, the Sun is much more obviously a variable in another sense.

Almost every one has either heard of, or seen, sunspots: dark areas which are associated with magnetic fields on the surface of the Sun. Both the total area and position of the spots changes with the well known cycle of about 11 years. How much is the total light of the Sun reduced when a large spot, or groups of spots crosses its surface? Of course, it depends on the size of the spot, or spot group, and its position, e.g. near the centre of the disc or near the limit, but certainly reductions in brightness of 0.25% (2.5 millimags) can occur. Therefore a sensitive enough instrument could, in principle, see a 2.5 millimag reduction in the brightness of our Sun from the passage of one of the larger sunspots across its surface.

At the moment we have no way of mapping the surface of a typical star like the Sun, nor even of determining either its rotational period or the length of its spot cycle. The exciting possibility exists of determining both the spot cycle length and the rotational period for solar type stars from accurate photometry alone. High-resolution spectroscopy can also do the same, but requires much larger telescopes and more expensive instrumentation. An ability to detect star spots would give us the chance of comparing the behaviour of our Sun with that of similar stars. By such means we might better understand how typical our Sun, and perhaps its family of planets, is. In turn, this might even give us some clue as to how we ourselves relate to the Universe at large. The crucial question is, therefore, whether it is realistic to search for such small variations in stars other than our star, looking at timescales of tens of days for the rotation period and years, or tens of years, for the spot cycles.

A start has already been made on doing it for a small number of stars by two American astronomers from Lowell Observatory, in Flagstaff, Arizona, Wes Lockwood and B A Skiff. Their results must rank among some of the most accurate photometry ever obtained and yet readers of this book will be pleased to know that their results were not obtained by the application of some new, multimillion dollar technique. Instead, they have resorted to the good practice recommended in this book and in fact have used rather old-fashioned equipment. Filters, photomultipliers, power supplies, etc, have all been left unchanged and whenever possible permanently switched on. They have always used two or more comparison stars of a similar spectral type to that of the star of interest and in so doing have found many of their comparison stars to be variable. When this has happened they have incorporated new comparison stars into the programme. They find that many of the stars vary slowly throughout each season and that some of those that seem constant within one year nevertheless show variation on a year to year basis.

Now, not many people will be able to observe in skies as good as those at Flagstaff, Arizona, so is there any hope of obtaining similar accuracies in poorer skies? Well, the answer is that the photometrically underprivileged have to be more subtle but the fibre-optic comparative photometer has shown what is possible on poor sites and it seems unequivocally certain that poor sites could be used for the same work given persistence and a comparative photometer. You are going to be looking for millimagnitude changes on timescales of days to years so it cannot be emphasized too much that you must work out your techniques early on in the programme and then stick to them. You must be prepared to use three comparison stars, rather than two. If there is any doubt as to the stability of any of your comparison stars and if you find (as you certainly will) that some of comparison stars are variable then you must use that as part of the data you are producing. You cannot, *a priori*, know the accuracy of your photometer. You might feel sure that it is good to one millimag but you cannot assume that everyone will believe you! The accuracy can be estimated in the following way. In any group of four stars, say, A, B, C and D, you can compare six pairs of stars: AB, AC, AD, BC, BD, CD. The most stable Δ mag of these gives you an upper limit to your accuracy; it might be better but you cannot demonstrate it. You can use that value to check for variability of the other two stars in the group. Note how the total number of pairs goes up faster than the numbers of stars in the group, e.g. 3 stars gives 3 pairs only, 4 stars gives 6 pairs, 5 stars gives 10 pairs and 6 stars give 15 pairs. Readers who live in good sites and who feel able to go round and round a group of say six solar type stars for perhaps one hour each night for years might be among the first to demonstrate the rotational period and star spot cycle of one of our Sun's close cousins.

There is one other type of variability for which you can search which is also seen in the Sun. It has been so recently discovered that its significance is not yet understood. It is only due to the launching of satellites designed to work for many years that it has been possible to discover the latter effect. Since the launch of the 'Solar Max' satellite in 1980 it has been found that there has been a tiny, slow decrease in the Sun's brightness by about 0.1% ($\sim$0.001 mag). How long this will continue, and whether it correlates with anything else, is not yet known. Several laboratory instruments are well able to work to the required accuracy and you might wonder how such a variation might have gone unnoticed until now. There are two reasons for this. One is the variable transparency of the Earth's atmosphere which renders accurate comparison of the Sun with laboratory standards impossible. The second reason is that, for this type of work, the great apparent brightness difference between the Sun and other star makes any direct comparison practically impossible. Neither of these restrictions applies to the intercomparison of several solar type stars in one group of the sky. In summary, the exciting possibility exists of determining both the rotation rates, spot cycles and long-term variation of solar type stars.

17.2 TESTS OF GENERAL RELATIVITY

We come secondly to a project which tests a very important piece of physics, and thereby has a very clear and useful aim. General relativity is fundamental to physics and some of its effects are not at all esoteric. For instance, the $O-C$ diagram for some eclipsing binaries can only be analysed by using general relativity. A case in point is DI Her; does it represent a good measurement project?

Most tests of general relativity are based on measurements in the Solar System. One test, the degree of gravitational redshift, can even be carried out on the surface of the Earth, though it is more accurate with rockets sent up from the Earth or in satellites. All agree, with the best precision being better than 1 in 1000. However, this only tests one of the axioms of general relativity, the principle of equivalence. There are other measurements which test the famous curvature of spacetime, an idea right at the heart of general relativity. These include the bending of starlight at eclipses, pioneered by Eddington, and now much improved in the form of the bending of quasar radio waves, measured using radio interferometers which have much higher precision than any optical telescope. Another test is the amount the perihelion point on the orbit of a planet swings round, due to the bending of space. The size of swing is higher for close compact objects and high eccentricity orbits, since, as with all general relativistic effects, it is proportional to the difference of gravitational potential involved. Mercury is the best bet in the Solar System, but it is not very good, since the predicted effect is only 0.43 seconds of arc per year.

It is easy to cite binary stellar systems which rate better in terms of speed and eccentricity. Many of the cataclysmic variables would win over Mercury. However they are, as stressed in Chapter 8, really three-body systems, with their internal motion substantially affected by tidal drag and/or mass transfer. These so complicate the motion that they are useless for testing general relativity. What is needed is a 'clean' system, where these complications are small, and also where the point of periastron can be measured accurately, and this means an eclipsing system. DI Her is thought to be such a system. Certainly it has beautifully sharp minima, and the $O-C$ diagram shows a very clear periodicity (figure 17.1(*c*)). (Incidentally, what is called perihelion advance for Mercury is called apsidal motion for stars, the two terms being equivalent.)

DI Her itself has been measured very well and there is not much hope that an amateur could compete. The result, 0.0065 ± 0.0018 degrees per year, is dramatically smaller than that predicted by general relativity assuming that the stars are small compact objects: the prediction is 0.0234 degrees per year. (Notice that these are degrees, compared with arcseconds for Mercury.) So does one say general relativity is wrong? Not immediately, because there are tidal effects and these must be estimated. That makes the disagreement worse, because they add to the general relativistic calculation for point objects, giving a total of 0.0427 ± 0.0030 degrees per year. The tidal effects can be, and in practice are, calculated using classical mechanics (an approximation to general relativity) so should one also ask whether

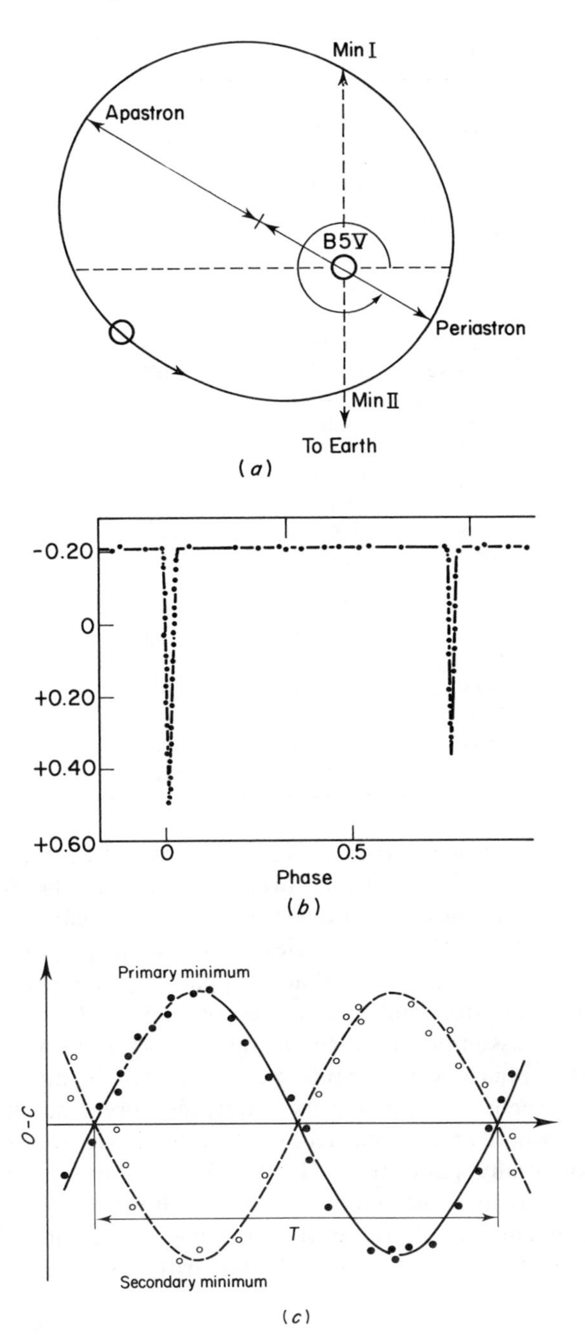

Figure 17.1 (*a*) The line of apsides. (*b*) The light curve for the eclipsing binary DI Her. (*c*) The measured $O - C$ diagram for DI Her. T is the time for a complete rotation of the line of apsides.

Table 17.1

Name	P (days)	Apparent visual magnitude
DI Her	10.550 17	8.3
BW Aqr	6.719 69	10.2
CD Aqr	4.837 72	10.1
V889 Aql	11.120 88	8.7
V459 Cas	8.458 29	10.3
EK Cep	4.427 80	8.2
TV Cet	9.103 29	8.7
V541 Cyg	15.337 95	10.2
V1143 Cyg	7.640 76	5.9
AI Hya	8.289 68	9.7
KM Hya	7.750 50	6.3
RW Lac	10.369 22	10.5
SS Lac	14.416 29	10.1
ES Lac	4.459 40	11.4
V345 Lac	7.491 86	11.1
RR Lyn	9.945 08	5.6
TZ Men	8.569 00	6.2
UX Men	4.181 10	8.2
EW Ori	6.936 80	10.4
GG Ori	6.631 47	10.8
VV Pyx	4.596 18	6.6
EO Vel	5.329 62	11.1
EQ Vul	9.297 16	11.5

classical physics is wrong? Now, it would not really be very sensible to discard all the accurate orbital measurements made in the Solar System in favour of a binary, however interesting, where one cannot see what is happening. DI Her, like all close binaries, is after all just a point of light, so it is difficult to reject the argument that its apparently anomalous behaviour arises from some invisible and undetected complexity of its structure.

Such an *ad hoc* assumption is also not satisfactory and it would be much better to find a binary which could be interpreted in an unambiguous manner. Where theory cannot sort itself out, measurements must lead the way and this is where an amateur can contribute. There are other binaries which are potentially good tests. Table 17.1 contains a list of them, compiled by a Spanish astronomer, Alvaro Gimenez. Notice that the periods are conveniently short, as they must be for a close binary, and the luminosities, though not very high, are within the scope of a small telescope.

Appendices

Appendix 1

Reduction procedures

There are various steps that have to be taken to convert raw data to magnitudes of luminosity. This appendix should provide enough background to carry out these reductions when you decide to do some photometry. The book by Hall and Genet and the article by Robert H Hardie in *Astronomical Techniques*, listed in appendix 5, contain much of use in this context.

We will assume as our starting point that you have been obtaining measures of the star + sky, and of the sky only, through an aperture which covers about one minute of arc on the sky. If you have a photon counting device you will have a count number equal to the total number of pulses, in your integration time. Alternatively, if you have a DC system you will have a number proportional to the total number of pulses generated by the analogue to digital converter in your system. Either way, your first task is to convert your total number of counts into a rate, e.g. counts/second. You then have to subtract the rate of your (sky only) reading from your rate of (star + sky) reading to give the total signal each second received from the star. Let us call this I which, ignoring spurious pulses, is the intensity of the star.

The equation to be used to convert intensity into magnitude is simply

$$m_i = 2.5 \log_{10} I(\text{counts/s}). \qquad \text{(A1.1)}$$

However, this number is not a very useful magnitude. It depends upon how large your telescope is, how clean the lenses and mirrors are, what filter you are using, the efficiency of your photomultiplier tube and a whole host of other factors. It needs to be referred to the standard stellar magnitude scale. The simplest equation which can be used to give a useful magnitude has a form something like

$$m = C + 2.5 \log_{10} I \qquad \text{(A1.2)}$$

where m is an observed magnitude and I is in counts/s. C is a constant under the following conditions: if you use stars near to the zenith and less than, say, one degree apart, through a filter which was known to give perfect agreement with one of the standard bands (U, B, V, R or I) and always use comparison stars of a very similar colour and spectral type to the variable.

If all these conditions are satisfied, then either equation (A1.1) or (A1.2) would suffice to allow you to publish magnitude *differences* on the standard stellar scale. Taking the difference subtracts out the effect of C. It will be seldom possible for you to satisfy all these criteria, and therefore it is normal to continue past equation (A1.2) and to apply corrections for spurious effects on C. Firstly, we will deal with the variable transmission of the sky.

We have already discussed in §10.3.4 how the amount of light absorbed by the Earth's atmosphere is not only intrinsically variable, but also varies as the secant of the zenith angle (sec Z). The sec Z approximation is derived from a very simple model of the Earth and its atmosphere and, as long ago as 1904, Bemporad derived more sophisticated models. However, Andrew T Young, an American photometric expert, has derived a rather simpler expression which suffices up to sec $Z = 4$ (i.e. zenith angle of 78°, or 12° above the horizon). Rather than just using sec Z he recommends that X, the air mass, should be derived from

$$X = \sec Z\,\{[1 - 0.0012\,(\sec Z - 1)]\}. \qquad \text{(A1.3)}$$

The absorption by the Earth's atmosphere is called the extinction coefficient and is usually referred to as k. We will describe below how blue light is absorbed more by the Earth's atmosphere than is red light—remember that the rising and setting sun or moon is manifestly redder than when higher in the sky, and therefore it is normal not just to write k for an extinction coefficient but to qualify this by wavelength, k_λ. Examples would be k_B or k_V, etc. The total extinction experienced by a star is, therefore, $k_\lambda X$ and the equation used is

$$m_0 = m - k_\lambda X \qquad \text{(A1.4)}$$

where m_0 is the apparent magnitude corrected for atmospheric *extinction*, m was derived from (A1.2), k_λ is determined empirically and X is derived from (A1.3). The equation for deriving sec Z is in §10.3.4. Note that the extinction correction is in magnitudes and is applied with a negative sign to the raw magnitude m. This has the effect of reducing the numerical value of m_0, which means the m_0 is less than m, i.e. the correct sense for a magnitude scale in which smaller numbers mean brighter stars.

It is now time to pause and to consider exactly what we are trying to do when we correct for the atmospheric extinction. What we would like to know is how bright the star would have appeared if the Earth had no atmosphere. Now sec Z equals 1 at the zenith, increases to two at $Z = 60°$, 4 at $Z = 78°$ and so on, and therefore we can only measure the extinction (from standard stars) for values of sec Z of 1 and upwards, yet what we really want to know is how bright the star would have been if there were no absorption, which is equivalent to sec Z = zero. Typically, the extinction is only determined at values of sec Z from 1 to 2, or sometimes 3, and yet we need to extrapolate backwards to sec $Z = 0$ (it does not matter that no physically possible value of Z gives sec $Z = 0$). This leads to what is sometimes called the 'lever arm' problem, in that we need to estimate the correction for extinction to a value for which we can never measure it. Figure A1.1 below shows the problem. This problem is inescapable in 'all-sky' photometry, and its avoidance is one of the powerful arguments that can be put forward in favour of only doing comparative photometry when the ultimate in accuracy is required.

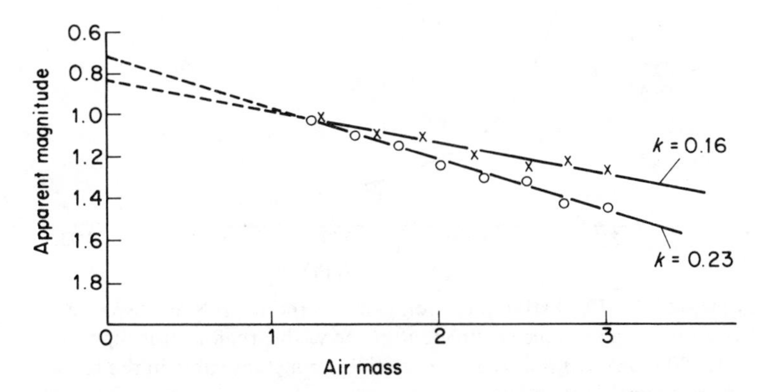

Figure A1.1 The extinction coefficient can only be determined over a range of air masses from 1 upwards. Extrapolating backwards to an 'above atmosphere' value can lead to uncertainties.

Thus far we have compensated for the various sizes of the telescopes and their overall efficiencies, the innate opacity of the Earth's atmosphere and, simplistically, the zenith distance of the observation. There is still a further complication. We have earlier drawn attention to the fact that the extinction coefficient varies with wavelength. The Earth's atmosphere is almost opaque below 300 nm and becomes increasingly transparent until about one micron, after which it varies in opacity all through the infrared region. The wavelengths at which the Earth's atmosphere is relatively transparent are referred to as 'windows', e.g. the visible window, the infrared windows, etc. Figure A1.2 shows the typical values for k at different wavelengths for the visible region at a good high-altitude site. Readers should not be surprised if their own site gives values significantly worse than this. A little study of figure A1.2 will give some clues as to the next complication. The problem arises because the light given off by stars does not have equal intensities at all colours—red stars have more red light then blue and blue stars have more blue light than red. The distribution of energy against wavelength is close to that for an ideal black body at a given temperature. However, due to the complexities of the absorption spectra of stars, the real situation

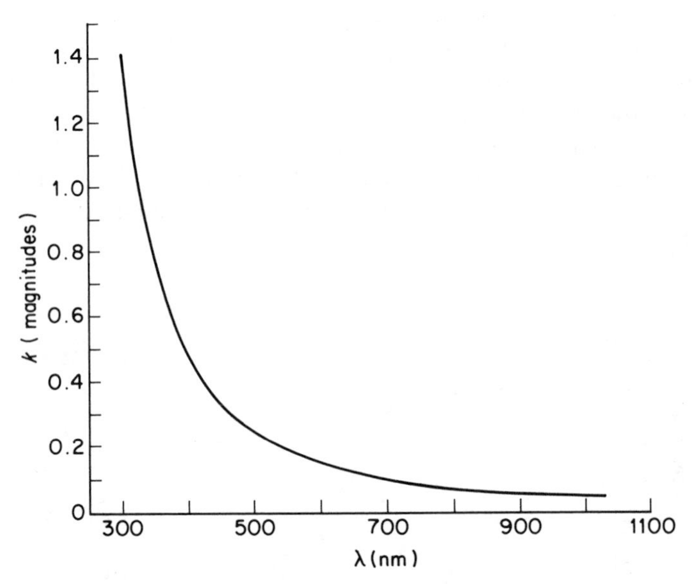

Figure A1.2 The extinction coefficient of the Earth's atmosphere is a strong function of colour throughout the visible region. It is high in the UV (~ 300 nm) and reduced to a relatively constant value in the red and infrared (up to 1 micron).

is more complicated than if the stars just had simple black body spectra. This has the effect of altering the effective wavelength of the various broad band filters. Figure A1.3 shows the effect schematically. What happens is that with a blue star the effective wavelength of any broad band filter is shifted to the blue and vice versa for red stars. Of course, if the curve of k against λ was as flat as the part of the curve from 700 to 1100 nm in figure A1.2 then a slight shift of effective wavelength of the filters would not much matter. Unfortunately, for the U + B bands the change in opacity with wavelength is so steep that even slight shifts in effective wavelength can lead to the need for an additional correction to be applied to the extinction coefficient and X (or sec Z) corrections (given in equation A1.3). The way that this is done is to apply another term to equation (A1.4) which is a function of the B–V colour of the star. The new equation has the form

$$m_0 = m - k'_\lambda X - k''_\lambda X(\mathrm{B{-}V})$$

where k' is that part of the extinction coefficient which is independent of the colour of the star, but is a strong function of wavelength (figure A1.2) and k'' is a coefficient to be multiplied by X and B–V to compensate for the shift in effective wavelength of the filters with colour. Typical values for k'' for the V band would be nearly zero, for the B band in the range – 0.02 to – 0.04 and for the U band 0.00 to – 0.04 mag. Unfortunately, the value of k'' for the U band is a very strong function of the spectrum of the star, as might be guessed from the steepness of the curve in figure A1.2 in that range. Equally unfortunate is the fact that when the UBV system was originally set up it was implicitly assumed that $k''_B = k''_U$. Therefore, if you wish to match the standard UBV system you will have to make this same assumption. You can have greater accuracy by determining k''_U empirically but you cannot then claim that your magnitudes are exact UBV ones.

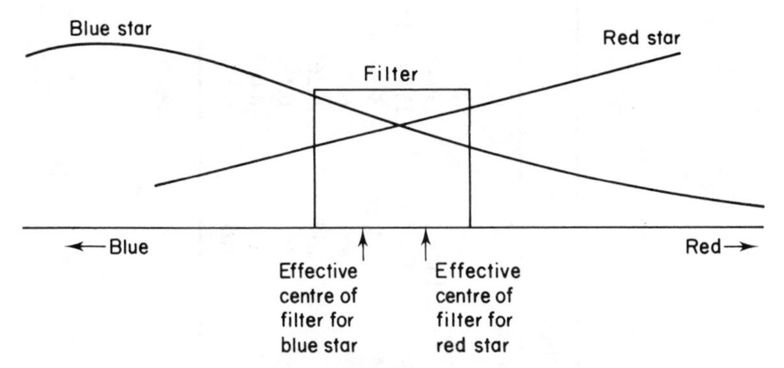

Figure A1.3 A schematic representation of how the effective wavelength of a filter changes depending upon whether a red star or a blue star is being observed.

The method that is used to determine this second-order correction (k''_X) is similar to that used to determine k'_X, in that standard stars are measured over as wide a range of X as is possible. However, for the k''_X term it is usual to choose a pair of stars of widely differing colours but which lie close together on the sky. Instead of plotting the change of magnitude against air mass, as for the first-order term, this time the plot is of colour index, e.g. B–V or U–B, against air mass. The value of the k''_X coefficient is determined by comparing the apparent change in colour (U–B or B–V) with changing X and calculating the gradient of this slope. It is marginally more accurate to use the observed colours rather than the reduced catalogue values for this exercise, as it is observed values which matter. However, in practice the errors in the procedure will probably outweigh any attempts at increased sophistication. Although this might seem a rather time consuming and tedious task, the good news is that once it has been done thoroughly it is unlikely to be necessary to repeat the exercise unless either the filters, photomultiplier tubes or some other part of the equipment is changed. Even deteriorating mirror surfaces probably only require the k''_X term to be determined once each year. Note how different this is from the k'_X term which not only changes from night to night but on many apparently photometric nights still changes by a few per cent through the night.

Table A1.1

	Pair 1		Pair 2		Pair 3		Pair 4	
Name			27 LMi	28 LMi				
HR		1534	4075	4081			8451	8453
HD	30544	30545	89904	90040	161261	161242	210419	210434
α 2000	$4^h48^m39^s$	$4^h48^m45^s$	$10^h23^m6^s$	$10^h24^m8^s$	$17^h44^m15^s$	$17^h44^m11^s$	$22^h10^m21^s$	$22^h10^m34^s$
δ 2000	3°38′56″	3°35′18″	35°54′29″	33°43′6″	5°42′48″	5°14′58″	−3°53′39″	−4°16′1″
V	7.32	6.02	5.878	5.500	8.315	7.805	6.27	6.005
B	7.265	7.22	6.028	6.68	8.365	9.085	6.265	6.985
U	6.955	8.37	—	—	8.23	10.19	6.195	7.82
B–V	−0.055	1.20	0.15	1.18	0.05	1.28	−0.005	0.98
U–B	−0.31	1.15	—	—	−0.135	1.105	−0.07	0.835

In table A1.1 we list four pairs of stars which are suitable for the determination of the k_X'' term. They are all close pairs of similar brightness and widely differing colour. They have been recommended by Douglas Hall (one of the authors of Hall and Genet's book) as suitable for this purpose.

Appendix 2

Cost and Types of Equipment

The cost of making interesting and useful measurements can be anything from zero (for the unaided eye) up to tens of thousands of pounds. Naturally the more expensive equipment allows a wider range of projects, but may require the resources of a society or research institution. The options could be summarized as follows.

		Rough Cost £	Limiting magnitude	Accuracy (mag)	
Visual	Unaided eye	0	6	0.1	Individual amateur astronomer
	Binoculars	50	8	0.1	Individual amateur astronomer
	Small telescope	500	12	0.1	Individual amateur astronomer
Photometric single-channel	Small telescope	2 000	12	0.01	Individual amateur astronomer; Amateur astronomical society
	Large telescope	10 000	15	0.01	Amateur astronomical society; Research institute
Photometric multi-channel	Small telescope	50 000	12	0.002	Research institute
	Large telescope	100 000	15	0.002	Research institute

It is quite possible to discover novae or supernovae, or follow the eclipses of bright binaries with the unaided eye, but we hope you would now like to do more. The next step is the use of either binoculars or a telescope. It will depend upon the brightness of the stars you choose to observe and whether or not they are in a crowded field as to whether binoculars or a telescope are most useful. 7 × 50 binoculars can cost from a few tens of pounds up to a few hundred pounds and will allow you to work on stars as faint as about eighth magnitude, depending upon your site. (Viewing from a city will obviously restrict you to brighter stars than if you were up a mountain far from the nearest street lights.) It is even possible to use B, V and R filters in front of the binoculars but it is then much better to use a telescope.

The size of the telescope will probably be less important than its mounting. Under no circumstances should you try to work with the type of telescopes which are often sold in department stores for a few tens of pounds. The mounting of these telescopes is generally totally inadequate for any type of serious use. If you wish to buy a telescope which has a mounting with an RA drive, i.e. capable of following the stars, then you will probably have to pay a few hundred pounds for a secondhand one, or approximately one thousand pounds or more, for a new one. Never buy a telescope without using it on stars first. If you choose to buy secondhand then if the vendor is confident in what he is trying to sell he will have no objections to demonstrating it to you. Make sure that you can move the telescope easily from star to star and that once the star is centred the telescope is not easily deflected by slight knocks or the wind. If financial constraints mean that it is impossible for you to spend several hundred pounds or more then remember that you can buy the mirrors or lenses and make your own telescope. (Some designs such as the Dobsonian allow you to construct a telescope from simple flat sheets of wood and do not require any great engineering skills. A set of mirrors and an eyepiece for a 15 cm (6 inch) Newtonian can be bought for about one hundred pounds. However a Dobsonian mount is not suitable for tracking a star.)

If you wish to do photoelectric photometry then you must have a telescope which is driven in Right Ascension and which is stable. The telescope must be capable of keeping the stellar image in the centre 10% of a one arc minute field for thirty seconds as an absolutely minimum requirement. Preferably it should be able to do this for several tens of minutes. It should have easily readable RA and Dec scales and if it is to be seriously usable it will almost certainly have to be equipped with an RA drive system with a diameter larger than 15 cm (6 inches). Very few of the telescopes intended for the amateur market will be suitable. You might have to pay several thousand pounds for a suitable new telescope but price alone is not a good criterion. A homemade telescope constructed with photometry in mind may be more suitable than a commercial product. If you possess the skills and the time then a suitable telescope could be constructed for under £1000. Such a project is probably best tackled using the combined skills and resources of an amateur society. The advice and encouragement of colleagues will be useful, even if you tackle it yourself. If you, or the Society, wants to buy a suitable *new* telescope then you will probably need to deal with a manufacturer of telescopes for professional use and even a 30 cm (12 inch) telescope and drive is likely to cost £5000 to £10 000.

At the present time your choice of commercially available photometers is strictly limited. Several dedicated amateurs have constructed their own. It requires a rare combination of optical, mechanical and electronic skills and some considerable dedication to discover the source of the various components. Your initial choice is between photometers which contain a solid state detector and those which have a photomultiplier tube. The solid state devices have the advantage of light weight and no need for a high tension supply, but will never give you the accuracy of which a photomultiplier is potentially capable. A solid state device is available from the USA for about $800 and if you wish to work on large amplitude red stars it could be the right choice for you. For really serious work there is still no choice but to use photomultiplier tubes.

The cheapest photomultiplier tubes which might be expected to produce useful results are the modern versions of the old 1 P 21 tube, used for most of the original stellar photoelectric photometry. Several manufacturers produce modern versions of the old design, which was originally intended to produce noise for radio transmission. Several amateurs have produced photometers based upon these tubes and they can certainly form the basis for a DC system. Tubes of this pattern from Thorn EMI and Hammamatsu will cost about £50. If cost is your main concern then perhaps a system using one of these tubes will be the one for you.

If you wish to undertake photon counting photometry then end window photomultiplier tubes with a specially designed dynode structure should be your choice. All the major manufacturers produce tubes suitable for single photon detection and those used in the larger professional observatories typically cost in the range of from £1000 to perhaps £5000 each. However the 9924, or its new version the 9124, produced by Thorn EMI, has been found to be an excellent small tube and has been chosen for the JEAP (see Chapter 10). It currently costs just under £200. (For a while EMI Gencom, in the USA, produced a photometer suitable for smaller observatories for a cost of just under $2000. Unfortunately various problems led to a discontinuation of this unit.) The JEAP costs about £1500 but must be used in conjunction with a personal computer with an RS232 interface. The advantage of the JEAP is that it uses exactly the same technology and components as the professionally orientated four-star system, facilitating the combination of results from small amateur observatories and professional observatories. Depending upon the degree of sophistication, and of automation required, the full four-channel comparative photometer costs several tens of thousands of pounds, restricting its use to research institutes or very large amateur societies. A single-channel photometer constructed from top quality commercially available components would cost about £10 000 plus labour and design. If you wish to work on a large number of nights to a higher accuracy, anywhere other than the very best sites, then you will have to use a comparative photometer or perhaps use two separate telescopes, each equipped with a photometer.

Is it necessary to use top quality components? After all, surplus photomultiplier tubes can be found from time to time in electronics stores specializing in surplus and secondhand items. Such tubes are often available for perhaps 10% of their new cost and several people have been tempted to build photometers around such tubes. You should be aware that not all photomultiplier tubes are suitable for the low light levels found in stellar photometry. Additionally remember that the photomultiplier tube is probably the most fragile component in your whole system. If too much light is allowed to enter the photometer then the tube will be permanently damaged. It might seem attractive at the start of the project to build a photometer around a £1000 tube purchased for £100. However this will seem much less desirable when it suddenly becomes necessary to replace the burnt-out tube with a new one at full price!

If you are a member of a small university team, rather than an amateur astronomer working alone, then it can often be advantageous to work in collaboration with your engineering or electronics groups. Construction of much of the equipment described here can easily be carried out by such groups at a much lower cost than purchase of commercial systems.

Appendix 3

Conceptual Diagram of Variable Star Astrophysics

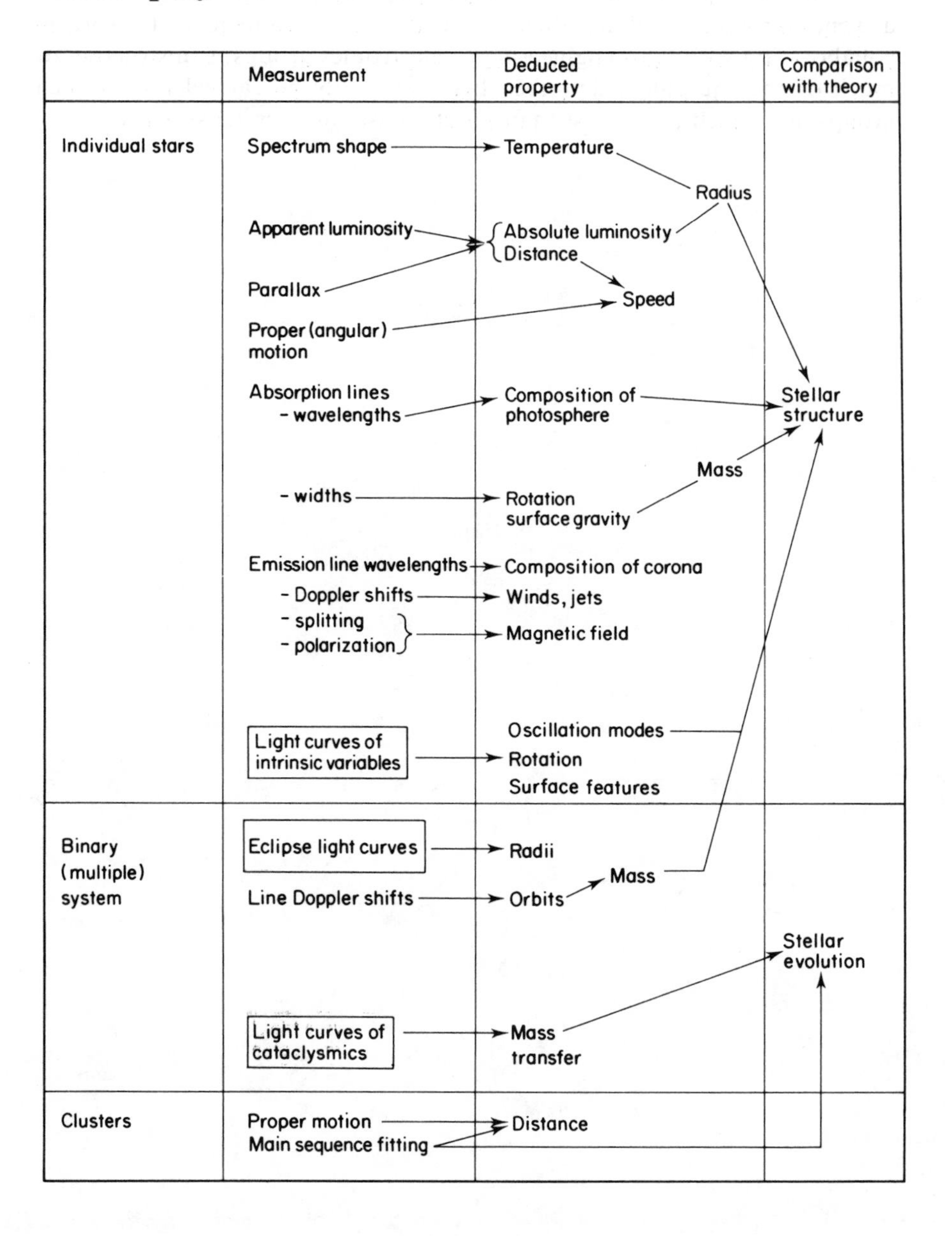

Appendix 4

Lists of Stars Suitable for Projects

Bright symbiotic stars (see Chapter 8)

	RA		Dec		Maximum magnitude	Period (days)
0 + VZ Cet	2 h	19 m	– 2°	59′	2.0	332
SS Lep	6	03	– 16	29	4.8	
AX Mon	6	28	5	54	6.6	
T Cr B	15	57	26	04	2.0	
RS Oph	17	48	– 6	42	4.3	
AR Pav	18	20	– 66	05	8.5	
V1017 Sgr	18	32	– 29	24		
FN Sgr	18	54	– 19	00		
CH Cyg	19	25	50	14	7.4	97
RR Tel	20	04	– 55	43		
AG Peg	21	51	12	38	6.0	800
Z And	23	34	48	49	8.0	694
R Aqr	23	44	– 15	17	5.8	387

Hypergiant stars (see §12.1)

Star	Spectral type	M_v	M_{bol}	V	B – V	Position
HD 224014 (ρ Cas)	F8pIa	– 9.5	– 9.4	4.59	1.26	23 54.4, 57 30
HR 8752	G0–G5 0–Ia	– 9.5	– 9.5	5.08	1.34	23 0, 56 57
HD 212466 (RW Cep)	K0 0–Ia	– 9.4	– 9.6	6.65	2.22	22 23, 55 58
HR 5171	G8 0–Ia	– 9.2	– 9.0	6.80	2.25	13 47, – 62 35
HD 96918	G0 Ia^+	– 9.2	– 9.2	3.93	1.25	11 08, 58 59
HD 74180	F0 Ia	– 9.0	– 8.9	3.85	0.69	8 41, 46 39
Cooler stars						
HD 206936 (μ Cep)	M2 Ia	– 8.0	– 9.5	4.02	2.38	21 43, 58 47
KY Cyg	M3 Ia	– 7.6	– 9.6	10.57	3.64	20 24, 38 11
BD + 24° 3902	M1 Ia	– 7.5	– 8.8	9.00	3.10	19 44, 24 34
HD 143183	M3 Ia	– 7.5	– 9.5	7.64	2.30	15 54, – 53 51

Some Wolf–Rayet stars (see §13.2)

	RA	Dec	Apparent magnitude	Period (days)
CV Ser	18 h 16 m	− 11° 39′	9	29.7
V1676 Cyg	20 04	35 39	6.75	
V444 Cyg	20 18	38 34	7.92	4.2
V1696 Cyg	20 40	52 24	10.3	4.3
CX Cep	22 08	57 28	12.1	2.13
GP Cep	22 17	55 52	9	6.7
CQ Cep	22 35	56 39	8.63	1.64

Some β CMa stars (see §13.3.1)

	RA	Dec	Apparent magnitude	Period (days)
γ Peg	0 h 13 m	15° 11′	2.83	0.1517
δ Cet (HR 779)	2 39	0 20	4.05	0.161
β CMa	6 23	− 17 57	1.98	0.25
δ Lup	15 21	− 40 39	3.22	0.165
σ Sco	16 21	− 25 36	2.89	0.2468
V986 Oph	18 02	1 55	6.12	0.323
BW Vul	20 55	28 31	6.2	0.201
β Cep	21 29	70 34	3.22	0.1905
12 Lac (DD Lac)	22 41	40 14	5.25	0.193
16 Lac (EN Lac)	22 56	41 36	5.59	0.1692

Some Mira stars which reach $V < 7$ (see §14.1)

	RA		Dec		Max apparent magnitude		Period (days)
KU And	0 h	07 m	43°	05′	6.5	10.5	750
T Cas	0	23.2	55	48	6.9	13	445
R And	0	24	38	35	5.8	14.9	409
W And	2	18	44	18	6.7	14.6	396
o Cet	2	19	−2	59	2.0	10.1	332
R Tri	2	37	34	16	5.4	12.6	266
R Aur	5	17	53	35	6.7	13.9	458
U Ori	5	56	20	10	4.8	12.6	372
R Gem	7	07	22	42	6.0	14	370
S CMi	7	33	8	19	6.6	13.2	333
R CnC	8	17	11	44	6.1	11.8	362
Y Dra	9	42	77	51	6.0	15	325
R LMi	9	46	34	31	6.0	13	372
R Leo	9	48	11	26	4.4	11.3	312
R UMa	10	45	68	47	6.7	13.4	302
SS Vir	12	25	0	48	6.0	9.6	355
T UMa	12	36	59	29	6.6	13.4	256
R Vir	12	39	6	59	6.0	12.1	145
R CVn	13	49	39	33	6.5	12.9	329
R Boo	14	37	26	44	6.2	13.1	223
S Cr B	15	21	31	22	5.8	14.1	360
V Cr B	15	50	39	34	6.9	12.6	358
R Ser	15	51	15	08	5.2	14.4	356
RU Her	16	10	25	04	6.8	14.3	485
U Her	16	26	18	54	6.5	13.4	406
R Dra	16	33	66	45	6.7	13.0	245
S Her	16	52	14	56	6.4	13.8	307
T Her	18	09	31	01	6.8	13.9	165
X Oph	18	38	8	50	5.9	9.2	334
R Aql	19	06	8	14	5.5	12.0	284
R Cyg	19	37	50	12	6.1	14.2	426
RT Cyg	19	44	48	47	6.4	12.7	190
χ Cyg	19	51	32	55	3.3	14.2	407
U Cyg	20	20	47	54	5.9	12.1	462
T Cep	21	10	68	29	5.2	11.3	388
R Peg	23	07	10	33	6.9	13.8	378
V Cas	23	12	59	42	6.9	13.4	229
R Cas	23	58	51	24	4.7	13.5	430

Classical Cepheids, $V_{max} < 7.5$ (see §14.2)

	RA	Dec	Mean apparent magnitude	Period (days)
Polaris, α UMi	2 h 32 m	89° 16′	2.0	3.97
SU Cas	2 52	68 53	5.9	1.95
RX Cam	4 05	58 40	7.7	7.91
SZ Tau	4 37	18 33	6.5	3.15
AW Per	4 48	36 43	7.5	6.46
RX Aur	5 01	39 58	7.7	11.62
T Mon	6 25	7 05	6.1	27.02
RT Aur	6 29	30 30	5.4	3.73
W Gem	6 35	15 20	7.0	7.91
ζ Gem	7 04	20 34	3.9	10.15
TT Aql	19 08	1 18	7.1	13.75
U Vul	19 37	20 20	6.2	7.99
SU Cyg	19 45	29 16	6.8	3.85
SV Vul	19 52	27 28	7.2	45.04
η Aql	19 53	1 00	3.9	7.18
S Sge	19 56	16 38	5.7	8.38
X Cyg	20 43	35 35	6.4	16.39
T Vul	20 52	28 15	5.8	4.44
DT Cyg	21 07	31 11	5.8	2.50
δ Cep	22 29	58 25	3.9	5.37

RR Lyr (see §14.3)

	RA	Dec	Mean magnitude	Period (days)
SW And	0 h 21 m	29° 08′	9.14	0.442
RR Gem	7 18	30 59	10.62	0.397
TT Cnc	8 30	13 22	10.72	0.563
RW Cnc	9 16	29 17	10.7	0.547
Z CVn	12 47	44 03	11.46	0.654
TV Boo	14 15	42 36	10.71	0.313
AR Ser	15 31	2 56	11.43	0.575
AR Her	15 59	47 04	10.59	0.47
RW Dra	16 35	57 56	11.05	0.443
DL Her	17 18	14 32	11.0	0.592
RR Lyr	19 24	42 41	7.06	0.567

Some Delta Scuti stars with $V < 6$ (see §14.4)

	RA		Dec		Mean apparent magnitude	Δm	Period(s) (days)
β Cas	0 h	09 m	59°	09′	2.28	0.06	0.1043
GN And	0	30	29	45	5.2	0.04	0.0696
CC And	0	41	42	00	9.2		0.1249, 0.0749
VX Psc	1	30	18	21	5.9	0.02	0.131
RV Ari	2	12	17	50	11.9	0.7	0.0931, 0.0720
UV Ari	2	45	12	27	5.2	0.04	0.0355
V376 Per	3	49	43	58	5.85	0.11	0.091
IM Tau	4	11	26	29	5.4	0.07	0.145
V483 Tau	4	20	14	02	5.6	0.03	0.054
V696 Tau	4	21	15	06	5.23	0.06	0.036
V480 Tau	4	51	18	50	5.1	0.02	0.042
V474 Mon	5	57	−9	23	5.9		0.136, 0.0826
V1004 Ori	5	58	1	50	5.9	0.01	0.054
FZ Vel	8	57	−47	02	5.1		0.065
υ UMa	9	51	59	02	3.8	0.09	0.133
DP UMa	12	02	43	03	5.2		0.06
AI CVn	12	24	42	33	6.0	0.26	0.2085
AO CVn	13	18	40	34	4.7	0.05	0.12168
K^2 Boo	14	14	51	47	4.5	0.08	0.07
δ Ser	15	35	10	32	4.2	0.05	0.134
γ Cr B	15	43	26	18	3.8	0.03	0.030
CL Dra	15	58	54	45	5.0	0.05	0.063
δ Sct	18	42	−9	03	5.1	0.18	0.194, 0.116
V1208 Aql	19	20	12	22	5.5	0.05	0.1497
δ Del	20	44	15	04	4.4	0.10	0.158
τ Cyg	21	15	38	03	3.7		0.083
BP Peg	21	31	22	31	11.7		0.1094, 0.0845
ε Cep	22	15	57	03	4.2	0.06	0.041
τ Peg	23	21	23	44	4.6	0.02	0.0543
HT Peg	23	53	10	57	5.3		0.06

ZZ Ceti stars (see §14.5)

	Apparent magnitude
MY Aps	13.75
ZZ Cet	14.1
RY LMi	15.5
VY Hor	15.0
PT Vul	15.5
VW Lyn	14.6
V470 Lyr	14.6
V411 Tau	15.0
AX Phe	15.3
TY Cr B	14.4
ZZ Psc	13.1
V468 Per	15.6
BG CVn	15.3

Dwarf novae (see §15.1)

	RA	Dec	Apparent magnitude	Orbital period (h)	Period(s) (days)
RX And	1 h 02 m	41° 02′	10.3		14, 60
WX Hyi	2 08	−63 33	9.6		14, 140
VW Hyi	4 09	−71 25	8.4	1.78	27.8, 180
U Gem	7 52	22 08	8.2	4.25	103.7, 198
YZ Cnc	8 08	28 18	10.2	2.21	11.5, 135
Z Cam	8 20	73 16	10.0	6.96	23.4, 106
AH Her	16 42	25 21	10.9	5.93	19.5, 62
SS Cyg	21 41	43 21	7.7	6.6	5.02, 118
RU Peg	22 12	12 27	9.0		69, 110

Massive x-ray binaries (see §15.2)

	RA	Dec	Apparent magnitude	Period
X Per 0352 + 309	3 h 52 m	30° 54′	6.0	835 days
GP Vel, Vela X-1, 4U 0900-40	9 00	−40 21	6.8	8.965 days
V779 Cen, Cen X-3	11 19	−60 18	13.4	2.1 days 4.8 s
V884 Sco, 4U 1700-377	17 04	−37 51	6.6	3.41 days
V1357 Cyg, Cyg X-1, 1956 + 350	19 56	35 04	8.9	5.6 days

RS CVn stars (see §15.3)

	RA	Dec	Maximum magnitude	Period (days)
RW UMa	11 h 38 m	52° 16′	10.2	7.328
RS CVn	13 08	36 12	7.9	4.798
SS Boo	15 12	38 45	10.3	7.603
RT Lac	22 00	43 39	8.8	5.074
AR Lac	22 07	45 30	6.1	1.983

Appendix 5

Further Reading

The ISBNs are for paperback/softcover where available. Prices subject to change.

Introductory/general interest

Hartmann W K 1985 *Astronomy: The Cosmic Journey* (Wadsworth) 3rd edn £38.95 ISBN 0-534-04011-X

Snow T 1983 *The Dynamic Universe* (West) £37.55 ISBN 0-314-69681-4

Ridpath I (ed) 1984 *Collins Guide to the Planets* (Collins) £5.95 ISBN 0-00-219067-2

Abell G 1978 *Drama of the Universe* (Holt, R&W) £11.50 ISBN 0-03-022401-2

Henbest N and Marten M 1983 *The New Astronomy* (CUP) £12.00 ISBN 0-521-25683-6

Astrophysics texts

Kitchin C 1987 *Stars, Nebulae and the Interstellar Medium* (Adam Hilger) £15.00 ISBN 0-85274-581-8

Shu F *The Physical Universe* (OUP) £30.00 ISBN 0-19-855706-X

Tayler R J 1970 *Stars, their Structure and Evolution* (Wykeham) £6.00 ISBN 0-85109-110-5

Abell G 1982 *Exploration of the Universe* (Holt, R&W) 4th edn £21.95 ISBN 0-03-058502-3

Pasachoff J *Astronomy Now* (Saunders) ISBN 0-7216-7100-4

Pasachoff J and Kutner *University Astronomy* (Saunders) ISBN 0-7216-7099-7

Variable Stars

Hoffmeister C, Richter G and Wenzel W 1985 *Variable Stars* (Springer) ISBN 0-387-13403

Petit M 1987 *Variable Stars* (Wiley) £35.60 ISBN 0-471-90920-3

Eggleton P and Pringle 1985 *Interacting Binaries* (Reidel) ISBN 90-277-1966-7

Pringle and Wade (ed) 1985 *Interacting Binary Stars* (CUP) £25.00 ISBN 0-521-26608-4

Percy 1986 *The Study of Variable Stars Using Small Telescopes* (CUP) ISBN 0-521-33300-8

Kenyon S J 1986 *The Symbiotic Stars* (CUP) ISBN 0-521-26807-9

Observational Techniques

Hardie R C 1962 *Astronomical Techniques* (University of Chicago Press)
Kitchin C 1984 *Astrophysical Techniques* (Adam Hilger) £15.00 ISBN 0-85274-484-6
Wood F B 1963 *Photoelectric Astronomy for Amateurs* (Macmillan) out of print
Walker G 1987 *Astronomical Observation* (CUP) £15.00 ISBN 0-521-33907-3
Hall D and Genet R *Photoelectric Photometry of Variable Stars*
International Amateur–Professional Photoelectric Photometry 1982
Henden and Kaitchuk 1982 *Astronomical Photometry* (Van Nostrand Reinhold) $30.00 ISBN 0-442-23647-6
New Zealand Symp. on Photoelectric Photometry vols I and II (Fairborn) $9.95 and $15.00
Wolpert and Genet R *Advances in Photoelectric Photometry* vols I and II
Photoelectric Photometry Handbook (Fairborn) $23.95 ISBN 0-911351-09-4
Hearnshaw and Cottrell 1986 *Instrumentation and Research Programmes for Small Telescopes* (Reidel) ISBN 90-277-2325-7
IAU 1988 *Proc. IAU Coll. 98. The Contributions of Amateurs to Astronomy* (Springer)

Astronomical Computing

Ghedini S 1982 *Software for Photometric Astronomy* (Willman-Bell) ISBN 0-943396-00-X
Duffett-Smith P 1981 *Practical Astronomy with your Calculator* (CUP) 2nd edn ISBN 0-521-28411-2
Automatic Photoelectric Telescopes (Fairborn) $23.95 ISBN 0-944389-00-7
Trueblood and Genet R *Microcomputer Control of Telescopes* (Willman-Bell) ISBN 0-943396-05-0

Catalogues

Hoffleit D and Jaschek C 1982 *Yale Bright Star Catalog* (Yale University Observatory) 4th edn
Hirshfeld and Sinnott 1982 *Catalogue 2000.0 vol 1: Stars to magnitude 8* (CUP) £18.00 ISBN 0-521-28913-0
Hirshfeld and Sinnott 1985 *Catalogue 2000.0 vol 2: Double Stars, Variable Stars and Non-stellar Objects* (CUP) £20.00 ISBN 0-521-27721-3
Wallenquist 1982 *Catalogue of Photoelectric Magnitudes and Colours of Visual Double and Multiple Systems* (Acta Universitatis Upsaliensis) ISBN 91-554-12165
Kholopov P N 1985 *General Catalogue of Variable Stars* Vols 1 and 2 (Nauka) 4th edn
Burnham R *Burnham's Celestial Handbook* vols I, II, III (Dover) £30.00 all three vols ISBN 0-486-23567-X, ISBN 0-486-23568-8, ISBN 0-486-23673-0
Webb 1962 *Celestial Objects for Common Telescopes* vol II (Dover) ISBN 0-486-20918-0
Kukarkin B V 1982 *New Catalogue of Suspected Variable Stars* Astronomical Council of the USSR Academy of Science

Huth H and Wenzel W 1981 *Bibliographic Catalogue of Variable Stars* (Centre de documentation stellaire, Strasbourg)

Draper H 1921 *Henry Draper Catalog* (Annals of the Astronomical Observatory of Harvard College)

Journals and Serials

Sky and Telescope

Sky Publishing Corporation, 49 Bay State Road, PO Box 9102, Cambridge, MA 02238-9102, USA $35.00 pa

Astronomy

Astromedia Kalmbach Publishing, 625 E St Pauls Ave, PO Box 92788, Milwaukee, WI 53202, USA $26.00 pa

Astronomy Now

Intra Press, 193 Uxbridge Road, London W12 9RA

The Astronomer

16 Westminster Close, Kempshott Rise, Basingstoke, Hampshire RG22 4PP £14.00 pa

Moore P (ed) 1988 *Yearbook of Astronomy* (Sidgwick and Jackson) £6.95 ISBN 0-283-99475-4

Appendix 6

Addresses

British Astronomical Association
Burlington House
Piccadilly
London W1N 9AG (Tel: 01 734 4145)

Royal Astronomical Society
Burlington House
Piccadilly
London W1N 9AG (Tel: 01 734 4582)

London Planetarium
Marylebone Road
London NW1 5LR (Tel: 01 486 1121)

Jodrell Bank Planetarium
Jodrell Bank
Macclesfield
Cheshire SK11 9DL (Tel: 0477 771339)

American Association of Variable Star Observers
25 Birch Street
Cambridge
MA 02138-1205
USA (Tel: 017 354 0484)

If you think you have the first sighting of a comet, nova, supernova, etc, the easiest way to report it in the UK is via the editor of '*The Astronomer*':

Guy Hurst
16 Westminster Close
Kempshott Rise
Basingstoke
Hampshire RG22 4PP

Tel: 0256 471074
Telex: 265 871 (MONREF G)

Appendix 7

Julian Dates

1988	244 7161.5
1989	7527.5
1990	7892.5
1991	8257.5
1992	8622.5
1993	8988.5
1994	9353.5
1995	9718.5
1996	245 0083.5
1997	0499.5
1998	0814.5
1999	1179.5
2000	1544.5

These dates apply to 0^h UT on 1 January in each year. For example, for 10 January add 9 to number.

Appendix 8

Summary Table of Variable Star Characteristics

This table gives details of the variable star types discussed.

Archetypal Star	GCVS type	Generally used name	Alternate generally used name	Names	Population	Luminosity class	Periodic	Irregular	Period range (typical) (days)
Radial pulsators									
β Cephei	BCEPS	β Cephei	β Cephei	β Cephei β CMa γ Peg α Vir σ Sco	I	V–III	Single + multiple		0.14–0.3
Do Eri	ACVO	Ap pulsator			I	V–IV	Multiple		6–12 min
δ Scuti	DSCT	δ Scuti		1 Mon 28 Aql SZ Lyn	I	V–III	Single + multiple		0.01–0.2 0.2
RR Lyrae	RR	RR Lyrae	Cluster variable	RR Lyrae AQ Leo RR Gem	II	IV–III	Single + multiple		0.2–2.4 0.5
δ Cephei	CEP	Cepheid		EV Sct UY Per SZ Cas	I	III–II	Single + multiple		1–100
o Ceti	M	Mira	Long period variables	χ Cyg T Cen BX Mon	I	I	Semi-periodic		90–1400
Non-radial pulsators									
ZZ Ceti	ZZ	ZZ Ceti	Variable white dwarf	MY Aps TY CrB ZZ Psc		DA	Multiple		100–1000 s
β Cephei	BCEPS	β Cephei	β Can Maj		I	V–III	Multiple		0.14–0.3
α Cyg	ACYG				I	I	Multiple		Days to tens of days
Irregular variables									
S Dor	SDOR	Supergiant variable		P Cyg η Car		I–II		Yes	Days to centuries
	GCAS			γ Cas					
	SR + SRA SRB, SRC SRC	Semi-regular	Long period variable	α Scu α Ori UU Her		III–I		Yes	~ 10 – ~ 1000
UV Ceti	UV	Flare star				V		Yes	

Absolute magnitude range (typical)	Amplitude of variation range (typical)	Spectral class or type	Number known	Mass $M_\odot$	Mechanism and comments
−3.5 to 4.5	0.01 to 0.3	B0–B3	~50	~5	Uncertain. See also non-radial pulsators as both types of variation may be present in these stars.
~ +2	~0.01	A7–F2	~12	~2	Uncertain
0 to +3	<0.1 to 0.7	A2–F4	~100+	~2–4	He II opacity
0 to +0.5	~1	A2–F6	Thousands	0.6	He II opacity. $\log_{10} P(\text{days}) = -0.85 \times \text{mag.}$
−1.5 to −6	~1	F6–K2	Hundreds	4–14	He II opacity. $\log_{10} P(\text{days}) = -0.394 \times \text{mag.}$
−1 to −2	2–4 mags	M, C and S	Thousands		This class of stars has variations which range from periodic to semi-regular. See irregular variables below.
13 to 16	0.01 to 0.3	DA	Hundreds		H, He, C + O ionization
−3.5 to −4.5	0.01 to 0.3	B0–B3	~50	~5	See also radial pulsators
−5	0.1 mag	A0	Few		There is still some doubt if this type of star is periodic or irregular.
−4 to −8	0.01 to several magnitudes	Variable	Tens		Extended envelopes and mass loss.
−2 to −6+	1 to 2	F–M	Thousands		Mass loss leading to white dwarf formation.
	1 to 6	K–M			

Appendix 8 *Continued*

Archetypal Star	GCVS type	Generally used name	Alternate generally used name	Names	Population	Luminosity class	Periodic	Irregular	Period range (typical) (days)
T Tau	Int	T Tauri		RW Aur T Cha RU Lup BO Cep		I	Quasi-periodic and irregular		Hours to years
	NA NB NC NL NR	Novae (fast–slow recurrent)		GK Per RR Pic RR Tel TCrB				Yes	
U Gem	UGSS UG SU UGZ	Dwarf novae		SS Cyg U Gem SU UMa Z Cam				Yes	Days to 1000s of days + seconds for the flickering
Z And	ZAND	Symbiotic		V1016 Cyg RX Pup RR Tel		II			1 to several years
	SNI SNII	Supernova					Single events		
Geometric variables									
	ELL	Ellipsoidal variable		b Per α Vir		All	Yes		Fractions of a day to tens of days
		Hot spot				All	Yes		Fractions of a day to tens of days
		Disc				All	Yes		
Algol β Lyrae WU Ma	EA EB EW	Eclipsing star				All	Yes		Minutes to years
U Gem	UG	Cataclysmic variables	Dwarf novae	SS Cyg U Gem SU UMa Z Cam			Yes		Minutes to days
α^2CVn	ACV	Ap star		α And τ^9Eri α Dor μ Lep		IV–IV	Yes		~0.5 day to years
RS CVn	RS	RS Can Ven stars		AR Lac SS Boo RW UMa RT Lac			Multiple		Days and years
BY Dra	BY	BY Drac stars		AV Mic		V	Variable		Fractions of a day to 120 days

Absolute magnitude range (typical)	Amplitude of variation range (typical)	Spectral class or type	Number known	Mass $M_\odot$	Mechanism and comments
	0.1 to 3	Ae⁻		0.3–3	Stars still contracting from interstellar gas and dust towards the main sequence. Associated with nebulosity. There are many different subgroups of these stars.
	3 + mag		Tens		Probably mass transfer in a binary system containing a degenerate star. As such they should be compared with the next class down (dwarf novae). The main difference between the two groups could be the mass and size of the donor star.
	2–6	WD+	Hundreds		Eruptions due to nuclear burning of transferred material This gives rise to the outburst and the flickering. (See also geometric variables below.)
−3 to −4	~4	Combined M+ early	~100		Mass transfer onto white dwarf within the envelope of the late type donor star.
−18 to −21 −16 to −18			Hundreds		Explosive events with a sudden release of more than 10^{42} J. Different mechanisms are involved, including the collapse of a white dwarf binary and photonuclear reactions.
All	⩽0.01 to ~0.1	All	Tens	All	Rotation of non-spherical stars in close binary systems.
All	⩽0.01 to ~0.1	All	Tens	All	Heating of the surfaces of the two stars in a binary system by the radiation from each other can cause their facing surfaces to have a higher temperature than the rest of the surfaces. Rotation of the star causes the high spots to move across our line of sight.
All	⩽0.01 to ~0.1	All	Hundreds	All	Material transfer between stars in binary systems can form a disc about the recipient star. Rotation of the binary causes the disc to present a variable area towards us.
All	⩽0.01 to magnitudes	All	Thousands	All	Binary systems in which the plane of the orbit lies close enough to our line of sight for one star to pass in front of the other.
	~0.5		Hundreds	0.5 to ~1.0	These interacting binary stars show regular eclipse variations as well as other variations (see irregular variables). Most of the orbital variation is due to eclipse of the hot spot in the system where infalling material from the donor star collides with the accretion ring around the white dwarf.
−2 to +2	⩽0.01 to 0.5	B2–F0	Hundreds	2–4	Rotation of stars which have chemically peculiar patches on their surfaces which have different temperatures to the rest of the photosphere.
	~0.1 to 0.2	G and K	Tens		Spotted binary stars. Some show eclipses. Rotation of the spotted stars causes photometric variations with the orbital period while drifting of the spots causes beat periods of years.
+6 to +8	⩽0.3	Ke–Me			Spotted late type stars. The spots cover perhaps 5–20% of the surface and rotation causes the photometric variation.

Appendix 9

Values of Constants

Fundamental physical constants

Gravitational constant (in $F = GMm/r^2$)

$$G = 6.67 \times 10^{-11}\ \mathrm{kg^{-1}\,m^3\,s^{-2}}$$

Speed of light

$$c = 299\,796\,458\ \mathrm{m\,s^{-1}}$$

Planck's constant (in $E = hf$)

$$h = 6.63 \times 10^{-34}\ \mathrm{J\,s}$$

Boltzmann's constant (in $E = \frac{7}{2}KT$)

$$k = 1.38 \times 10^{-23}\ \mathrm{J\,K^{-1}}$$

Stefan–Boltzmann constant (in $E = \sigma T^4$)

$$\sigma = 5.67 \times 10^{-8}\ \mathrm{W\,m^{-2}\,K^{-4}}$$

Useful values

Solar mass $M_\odot = 1.99 \times 10^{30}$ kg

Solar radius $R_\odot = 6.96 \times 10^{8}$ m

Solar total luminosity $L_\odot = 3.86 \times 10^{26}$ W

1 light year $= 0.95 \times 10^{16}$ m

1 parsec $= 3.09 \times 10^{16}$ m

Glossary

Absorption spectrum Variation of intensity with wavelength showing dips due to selective absorption of photons.
Accretion Capture of gas and dust by a (compact) star.
Accretion disc A disc of gas (and dust), which forms about a star as it gravitationally attracts material from a companion star (or the interstellar medium). This 'accreted' material has enough angular momentum to go into orbit about the accreting star, rather than falling directly onto it but eventually it will spiral down into the capturing star.
Accuracy of a count In any physical phenomenon it is not possible to make a perfectly accurate measurement. Photon counting has this same characteristic. The statistical accuracy of a count is the square root of the counts; e.g. 10^6 counts give a statistical error of 10^3 (an accuracy of 10^{-3} or 0.1%), 10^4 counts gives an error of 10^2 (an accuracy of 10^{-2} or 1%), and so on.
Alpha Centauri The brightest member of a group of three stars which are the closest to the Earth.
Anomaly Said of an element whose abundance in a star departs (markedly) from the standard composition for that type of star.
Assigning harmonics Choosing a set of waves which when combined closely reproduce a given light curve.
Asteroid collisions A typical asteroid is thought to collide with another every few tens of millions of years.
Asteroid spin periods Periods of rotation about the one or more axes of rotation of an asteroid.
Autocorrelation A method of comparing a curve (e.g. the mean light curve of a variable star) with subsets of the data in order to find an optimum fit.
Axes of rotation Lines through the centre of mass of a body about which a body is rotating.
Background radiation Radiation from the Big Bang, now pervading the Universe in the form of radio waves.
Balance of forces For a star in a steady state the net force at any point must be zero.
Beats If two periodic phenomena have similar, but not identical periods,

then there will be some times when they are in phase, i.e. their maxima coincide, and other times when the maximum of one component coincides with the minimum of the other. A combination of the two signals will therefore sometimes be enhanced, when the components are in phase, and sometimes will be reduced when the components are out of phase. This slower wave is known as a beat.

Betelgeuse A giant star in Orion.

Big Bang The explosive start of the Universe.

Binary Two stars in mutual orbits about their common centre of mass.

Binary asteroid Two minor planets orbiting each other, as together they orbit the Sun. (None has so far been discovered.)

Bipolar jets in star formation Back to back jets of matter commonly seen issuing from a site presumed to be that of star formation.

Black body An idealized body which absorbs all radiation at all wavelengths.

Black hole Hypothetical compact objects which are most easily visualized as having an escape speed that is greater than the speed of light. Thus, they are black in the sense that photons cannot escape, but are expected to be detectable by the intense gravitational perturbation they cause. To calculate the effects of black holes, the theory of general relativity must be used to find how they curve space and time.

Blazhko effect A distortion near to the maximum in the light curves of RR Lyrae stars which changes its amplitude and phase over periods much longer than the pulsational period of the star.

Blue edge The edge on the high temperature side of the instability strip (the left side of a HR diagram).

Brown dwarfs Low mass stars which do not heat up enough in the formation process to start a thermonuclear reactions.

Carbon A atom with six protons in its nucleus, formed in red giant cores.

Cepheids Pulsating variable giant stars.

Classical nova Single stellar outburst of large amplitude, but less than that of a supernova.

Clusters Localized groups of stars, which separate into two types: open clusters with hundreds or thousands of young stars, and globular clusters with tens of thousands of old stars.

Coefficient Strength of a component.

Component In our context, a particular mode or frequency of pulsation.

Convection Displacement of less dense (usually hotter) material in a gravitational field causing turbulence within a star.

Core pressure The pressure in the central part of a star where thermonuclear reactions take place.

Cosmogony Theories of the origin of astronomical bodies, especially the Solar System.

Couple The turning effect of a force whose line of action does not pass through the centre of mass of a body.
Dark count A term used in photon counting to indicate the amount of signal being generated by the photomultiplier tube and its ancillary electronics when no light is falling upon the tube. See also 'dark noise'.
Dark noise The random fluctuations in the dark current of a photomultiplier tube. See also 'dark count'.
DC photometry A method for the measurement of luminous intensity which uses the amplified current passed by a photomultiplier tube and does not have sufficient time resolution to distinguish the arrival of individual photons. Intensity is not proportional to photon count, except at a single given wavelength.
Dead time The speed of response of any photomultiplier and its associated electronics is finite and occasionally photons will arrive so closely together that two, or more, will be detected as a single event. This effect has to be corrected for at high count rates and it has inherited the term 'dead time' from some early electronic devices which really did have a period of insensitivity following a count.
Detectors Devices which give an electrical pulse when traversed by particles.
Detector sensitivity The minimum particle number or energy required to produce a recognizable pulse.
Diffraction The spreading out of light as it passes by any object. Most easily understood in the wave model of light. It sets a lower limit on the angular size of an object which can be resolved by a given telescope.
Diode A device which only transmits an electric current in one direction.
Dipole pulsation A periodic distortion of a star in which there are only two, diametrically opposite, bulges at any one time.
Driving region The shell in a variable star which generates pulsation, normally by modulating the outflowing radiation.
Dwarf nova A star which undergoes irregularly repeated outbursts (also called U Gem stars).
Dynode Accelerating electrode inside a photomultiplier.
Eclipsing binary A binary whose orbital plane passes close enough to the Earth to produce eclipses.
Ejecta from collision Fragments dislodged by a collision which may escape from the parent body, in particular an asteroid.
End window In some photomultiplier tubes the photocathode is coated onto the end of the cylindrical glass container and thus light has to enter through this end in order to be absorbed by the photocathode. Such photomultiplier tubes are said to have an 'end window'.
Epoch A term used to express the time at which some observation (minimum of an eclipse, etc) occurs.

Extinction Absorption in the Earth's atmosphere caused by both the molecules of air themselves and floating grains of dust, pollen, etc. It has the effect of making the Sun and stars appear both fainter and redder when they are viewed at low elevations, e.g. rising or setting.

Fabry lens A lens, the purpose of which is to avoid the movement of an image in the focal plane of some optical system, e.g. a telescope, from causing loss of light off a detector. To be a true 'Fabry' lens, it must image the entrance pupil of the telescope onto the detector and additionally turn light from the focal plane into parallel light.

Filter A device, generally of glass or gelatine for optical wavelengths, which has the property of absorbing some wavelengths of light and transmitting others.

Flip mirror A mirror which is pivoted at one end and can rotate through 45° or more. Its purpose is either to reflect light to an eyepiece for direct viewing or to be 'flipped' or pivoted out of the way so that the light can continue unobstructed.

Folded data In order to show the average light curve of some variables it is often necessary to plot the data from many cycles on top of each other. These data are said to have been folded with the mean period and are referred to as folded data.

Fourier analysis The method of analysing any light curve and breaking it down into discrete sine waves. Named after the French scientist Fourier who first discovered the method.

Fourier decomposition Analysis of a signal into sine waves with an arithmetical sequence of frequencies.

Free precession Wobbling of a non-sperical rotating body with no external torque.

g wave Non-radial wave, in which bouyancy acts as a restoring force.

Gamma rays Radiation from a nucleus, with photon energies typically greater than those of x rays.

Glancing collision A collision with little energy transfer which is therefore practically elastic. Correspondingly it is a collision where little energy is dissipated by friction.

Harmonics Waves with frequencies which are a simple multiple of the fundamental.

Heliocentric correction (to JD) As the Earth orbits the Sun it sweeps out an approximately circular path which has a diameter of about 17 light minutes (3×10^9 km). This means that the light from any object which is not at one of the poles of the ecliptic will arrive earlier or later than at the Sun according to the position in the orbit. Attempts to find accurate periods for any short period variables will be subject to some difficulties if this extra variation is not compensated for. It is, therefore, normal to apply a correction to all observational epochs to correct for this variable distance from the Sun. The text describes the equation which must be used and this correction for light travel time is known as the heliocentric correction.

Helium An atom with two protons in the nucleus.

Hertzsprung–Russell diagram A diagram with axes of luminosity and either spectral class or temperature, on which each star is represented as a point.

Hot spot A small region in the photosphere of a star or on an accretion disc which is markedly hotter than its surroundings, usually because of heating by infalling matter.

HR diagram We use the abbreviated form to mean an absolute luminosity–temperature diagram.

Hypergiant The terms giant, supergiant and hypergiant are used to mean stars of roughly 10,100 and 1000 times a solar radius, respectively.

Instability strip The narrow band on a Hertzsprung–Russell diagram passing from top right to bottom left and containing many types of variable star.

Integration time The length of time during which counts are taken (mean) or current is measured. Depending upon the project the integration times in photoelectric photometry would typically be from 1 second up to 2 minutes.

Interference filters These are composed of several very thin layers, the thickness of which is designed to cause the constructive interference of required wavelengths and the destructive interference of others. They can be 'tuned' to have very narrow and very precise transmission bands.

Inversion layer A level in the atmosphere of the Earth at which the change of temperature with height (typically $-1\,^{\circ}C$ per 100 m at low altitude) reverses. This has the effect of stopping the convective rise of clouds and dust and can often be seen from mountains or aeroplanes as a sharp change in the transparency of the air. Telescopes perform better above this level.

Ionization mechanism A mechanism in which variation of opacity with temperature creates a region which excites pulsations.

Julian date This is the number of days since 1 January 4713 BC. It is used to provide a continuous reference frame for the epochs of astronomical phenomena. The start of each Julian day is at 1200^{h} GMT, i.e. 0.5 days UT.

Kirkwood gaps Ranges of semi-major axis in which few asteroids occur. They are numerically related to the semi-major axis of Jupiter.

Least squares A method of discovering the best fit to some data by minimizing the sum of the squares of the residuals from a calculated line or curve.

Limiting magnitude The largest magnitude (i.e. faintest) star which can be recorded, or which can be measured (according to the context).

Linear scale Any scale in which the distance between each pair of marks represents equal increments in value.

Luminosity The energy radiated from a star. It should be specified whether one is referring to emitted radiation ('absolute luminosity') or received radiation ('apparent luminosity'), and the wavelength range should be given (visual, infrared, etc).

Magnitude A unit for the measurement of stellar brightness. The early Greeks called the brightest stars 'first magnitude' and the faintest stars visible to the unaided eye 'sixth magnitude'. The concept has now been put into a quantitative system wherein a magnitude difference of 5 exactly equals a brightness ratio of 100. Thus a difference of one magnitude equals a brightness ratio of the fifth root of $100 = 2.512$.

Mass transfer Flow of matter from (the atmosphere of) one star to another.

Median value In principle, the best estimate of the true value of a quantity based on a set of measurements is the average (mean) value. When only a small number of measurements are available there are however advantages in calculating the median of the distribution rather than the mean. This is done by finding that value having as many measurements above it as below it. For example, with seven samples, the median would be the value of the point fourth from the top or bottom. With eight samples it would be midway between the fourth and fifth points from the bottom, or top. The main advantage is that the median is less affected by an anomalously high or low reading, such as can easily occur in photometry. It is also quicker when doing hand calculations.

Mira A large, red variable star. Its long period irregular pulsations have made it the archetype for the class.

Mode Used in this book to mean type of wave, but elsewhere used more loosely, interchangeably with 'harmonic'.

Mode switch Change from one pulsation pattern to another.

Multiple stars Two or more stars in mutual orbits.

Neutrino Particle with zero mass, but unlike the photon, no electromagnetic interaction.

Node A point in a standing wave where displacement is zero.

Non-radial mode Waves running around a star.

Nova A sudden outburst of radiation from a star, often caused by mass transfer.

Nuclear reactions Collisions which are sufficiently energetic to change the number of protons and/or neutrons in a nucleus.

Oblique rotator model Many stars seem, like the Earth, to have their magnetic axes angled or displaced from their rotational axes. The rotation of the star causes the component of the magnetic field seen from the Earth to change its intensity.

$O - C$ diagram A diagram with time along the horizontal axis and the difference between the observed and calculated ($O - C$) times for some phenomenon (maxima, minima, mid-eclipse, etc) along the vertical axis.

Oort cloud A hypothetical cloud of comets, perhaps orbiting the sun at a distance of 50 000 AU and believed to contain remnants from the original formation of the Solar System. Named after Jan Oort, a famous Dutch astronomer who postulated the existence of this cloud.
Opacity The degree of absorption of radiation.
Period–density relation The correlation between period and density in (most) variable stars.
Period–luminosity relation The correlation between period and luminosity, arising from the period–density relation.
Phase The fraction of a cycle (of a given periodic phenomenon) completed at a given time.
Photon counting A method for the measurement of radiation which has sufficient sensitivity and temporal resolution that it is able to detect and resolve the arrival of individual photons.
Photomultiplier A device which is able to absorb photons of light and release electrons in one part of it (the photocathode) and then multiply the number of electrons by moving them past a series of charged plates (the dynodes). Typically of the order of a million electrons are collected for each one liberated from the photocathode.
Photon Radiation consists of a stream of packets of energy called photons. Photon energy is proportional to frequency. Intensity is the sum of the energies of all the photons arriving in unit time.
Plasma A highly ionized gas.
Proton A positively charged fundamental particle. Every nucleus contains several protons, except hydrogen which has only one.
Planck's constant The ratio of the energy to the frequency of all photons is the same and known as Planck's constant.
Planetary systems Systems like the Solar System, assumed to exist round other stars.
Positive feedback A system which reacts on itself to increase the effect of any disturbance is said to have positive feedback.
Power spectrum A graph used, in our context, to display, and reveal, which oscillation frequencies are present in a set of data. The horizontal axis is frequency and the vertical axis the square of the intensity of the component at a given frequency. Sharp peaks indicate the major components present.
Precession Rotation of an axis of rotation.
Przybylski's star A striking example of a cool Ap star.
Pulsar A compact star emitting a beam of (mainly radio) radiation intercepting the Earth as pulses.
P wave A compressional longitudinal wave, such as sound.
Quadrupole pulsation A pulsation with four regions of simultaneous displacement—see dipole pulsation.

Radial mode A wave pattern along a radial direction (as opposed to round a circumference).
Radiation transfer Radiation flow outwards in a star can be pictured as a slow diffusion of photons with innumerable scatterings in which they share energy. (Energy may also be transferred by convection and/or conduction.)
Radius mechanism A driving mechanism due purely to changes in pressure and temperature.
Recurrent nova Star with several recorded outbursts.
Red edge The right-hand (on a HR diagram) edge of the instability strip, in the region of a given type of star.
Red giant A large cool star, after hydrogen fuel is depleted.
RR Lyrae A type of variable star whose pulsational frequency varies little with luminosity. They are generally smaller and less massive than the Sun, and so have lower luminosity.
Scattering Used in its normal sense, but for photons as well as particles.
Semi-major axis Half the longer axis of symmetry of an ellipse, in particular of an orbit.
Shock wave A sharp disturbance which can arrive in conditions where wave speed depends on amplitude.
Side window In some photomultiplier tubes the light enters through the side wall of the glass envelope to strike the photocathode. Such photomultiplier tubes are said to have a 'side window'.
Sinusoidal components In principle, any wave form, no matter how complex, can be produced by the superimposing of some, or many, sine waves. The frequency, amplitude and phase of the various sine waves have to be adjusted to optimize the fit and the individual sine waves are known as the 'sinusoidal components' of the composite wave form.
Spectral class A categorization of spectrum types, closely related to temperature.
Spectral line A frequency at which radiation is either strongly emitted or strongly absorbed.
Spectral resolution The smallest change in wavelength that is detectable with given equipment.
Spectroscopic binary A binary in which the members are too close to resolve, but whose nature is revealed by periodic spectral changes.
Spectroscopy The identification of frequency components, in emission or absorption.
Stellar neutrinos Neutrinos produced by stellar thermonuclear reactions. Solar neutrinos have been detected on the Earth, but in less than the expected numbers.
String length A term used in period finding wherein a series of straight lines (an 'imaginary piece of string') is used to connect all the consecutive points in a phase diagram. When the data have been fitted to a period which

is the best fit then the phase diagram should be the least scattered and the length of the 'piece of string' should be a minimum.

Strong nuclear force The force holding nucleons in a nucleus, against Coulomb repulsion between protons.

Sunspots Spots of temperature ~ 1000 K less than the surroundings on the surface of the Sun (and presumably on other stars).

Supernova A stellar explosion with energy release $> 10^{40}$ J. There are several quite different reasons for such explosions.

Surface acceleration Acceleration due to gravity at the surface of a body.

Symbiotic star A close binary in which the more compact star is within the envelope of the donor (giant or supergiant) star.

Thermal radiation Radiation, with a distinctive continuous spectrum, arising from thermal agitation in a body. Total intensity is proportional to T^4 but total photon counting is proportional to T^3.

UBV filters Coloured glass filters, each of which transmits wavelengths over a band about 100 nm wide and which have respectively central wavelengths near 350 nm in the U (ultraviolet), 420 nm in the B (blue) and 500 nm in the V (visible).

Universal time This is effectively Greenwich Mean Time, but expressed in decimals of a day, rather than with hours, minutes and seconds.

Visual binary A binary with resolved separation.

Wave In general, any sort of periodic variation, but in our context a periodic variation of density, like huge sound waves, travelling outwards or around stars.

X rays Photons with wavelength between UV and gamma rays (typically ~ 1–10 nm).

X-ray bursters (Binary) stars which show sudden emission of x rays, lasting a few minutes, usually repeated at fairly regular intervals.

X-ray pulsars (Binary) stars with fast pulses of x rays, of great regularity.

ZZ Ceti stars Variable white dwarf stars which typically vary with many different periods and which are probably non-radial pulsators.

Index